The Metaphysics of the
Pythagorean Theorem

SUNY series in Ancient Greek Philosophy

Anthony Preus, editor

The Metaphysics of the Pythagorean Theorem

Thales, Pythagoras, Engineering, Diagrams, and the
Construction of the Cosmos out of Right Triangles

Robert Hahn

Published by State University of New York Press, Albany

© 2017 State University of New York

All rights reserved

Printed in the United States of America

No part of this book may be used or reproduced in any manner whatsoever without written permission. No part of this book may be stored in a retrieval system or transmitted in any form or by any means including electronic, electrostatic, magnetic tape, mechanical, photocopying, recording, or otherwise without the prior permission in writing of the publisher.

For information, contact State University of New York Press, Albany, NY
www.sunypress.edu

Production, Ryan Morris
Marketing, Anne M. Valentine

Library of Congress Cataloging-in-Publication Data

Names: Hahn, Robert, 1952– author.
Title: The metaphysics of the Pythagorean theorem : Thales, Pythagoras, engineering, diagrams, and the construction of the cosmos out of right triangles / by Robert Hahn.
Description: Albany, NY : State University of New York, 2017. | Series: SUNY series in ancient Greek philosophy | Includes bibliographical references and index.
Identifiers: LCCN 2016031414 | ISBN 9781438464893 (hardcover : alk. paper) | ISBN 9781438464909 (pbk. : alk. paper) | ISBN 9781438464916 (ebook)
Subjects: LCSH: Philosophy, Ancient. | Pythagoras. | Thales, approximately 634 B.C.–approximately 546 B.C. | Euclid. | Mathematics, Greek. | Pythagorean theorem.
Classification: LCC B188 .H185 2017 | DDC 182—dc23
LC record available at https://lccn.loc.gov/2016031414

10 9 8 7 6 5 4 3 2 1

पुस्तकं समर्पयामि श्रीगुरुचरणकमलेभ्यो नमः

At the Pythagoras statue in Samos.

Contents

Preface	ix
Acknowledgments	xv

Introduction
Metaphysics, Geometry, and the Problems with Diagrams — 1

- A. The Missed Connection between the Origins of Philosophy-Science and Geometry: Metaphysics and Geometrical Diagrams — 1
- B. The Problems Concerning Geometrical Diagrams — 4
- C. Diagrams and Geometric Algebra: Babylonian Mathematics — 7
- D. Diagrams and Ancient Egyptian Mathematics: What Geometrical Knowledge Could Thales have Learned in Egypt? — 12
- E. Thales's Advance in Diagrams Beyond Egyptian Geometry — 25
- F. The Earliest Geometrical Diagrams Were Practical: The Archaic Evidence for Greek Geometrical Diagrams and Lettered Diagrams — 32
- G. Summary — 41

Chapter 1
The Pythagorean Theorem: Euclid I.47 and VI.31 — 45

- A. Euclid: The Pythagorean Theorem I.47 — 46
- B. The "Enlargement" of the Pythagorean Theorem: Euclid VI.31 — 66
- C. Ratio, Proportion, and the Mean Proportional (μέση ἀνάλογον) — 70
- D. Arithmetic and Geometric Means — 72
- E. Overview and Summary: The Metaphysics of the Pythagorean Theorem — 81

Chapter 2
Thales and Geometry: Egypt, Miletus, and Beyond — 91

- A. Thales: Geometry in the Big Picture — 92
- B. What Geometry Could Thales Have Learned in Egypt? — 97
  - B.1 Thales's Measurement of the Height of a Pyramid — 97
  - B.2 Thales's Measurement of the Height of a Pyramid — 107
- C. Thales' Lines of Thought to the Hypotenuse Theorem — 116

Chapter 3
Pythagoras and the Famous Theorems — 135

- A. The Problems of Connecting Pythagoras with the Famous Theorem — 135
- B. Hippocrates and the Squaring of the Lunes — 137
- C. Hippasus and the Proof of Incommensurability — 141
- D. Lines, Shapes, and Numbers: Figurate Numbers — 148
- E. Line Lengths, Numbers, Musical Intervals, Microcosmic-Macrocosmic Arguments, and the Harmony of the Circles — 153
- F. Pythagoras and the Theorem: Geometry and the Tunnel of Eupalinos on Samos — 157
- G. Pythagoras, the Hypotenuse Theorem, and the μέση ἀνάλογος (Mean Proportional) — 168
- H. The "Other" Proof of the Mean Proportional: The Pythagoreans and Euclid Book II — 182
- I. Pythagoras's Other Theorem: The Application of Areas — 189
- J. Pythagoras's Other Theorem in the Bigger Metaphysical Picture: Plato's *Timaeus* 53Cff — 195
- K. Pythagoras and the Regular Solids: Building the Elements and the Cosmos Out of Right Triangles — 198

Chapter 4
Epilogue: From the Pythagorean Theorem to the Construction of the Cosmos Out of Right Triangles — 213

Notes — 241

Bibliography — 263

Image Credits — 271

Index — 273

Preface

In this Preface I have three objectives. I want to (A) explain how I came to write this book, (B) describe the audiences for whom it has been written, and (C) provide a broad overview of the organization of the book.

A

Almost a decade ago, I stumbled into this project. My research for a long time has focused on the origins of philosophy, trying to understand why, when, and how it began in eastern Greece. My approach had been to explore ancient architecture and building technologies as underappreciated sources that contributed to that story, specifically, to Anaximander's cosmology. Anaximander identified the shape of the earth as a column drum, and reckoned the distances to the heavenly bodies in earthly/column-drum proportions, just as the architects set out the dimensions of their building in column-drum proportions. I argued that Anaximander, who with Thales was the first philosopher, came to see the cosmos in architectural terms because he imagined it as cosmic architecture; just as the great stone temples were built in stages, the cosmos too was built in stages. My last book, *Archaeology and the Origins of Philosophy*, tried to show that since we learned about ancient architecture from archaeological artifacts and reports, research in ancient archaeology could illuminate the abstract ideas of philosophy. Philosophers rarely appeal to archaeological evidence and artifacts to illuminate the ideas of the ancient thinkers; this last project showed how archaeology is relevant and how it may be used in future research. Since my work had focused on eastern Greece—Samos, Didyma, and Ephesus—I traveled to southern Italy to explore how my work might continue there, in Crotona and Tarento where Pythagoras emigrated after leaving Samos. As I sat at the temple to Athena in Metapontum, a place where, according to one legend, Pythagoras fled as he was pursued by his enemies and where he took refuge until he died in a siege, I recalled a book on Pythagoras I read with great admiration as a graduate student, Walter Burkert's *Lore and Wisdom in Ancient Pythagoreanism*. Burkert had argued that the evidence connecting Pythagoras with the famous theorem that bears his name was too late and unsecure, and that Pythagoras probably had nothing to do with mathematics at all. That book consolidated an avalanche of consensus among scholars discrediting "Pythagoras the mathematician." I wondered if that judgment merited a review, since my research on ancient architecture and building technologies supplied so many contemporaneous examples of applied geometry. Might Burkert have gotten it wrong? And as I started to look into this matter, I tried to recall what the Pythagorean theorem—the hypotenuse theorem—was all about, and whether it might have been known in some different way, as yet unacknowledged. Late reports credit Pythagoras with the theorem, and no other name is connected to it. Although the evidence connecting Pythagoras to geometry was very contentious, the evidence connecting Thales, Pythagoras's much older contemporary, to geometry was less contentious. So, I started to see what Thales might have known geometrically, and thus what geometrical knowledge might have been current to young Pythagoras. This is how I stumbled into my project.

I tried first to figure out what the Pythagorean theorem means. I consulted dozens and dozens of books and articles, high school and college math textbooks, and scoured whatever sources I could Google on the Internet. For a long time all the accounts I found read like reports from a math seminar, and almost all explained the hypotenuse theorem in a forward-looking way to later developments in mathematics. But what I wanted to figure out was what it could have meant had it been known by the Greeks of the sixth century BCE, the Archaic period, when geometry was in its infancy and there were no geometry texts in Greece. Had Thales known the hypotenuse theorem, what could it have meant to him? What does anyone know who knows it?

When I reviewed all the distinguished studies in the history of ancient philosophy I could hardly find any study of Thales that showed the diagrams of geometrical theorems with which he is credited. And so I started making diagrams to begin to explore what he arguably knew. I placed the diagrams for theorems next to the ones that would have become evident had he tried to measure the height of a pyramid in Egypt and the distance of a ship at sea, as the ancient reports say he did. Concurrently, I sought to understand the geometrical diagrams connected to the proof of the hypotenuse theorem. At first, I began by thinking in terms of a formula I had learned in high school: $a^2 + b^2 = c^2$. Given the lengths of two sides of a right triangle, one can apply the formula and determine the length of the missing side. But it quickly became apparent to me that this was a false start. Here was an algebraic equation, but the Greeks did not have algebra, so I focused more directly on Euclid to remind myself what the geometrical proof was. That proof I recalled distantly from my sophomore year of high school, Euclid Book I.47. According to an ancient commentator, this proof traces to Euclid himself, and hence to circa 300 BCE. Even if Thales and Pythagoras knew the hypotenuse theorem, they did not know *this* proof. But the evidence for measurements with which Thales is credited showed he was familiar with ratios and proportions, and with similar triangles.

And then, by chance, I picked up yet another book, by Tobias Dantzig, *Mathematics in Ancient Greece*, and although he touches on the focus of my project on only a few pages, all at once the whole picture became clear to me. Dantzig drew my attention to the second proof of the hypotenuse theorem, known as the *enlargement* of the Pythagorean theorem. It appears in Euclid Book VI.31, a proof by ratios, proportions, and similar triangles. And as I studied that proof I realized that behind it lay, in nascent stages, the lines of thought by which Thales plausibly knew an areal interpretation of the theorem. And if Thales had such insights, the knowledge was current in Ionia already by the middle of the sixth century, and so there could be no reason to deny a knowledge of it, and perhaps a more refined investigation and proof of it, to Pythagoras and his retinue.

Before this breakthrough, I had taken a keen interest in the evidence from Babylonian and Egyptian mathematics. Since we have strong evidence, including geometrical diagrams, that the Babylonians knew an interpretation of the hypotenuse theorem at least a millennium before the time of Thales, and we have geometrical diagrams from Egypt of comparable antiquity, perhaps Thales, Pythagoras, and the Greeks of the sixth century got their insights from these ancient traditions? As I explored this further, however, I came to see a whole different approach to the question. For all we know, Babylonian and Egyptian knowledge of geometry, and perhaps the famous theorem, were part and parcel of what the Greeks of the sixth century knew through travels and transmission from and with them, but Thales's understanding emerged within a very different context. What different context was that? It was the one that showed that geometry was originally a part of metaphysical inquiries, and this had been missed, so far as I could tell from my research in the field, I had come to the conclusion that, despite the long and heated debates, Thales and the Milesians were metaphysical monists; by that I mean that Aristotle had Thales right in general when he came to identify him as the first philosopher-scientist. Thales came to realize through his investigations that there was a single unity underlying all things that he called ὕδωρ ("water"): all things come from this unity, all appearances are only alterations of this unity, and all divergent appearances return ultimately back into this original unity. There is no change in Thales's metaphysics of the cosmos, only alteration. In a range of well-known scholarly studies of the history of ancient philosophy, this view of Thales and the origins of philosophy is broadly conventional. But I began to wonder, having traveled with my students to Greece each year for decades, how my Greek friends would have responded had some "Thales" stood up at the local Σύσκεψη του Συμβουλίου—council meeting—and announced that although it seems that our experiences are full of different things that change, in reality, there is only a single underlying unity that alters without ever changing. Surely, I reasoned, knowing my inquisitive and skeptical Greek friends, someone would have stood up and asked: "Very well then, best of friends, just *how* does this happen?" How does a single underlying unity alter without changing to produce the extraordinary experience we seem to have of a world full of so many different things and full of change? And I thought about this question for a long time. And when I placed the few suggestions I got from Dantzig together with the diagrams I had made of Thales's theorems and measurements, and placed them alongside the ones connected with the enlargement of the hypotenuse theorem by ratios, proportions, and similar triangles, the whole story came to me in a flash.

The answer I realized was this: just as Thales had come to the conclusion that, despite appearances to the contrary, there was only a single underlying unity, he sought to figure out the structure of that basic unity that, repackaged and recombined, was the building block of the whole cosmos. Thus, Thales turned to investigate geometry because he was looking for the fundamental geometrical figure into which all other rectilinear figures reduce, and he came to realize that it was the right triangle. This discovery is encapsulated in what is preserved as Euclid VI.31, the enlargement of the Pythagorean theorem. And from this discovery, a new metaphysical project was created to explain all divergent appearances as alterations of the right triangle, and triangles into which all other rectilinear figures dissect. The building the cosmos out of right triangles is a narrative preserved in Plato's *Timaeus* 53Cff. This metaphysical project begins with Thales and is taken up by Pythagoras, who is also credited with the "application of areas" theorem, the construction of all rectilinear figures out of triangles in any angle, and the "putting together" of the regular solids, later called "Platonic solids," that are the molecules, created from right triangles, out of which all other appearances are constructed.

But let me add a word of caution: there's no doubt about it, this must be a speculative thesis, a circumstantial case at best. Some scholars think the evidence concerning almost anything about Thales, and even more so Pythagoras after him, is so meager and unsecure that we cannot make a solid start about their metaphysical views, much less their knowledge of geometry. It is unlikely that any of those doubters will be convinced of my thesis. But if we notice the evidence from architecture and engineering, contemporaneous with and earlier than the sixth century, and add to it what we know of geometrical knowledge and diagrams from Egypt and Mesopotamia, the reader can come to see the plausibility of my proposal regarding how the case I propose unfolded, underlying Euclid's Book VI. From what sources did the idea germinate originally for building the cosmos out of right triangles? It plausibly came from Thales, Pythagoras, and the Pythagoreans after him.

B

I have written this book with a few different audiences in mind. My first audience consists of scholars, students, and others with interests in early Greek philosophy and the origins of philosophy. How did it begin? Why there and then? I have broadly explored this central theme for most of my career. While the ideas of Milesian philosophy and metaphysical monism are familiar to those already aware of early Greek philosophy, it is my perception that the early stages of geometry are not part of that familiar understanding. Indeed, some scholars of ancient philosophy might at first regard this project as "ancient mathematics" and not "ancient philosophy," but I hope they will realize such a view is wrongheaded after reading the book with care.

The second audience I hope to reach with this book is advanced seminars, probably at the graduate level, in philosophy and also in the history of mathematics. The book was written to be teachable; certainly separate seminars could be devoted to Euclid I.47 and then VI.31. We could have a seminar meeting or two on Thales's measurements and theorems, and a series of seminars on Pythagoras and the Pythagoreans' hypotenuse theorem. Discussions could also be devoted to musical intervals, figurate numbers, the squaring of the rectangle and the mean proportional, the application of areas theorems, and the putting together of the regular solids (i.e., the Platonic solids).

The third audience I hope to reach is high school students and math clubs, with two purposes in mind. Since many mathematicians regard the Pythagorean theorem as among the most important theorems in all of mathematics, students can learn *both* proofs diagrammatically, and consider, perhaps for the first time along with their teachers, the metaphysical possibilities of geometrical propositions.

C

The book begins with an Introduction that deals broadly with problems concerning, and the evidence for, early geometrical diagrams. First, in clarifying how I am using the term "metaphysics" in this book, we place Thales as first-philosopher and scientist for his metaphysical monism, and then as first geometer for his exploration of spatial relations through diagrams. The compelling context in which Thales's diagram-making finds a place is composed of Babylonian cuneiform tablets, the Egyptian Rhind and Moscow mathematical papyri, the Egyptian techniques of land surveying and square grids for tomb painting, the Archaic Greek evidence of mathematically exact diagrams from roof tiles of the temple of Artemis at Ephesus, and the scaled-measured diagram inferred from the Eupalinion tunnel in Samos, where the letters painted in red are still visible on the western wall. Thales's making of geometrical diagrams gains a clarity when placed within this context.

Chapter 1 leads the reader through two proofs of the "Pythagorean theorem" in Euclid, the famous one that most high school students learn, Book I.47, and also the so-called "enlargement" of the theorem at VI.31. In both cases, I lead the readers through the suppositions—earlier proofs—that are presupposed in each proof separately. Finally, the reader gets some sense of the metaphysical implications that will be explored in the next chapters.

In chapter 2, now that the reader has been alerted to the proof of the hypotenuse theorem by ratios, proportions, and similar triangles, we explore Thales's measurement of the height of the pyramid and the distance of the ship at sea, and place those diagrams next to the diagrams of the theorems with which he is credited, most importantly what turns out to be that "every triangle in a (semi-)circle is right angled." After reviewing the many diagrams, and placing together the project of metaphysical monism, I make a plausible case that Thales knew an areal interpretation of the hypotenuse theorem along the lines of VI.31.

In chapter 3, I continue directly from chapter 2: if Thales knew an areal interpretation of the hypotenuse theorem, then that knowledge was current by the middle of the sixth century, and there is no reason to deny a knowledge of it to Pythagoras and others in eastern Greece; it was part and parcel of a metaphysical project to explain the *structure* of some basic underlying unity that alters without changing. That structure was the right triangle—this is part of what the hypotenuse theorem meant to Thales and his contemporaries in the middle of the sixth century BCE, and from it the project was created to explain how all appearances could be imagined in terms of rectilinear figures all constructed out of right triangles. Pythagoras is credited, along with Thales, with the knowledge/discovery that every triangle in a (semi-)circle is right angled. Pythagoras worked along the same lines as Thales because he inherited and embraced this part of his project. After reviewing the evidence for the knowledge of the hypotenuse theorem in the fifth century, the thesis that Thales knew it in the middle of the sixth century is proposed again. Then, the evidence for Pythagoras's connection to figurate numbers, lengths, numbers, and musical intervals are investigated diagrammatically. Subsequently, as a practical application of geometrical knowledge, I review the details of the digging of the tunnel of Eupalinos, contemporaneous with young Pythagoras. I next detail a plausible hypothesis for the development of Thales's project by Pythagoras. The Pythagoreans are credited with a proof for squaring the rectangle—making a rectangle equal in area to a square. This is the construction of a mean proportional, the proof of a continuous proportion, the pattern by which the right triangle expands or collapses, and the pattern by which the right triangle grows; this is the other part of what the hypotenuse theorem meant to our old friends of the sixth century. Then, we turn to investigate Pythagoras's *other* theorem; he is credited with the application of areas theorem, which is to construct a rectangle equal in area to a triangle, attached to a given line, and another triangle (of different size but) similar to the first. This theorem shows the project of what I call *transformational equivalences*, how a basic unity alters without changing, how one and the same unity appears differently in different shapes yet sharing the *same* area. And finally, we explore how Pythagoras "put together" the regular solids—later called the "Platonic solids"—the molecules built out of right triangles from which all other appearances are constructed. While space is imagined as fundamentally flat, the four elements are constructed by folding up the patterns of the regular solids.

In the last chapter, chapter 4, the Epilogue, the diagrams that form the argument of the book—the construction of the cosmos out of right triangles—are presented in sequence succinctly, from the mathematical intuitions that underlie the Pythagorean theorem by ratios, proportions, and similar triangles at VI.31, to the application of areas theorem, and through to the "putting together" of the regular solids from right triangles.

In closing, let me pose to the reader this question. In Plato's *Timaeus* we get an account of the construction of the cosmos out of right triangles. Where did that idea come from? We find no explicit evidence for this theory anywhere before Plato, and yet he places it in the mouth of a Timaeus of Locri, a man from southern Italy, in a dialogue that contains Pythagorean ideas, among others. It is the thesis of this book that in nascent form, the idea is contained already in the hypotenuse theorem along the lines of VI.31. It was plausibly discovered and understood by Thales, who was looking for the fundamental geometrical figure (= right triangle) that was to ὕδωρ as ὕδωρ was to all other appearances, altering without changing. This is what I am calling the "lost narrative" of Thales, and the argument for it is delivered, for the first time, by following the geometrical diagrams.

Acknowledgments

Over the years in which this book took shape I received valuable assistance. In the early stages, I was fortunate enough to have help making diagrams from Richard Schuler, an astronomer-engineer who traveled with our groups to both Greece and Egypt, and to receive very helpful lessons in geometry from mathematicians Jason Taff and Michael Lockhart. I also benefitted greatly by consulting two websites, both of which present dynamically illustrated editions of Euclid's *Elements*: David E. Joyce's *Euclid's Elements*, and (ii) Ralph Abraham's website *The Visual Elements of Euclid*. I would also like to appreciatively recognize clarifications I received from mathematicians Travis Schedler and Jerzy Kocik. I had the chance to get useful help from mathematician Laurence Barker of Bilkent University in Ankara, Turkey. I'd like to express my appreciation to Leonid Zhmud, who was kind enough to read my chapter on Pythagoras and to make thoughtful comments and suggestions about it, and to Dirk Couprie, who was generous enough to read the whole manuscript and who saved me from making some errors. I gratefully acknowledge classicist Rick Williams for reviewing the Greek in the manuscript, and astronomer Bruce Wilking for discussing and reviewing the details of the pyramid measurement. And after a series of correspondences over a few years, mathematician Michael Fried graciously agreed to review my whole manuscript in the hopes that, even if I were not able to persuade him of my thesis, he might save me from careless mathematical errors. I want to thank him so very much for this assistance. And I wish to acknowledge the useful comments and suggestions of the anonymous readers who reviewed my manuscript. The responsibility for whatever errors, ineptitudes, and misrepresentations remain in the book is mine alone.

    I have given presentations on parts of this book to the Society for Ancient Greek Philosophy, to the University of Prague, to the University of West Bohemia in the Czech Republic, to the Mathematics Seminar at Bilkent University in Ankara, to the Deutsches Archäologisches Institut in Athens, to the Ancient Science Conference at the University of London, and to the Illinois Classics Conference; I benefitted from many of the comments and suggestions I received in these forums.

    I wish to record with special gratitude the support and encouragement I received from then-dean of the College of Liberal Arts, Dr. Kimberly Leonard, and to the administration at Southern Illinois University Carbondale for approving yet another sabbatical leave that allowed me to complete this manuscript.

    I would like to acknowledge with enormous gratitude my continued collaboration with Cynthia Graeff, who made the lion's share of the figures for this book; she also supplied the illustrations for my previous three research books. I could not have presented my case as well without her expertise and her enormous patience.

And again, in a class all its own, I wish to record my deepest appreciation for the support and encouragement that I have received over the decades from Tony Preus, without which I could scarcely have imagined the honor of delivering my fourth book to his estimable series on ancient Greek philosophy.

Introduction

Metaphysics, Geometry, and the Problems with Diagrams

This Introduction has two purposes. In section A, I want to clarify that Thales's metaphysics committed him to a doctrine of monism, thereby clarifying the senses in which I am using the term "metaphysics." In sections B, C, D, E, and F I contend that there is evidence for geometrical diagrams in Mesopotamian sources had they been transmitted to Ionia, that Thales could have seen geometrical diagrams in Egypt when he traveled there, and that there is evidence of geometrical diagrams from architectural and engineering projects in nearby Ephesus and Samos, that is, in eastern Greece, earlier and contemporaneously. Thales's geometrical diagram-making needs to be reviewed within this context.

A

The Missed Connection between the Origins of Philosophy-Science and Geometry: Metaphysics and Geometrical Diagrams

Aristotle[1] identifies Thales as the first scientist and first philosopher; he is the first among the "inquirers into nature"[2] and the first to propose an underlying ἀρχή—a "principle," or "cause" (αἰτία)—from which all things arise,[3] the first Greek of whom we know to offer a rational account of the universe and how it began, foregoing traditional mythologies that offered divine explanations. While it is true that Aristotle identifies Thales as first in a line of *physical* philosophers, and not as the first of all philosophers tout court, the fact that Aristotle can identify no one earlier than Thales who posits a first principle or first cause entitles him to the title of the first philosopher. According to Aristotle, not only do all things arise from this single underlying unity that Thales called ὕδωρ ("water"), all successive appearances are only ὕδωρ in altered forms, and upon dissolution all different appearances return to ὕδωρ. Thales and the early philosophers adopted a metaphysics that rejected the idea of real change; while we do perceive qualitative differences—at one moment some thing looks fiery and at another hard and solid—these are only modifications of this one underlying unity. The credit for inquiring into nature rationally earns Thales the renown as the first scientist; the credit for positing an ἀρχή or principle of an underlying unity earns Thales the title of the first philosopher. Aristotle's pupil, Eudemus, who wrote the first *History of Mathematics*, identifies Thales as the first to introduce geometry into Greece, crediting him with theorems some of which he approached more empirically and some more generally. Thus, Thales stands as the first Greek philosopher and scientist because of his metaphysics; he stands as the first Greek geometer for his explorations of spatial relations through diagrams. What has been missed is the place of geometrical investigations and diagrams as part of his metaphysics. It is a central thesis of this book that, in its nascent state, geometry offered Thales a way to express structurally how a fundamental unity could appear so divergently without changing, merely altering its form, if only Thales could discover the fundamental geometrical figure. The discovery that it was the right triangle turned out to be the discovery of the "Pythagorean theorem"—the hypotenuse theorem—as metaphysics. It is the thesis of this book that Thales plausibly knew an interpretation of it, and hence Pythagoras and the Pythagoreans after him, because this is what he was looking for. Once discovered—the right triangle—geometry offered a way to express how this fundamental figure, repackaged and recombined, was the building block of all things. The whole cosmos was imagined to be built out of right triangles; geometry shows how all appearances can be transformed into one another, altering without changing, and this is what I am calling the "lost narrative" traceable to Thales that is echoed finally in Plato's *Timaeus* at 53Cff.[4]

The narrative I propose to relate shows the link between these activities in a manner that has not been appreciated, speculative as I acknowledge it must be. Let there be no doubt that the evidence about Thales's views is meager, long contested, and much debated, and Aristotle has placed Thales within a system of his own, for his own exegetical purposes. But when these ancient doxographies are reviewed against a background of the reports of his two successors in the Milesian school, Anaximander and Anaximenes, who embraced the same project to explain how the cosmos arose from a single underlying ἀρχή but named that principle differently—ἄπειρον (unbounded) and ἀήρ (mist-air)—and how that thematic project blossomed subsequently through the developed systems of Plato and Aristotle, we should be able to see Thales's rightful place as the first scientist, first philosopher, and the first geometer.

The convergence of Thales-the-scientist and Thales-the-philosopher is the view that reality is not as it seems. Despite the obvious appearance of diversity and change, there *really* is none—this is the metaphysical doctrine of *monism*. Philosophers sometimes distinguish between "source monism" and "substance monism": in the case of the former, there is an ultimate source from which all things come; in the case of the latter, there is a single substance to which all things reduce.[5] While such a distinction deserves to be noted conceptually, the two different senses cannot be separated in Thales's case. There is an underlying unity, a source from which all appearances come, and all appearances are ultimately the same underlying unity; despite the appearance of change and diversity, there is an underlying permanence. Thus, the argument here is that Thales reached the conclusion that the ultimate *structure* of ὕδωρ, the source and substance of all things, is the right triangle, and by setting out the diagrams that constitute the argument for it, we shall see the metaphysical project as the construction of the cosmos out of right triangles. By stating the matter this way, I mean to clarify how I am using the term "metaphysics." While it is certainly true that the term "metaphysics" can be applied appropriately to the inquiries of other early Greek philosophers who embraced various forms of pluralism, not monism, when I use the term "metaphysics" in this study I mean monistic metaphysics, specifically, source and substance monism.

Thales, and the members of his Milesian school, embraced a metaphysics of source monism. And with hardly a scholarly dissenter,[6] and even given the hand-wringing about the paucity of the surviving evidence, there is a general consensus that Thales and the Milesian school held that there was only a single underlying reality altered but unchanged. Why exactly Thales identified this underlying unity as "water" while his younger contemporary, Anaximenes, named it "mist-air," the evidence will not allow us to answer with confidence. Perhaps Thales noticed that while water flows without its own shape at room temperature, it turns to solid ice on freezing, turns to steam when heated, and evaporates into misty air after a rainstorm on a hot day. Perhaps. Exactly why Thales identified the source and substance of the underlying permanent as ὕδωρ cannot be determined by the evidence. Aristotle speculates that Thales might have meant "moisture" when he referred to ὕδωρ since semen is the procreative source of animal life, and things sprout and grow from moisture throughout nature, but Aristotle always prefaces these speculative remarks by "perhaps." Clearly, he did not have access to any reports that supplied an answer for Thales's choice.

The reader may wonder, understandably, how Thales got from the elemental theory of water to a theory of elemental triangles. The argument of this book is that beginning with a source and substance monism, he turned to geometry to reveal the underlying fundamental *structure* of this permanent from which all appearances come; or, of course, it might have been just the reverse. Having explored geometrical figures, inspired in Egypt, and having concluded that the right triangle was the fundamental figure out of which all other shapes could be explained transformationally, Thales turned to look for the element fundamental to all others. This ingenious insight, whichever way it dawned, led to a project to show how from the right triangle, repackaged and recombined, all appearances could be explained. If one studies Euclid, as we do in this book, and then looks backward more than two centuries, as it were, to try to detect the earlier stages as the geometers first ventured and experimented, we see the formulation of just these rules for the transformational equivalences of all appearances. The reader can judge whether a plausible case has been made that geometry showed how to account for the transformational equivalences of all appearances from triangles in the formulation of the theorems of the application of areas that transform all triangles into different shapes, and the putting together of the regular solids, the elements out of which all appearances are constructed. As the readers will see, all triangles dissect into right triangles, and the elements—fire, air, water, earth—are all constructed out of right triangles. Thales got to the theory of elemental triangles from the elemental theory of a source monism because he searched for the structure of all appearances, and geometry revealed the structure that could alter without changing.

The evidence that is less in doubt allows us to take as our starting platform that Thales posited a single underlying unity—a source and substance monism. In our language of modern science today, we might say that all the diverse things we see have an underlying atomic structure that can be organized in a myriad of ways to produce the world we experience. At first we cannot see this underlying atomic unity; its reality is to be inferred. *This kind of vision of the cosmos, the questioning that leads to and posits an underlying unity, and then tries to explain how multifarious appearances emerge through alteration without change, is what I mean by "metaphysics" in this book.* Since the reader will discover that this is part of what the Pythagorean theorem—the hypotenuse theorem—meant to Thales and Pythagoras in the sixth century, the title of this book *The Metaphysics of the Pythagorean Theorem* captures just this story.

B

The Problems Concerning Geometrical Diagrams

To investigate this missed connection between geometry and metaphysics, we shall now explore the background of geometrical diagram-making in the context of which Thales's innovations are illuminated. Because there are no surviving diagrams by Thales, only references to theorems, the argument for the plausibility that Thales made them is to show that we do have evidence of diagram-making in Mesopotamia whose influence may have been communicated to Ionia, from Egypt where Thales measured the height of pyramid, and also evidence in eastern Greece—Ionia—where Thales lived. But let me try to be very clear on this point: it makes no sense to credit Thales with the isosceles triangle theorem, for example—that if a triangle has two sides of equal lengths, then the angles opposite to those sides are equal—without crediting him with making diagrams! And there are no good reasons to deny Thales credit for such a theorem. Furthermore, I will argue that Thales plausibly knew an interpretation of the Pythagorean theorem, and this is because, together with the doxographical reports about Thales's efforts in geometry, the knowledge of that theorem turned out to be fundamental to his metaphysics. An interpretation of it was certainly known at least a millennium before Thales's time in Mesopotamia, and it is possible that some interpretation of it was known in Egypt, but my argument is that the case for Thales's knowledge of it, and Pythagoras's and the Pythagoreans' in turn, emerged from a completely different context of inquiries.

The origins of geometry in ancient Greece have been the subject of endless controversies, but we can shine a new light into the early chapters by taking a new approach to understanding the Pythagorean theorem. Although a consensus among scholars of ancient philosophy has consolidated around Burkert's claim that Pythagoras himself probably had no connection with the famous theorem,[7] the knowledge of the hypotenuse theorem belongs to an early insight in this unfolding story, customarily placed in the fifth century. I will argue that its discovery among the Greeks[8] is plausibly earlier, as early as the middle of the sixth century BCE, regardless of whose name we attach to it, though the case I argue for is that Thales plausibly knew an interpretation of it. If Thales knew it, then the members of his school knew it, and if they knew it, it makes no sense to deny Pythagoras knowledge of it, whether he learned it from Thales, Anaximander, or a member of their retinue. When we see why and how it was plausibly discovered in the Greek tradition at such an early date, we get a new insight into the origins of geometry there. To explore this matter, we first have to grasp what the famous theorem means, what it could have meant to the Greeks of the Archaic period had they known it then, and to recognize that these answers might be quite different. For from the standpoint of trained mathematicians, we might get some kinds of answers, and from the context of the Greeks of the sixth century when there were no geometry texts and geometrical knowledge was for them in its infancy, we might get some other kinds of answers. We have got to figure out what we are looking for.

It is my contention, and the argument of this book, that the theorem was discovered by the Greeks of the sixth century who were trying to resolve a metaphysical problem. It was not originally a theorem discovered in the context of a mathematics seminar. The whole picture of understanding the hypotenuse theorem, and hence the origins of Greek geometry, has been misconceived. Geometry was originally the handmaiden to metaphysics.

Recent studies by Netz, for example, have directed us to see the origins of geometry in the context of geometrical diagrams, and specifically lettered diagrams, and the prose writing that is connected inextricably with them.[9] At one point Netz argues that "Our earliest evidence for the lettered diagram comes from outside mathematics proper, namely from Aristotle."[10] But, the dating for the earliest geometrical diagrams Netz fixes by speculative inference just after the middle of the fifth century BCE with the "rolls" of Anaxagoras, Hippocrates, and Oinopides; these are the earliest "publications" in which later writers referred to diagrams, though none of them survive from the fifth century. When Netz tries to disentangle the testimonies of Simplicius (sixth century CE), who refers to Eudemus's testimony (fourth century BCE) about Hippocrates's efforts in squaring the *lunulae* (fifth century BCE), he cautions and advises us about "layering" in these testimonies to sort out who said what: Simplicius had his own agenda in selecting the pieces in Eudemus's report about Hippocrates, as did Eudemus in presenting and discussing them.[11]

Netz's general conclusion is that our view of even the fifth century is distorted as a consequence of disciplinary boundaries set by Aristotle and others in the fourth century upon whom the doxographical reports rely, who separate the world into disciplinary divides of their own choosing and for their own purposes. Ironically, Netz has done much the same thing without grasping adequately that he has done so. His case about the origins and early development of geometry focuses on geometrical diagrams, and specifically lettered diagrams, to identify the earliest episodes in Greek mathematics, but his approach defines away evidence that changes the story itself. Netz focuses on diagrams in a mathematically "pure" sense, not applied. Netz's mathematician wants to know when the diagram has a reality all of its own, not as an architectural illustration of a building but as the building itself. Let me grant out of hand that regarding the diagram as an object onto itself—like grasping a Platonic form—reflects a different, special kind of consciousness; it is the mathematical object per se within the disciplinary guidelines for the mathematician, and exists in a realm all its own. And Netz is certainly entitled to focus on this moment and tell a story that springs from it. But it just so happens that we have evidence—that we shall consider in part D here—of lettered diagrams almost a century earlier than Anaxagoras and Hippocrates, and two centuries earlier if we fix on Netz's claim that it is Aristotle who provides the earliest evidence; the evidence for a lettered diagram traces to the middle of the sixth century in Pythagoras's backyard itself—from the tunnel of Eupalinos of Samos—and evidence from other building sites, such as the nearby monumental temple of Artemis at Ephesus, which dates even earlier, to the middle of the seventh century BCE, which suggest that thinking and reflecting on geometrical diagrams was probably widely acknowledged and practiced much earlier when architects/engineers sought to solve their problems, guide the workers producing the architectural elements, or explain their solutions to others. Let us consider how this evidence changes the story of the origins of Greek geometry.

Netz's argument is that "the mathematical diagram did not evolve as a modification of other practical diagrams, becoming more theoretical until finally the abstract geometrical diagram was drawn. Mathematical diagrams may well have been the first diagrams . . . not as a representation of something else; it is the thing itself."[12] Now it should be pointed out that Netz acknowledges "but at first, some contamination with craftsman-like, the 'banausic,' must be hypothesized. I am not saying that the first Greek mathematicians were e.g. carpenters. I am quite certain that they were not. But they may have felt uneasily close to the banausic."[13] Those fifth-century "mathematicians" to whom Netz is referring might well have felt uneasily close to the banausic, but probably not as much as modern mathematicians do who distinguish sharply between pure and applied mathematics by their own modern disciplinary divides. And when we see that geometrical diagrams were explored long before the middle of the sixth century, it should be clearer yet that the distinction of "pure" vs. "applied" is perfectly misleading to how these investigations started. The point is that the evidence we do have shows that early forays in Greek geometry in the Archaic period were connected to architectural and building techniques, engineering problems, and practical, technical activities such as measuring the height of a pyramid by its shadows or measuring the distance of a ship at sea. I argue that these practical applications of geometrical techniques led to and involved the making of geometrical diagrams and reflecting further on them, at least as early as the time of Thales; these practical applications urged and invited Greeks of the Archaic period to think about the relations among rectangles, squares, triangles, and circles. Netz contends that "What is made clear by [his] brief survey is that Greek geometry did not evolve as a reflection upon, say, Greek architecture."[14] At all events, I contend that the mathematical diagram did indeed evolve as a modification of other practical diagrams, especially from architecture and engineering projects, since these diagrams proliferated for more than a century before the time of Anaxagoras and Hippocrates, and more than two centuries before the time of Aristotle. This is how the evidence changes the story.

The main theme missing from the scholarly literature in the narrative I am going to unfold is that the connection that made all this possible is the principle that underlies the *practical diagram*—and I will argue that we can begin with Thales and his contemporaries. The practical diagram was an early form of "proof," a making visible of a set of connections that was being claimed. The diagram and its explanation may well fall short of the kinds of rigorous deductions we find a century or more later, but as von Fritz expressed it,[15] it was persuasive for an audience of the Greeks by the standards of that time. The revealing point of the practical diagram was that the cosmos could be imagined in such a way that an underlying, orderly structure—a nonchanging frame of relations—was discoverable, and could illuminate our everyday experience even though it wasn't directly derived from that experience. The limestone hill through which Eupalinos's tunnel is cut has no perfectly straight lines, and yet by imagining and planning a geometry of straight lines, the tunnel was successfully dug from two sides. I will get to this case by the end of the chapter, that Eupalinos had to have made not only a lettered diagram but also a scaled-measured diagram to achieve his result, and the evidence for it is still on the tunnel walls! It is this sort of evidence that undermines the kind of narrative Netz embraces so far as it reveals the origins of geometrical thinking for the Greeks; by focusing on the diagram as a thing-in-itself, what is missed is the fact that lettered diagrams did not start this way, but rather began by presupposing that an intelligible, nonchanging structure underlies and is relevant to the world of altering appearances that we experience, that the diagram contains an insight into the world of appearances. Deep truths lie behind appearances, and geometry offered a way to reveal and express them. It is only because the architectural diagram—the practical diagram—"represents the building" that geometry came to be seen as offering a new and special window into the structure that underlies it. Thus, because the practical diagram pointed to a hidden underlying structure that nevertheless found application to our world of ever-changing experiences, it came to acquire a metaphysical relevance.

C

Diagrams and Geometric Algebra: Babylonian Mathematics[16]

The thesis that Babylonian mathematics informed, stimulated, and influenced Greek geometry has had supporters and currency. One way to see this discussion is to consider the theme of "geometric algebra," that Mesopotamian mathematics from Cuneiform tablets—from the early days of the Sumerians to the fall of Babylon in 539 BCE—supplied evidence for what the historian of mathematics Otto Neugebauer described as "quadratic algebra."[17] And when scholars reviewed Euclid's geometry they came to the hypothesis that some of the *Elements* were geometric solutions to these very same problems in quadratic algebra. These geometrical solutions came to be termed "geometric algebra," and this was the basis for suggesting Mesopotamian influence and stimulation. While there is no evidence to discount possible exchange of mathematical knowledge and stimulus, there is also no direct evidence for it. The argument for substantial Babylonian influence on the character of Greek *geometry* seems to me to be tenuous; but positively and more importantly for our project, what the evidence does show is that had Thales and his sixth-century compatriots been aware of what the cuneiform tablets supply, it would have provided robust evidence for thinking about spatial relations through diagrams.

In the recent historiography of ancient Greek mathematics, the term "geometrical algebra" was introduced at the end of the nineteenth century by Tannery[18] and Zeuthen[19] and became current after the publication of Heath's edition of Euclid's *Elements*, where he made extensive references to it. The phrase "geometrical algebra" suggested an interpretative approach for reading a number of propositions in Euclid's *Elements*, especially Book II, a book customarily credited to the Pythagoreans.[20] According to those who regarded that the Greeks thought algebraically but decided to systematically organize this thought geometrically, in Book II.4, for example, the algebraic identity $(a+b)^2 = a^2 + 2ab + b^2$ is expressed geometrically (below, left, where HD is 'a' and AH is 'b'); and II.11 (below, right) is considered to be a geometrical solution of the quadratic equation $a(a-x) = x^2$ (where 'x' is AF, and 'a' is AC).

8　The Metaphysics of the Pythagorean Theorem

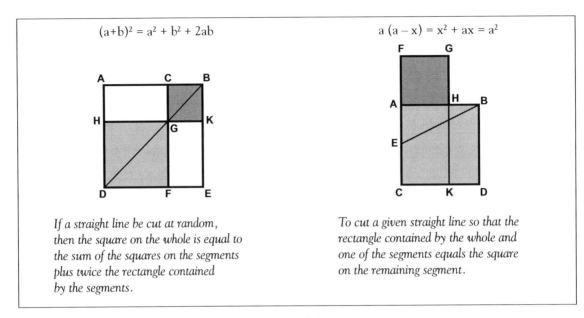

Figure I.1.

In the 1930s, Neugebauer's work on ancient algebra supplied a picture of Babylonian arithmetical rules for the solution of quadratic equations, also known as "second-degree problems,"[21] and the earlier hypothesis by Tannery and Zeuthen of geometrical algebra of the Greeks came to be seen as little more than the geometrical formulation and proofs of the Babylonian rules, a point that Neugebauer himself also advocated.[22] This argument, trumpeted by van der Waerden,[23] was that the Greeks were driven to geometrize arithmetic in light of the discovery of incommensurability.

Árpád Szabó was the first to break ranks with the hypothesis of geometrical algebra, and hence the view that some parts of Greek geometry were driven to geometrize Babylonian arithmetical rules, by arguing that the propositions of Book II belong to a pre-Euclidean stage of development of Greek geometry that showed a disregard for both proportions and the problem of incommensurability.[24] In the 1970s the debate was propelled further by Sabetai Unguru, who attacked the historiographic position on geometrical algebra; Unguru argued that it was a mistake to apply modern mathematical techniques to ancient ones to discover better how the ancient mathematicians thought through their problems: *Geometry is not algebra*. As Unguru stated the matter:

> Geometry *is thinking about space and its properties*. Geometrical thinking is embodied in diagrammatic representation accompanied by a rhetorical component, the proof. Algebraic thinking is characterized by operational symbolism, by the preoccupation with mathematical relations rather than with mathematical objects, by freedom from any ontological commitments, and by supreme

abstractedness. "Geometrical algebra" is not only a logical impossibility it is also a historical impossibility.[25]

Because the centerpiece of the argument for geometric algebra was Euclid's Book II, van der Waerden's reply to Unguru centered on it. He insisted that (i) *There is no interesting geometrical problem that would justify some of the theorems in Euclid Book II*, (ii) there is a step-by-step correspondence between the arithmetical methods of the Babylonians and certain theorems of Euclid Book II, and (iii) there are, generally, many points in common to Babylonian and Greek mathematics that indicate a transfer of knowledge from the former to the latter.[26]

Fowler countered van der Waerden's assessment of Euclid Book II by arguing that the propositions display techniques for working out ratios using the *anthyphairetic* method, a pre-Eudoxean method of proportions.[27] Mueller challenged further the claim that the lines of thought in Book II were algebraic by arguing that since Book II explores the function of geometrical properties of specific figures and not abstract relationships between quantities or as formal relationships between expressions, there is no reason to assume that Euclid is thinking in an algebraic manner.[28] And Berggren argued that in the absence of evidence showing that the Babylonian knowledge was transferred to the Greeks, certain coincidences cannot substantiate such an inference.[29] Knorr responded to the debate trying to find some areas of compromise. He is clear that the Greeks never had access to algebra, in the modern sense, but they did introduce diagrams not only to explore visually geometric relations but also to apply the results in other contexts. And so, Knorr concluded that "the term 'geometric algebra' can be useful for alerting us to the fact that, in these instances, diagrams fulfill this function; they are not of intrinsic geometric interest here, but serve only as auxiliaries to other propositions."[30]

My position is to acknowledge the use of the term "geometric algebra" but to caution about its usefulness; it misleads us in understanding how the earliest Greek geometers worked. In my estimation, the Greeks did not have algebra, nor was the pattern of their geometrical thought algebraic, though it might well be amenable to algebraic expression. My focus is not on Euclid Book II at the end of the fourth century, but rather on the earliest forays into geometry in the sixth century that eventually led to it. Szabó and Fowler thought that Book II was a pre-Eudoxean exploration, and it might well be that by the middle of the fifth century the whole form of thought in what became Book II displayed such a character. But it is my argument that the earliest problem that underlies what later became Book II—as I discuss at length in chapter 3—is the squaring of the rectangle, the realization of areal equivalence between a square and a rectangle. This is theorem 14, the last one in the shortest book of Euclid, and I shall detail the lines of thought that constitute the proof later in this book. This is the "interesting geometrical problem" that escaped the notice of van der Waerden; this was the key to grasping continuous proportions, the mean proportional or geometric mean between two line lengths without arguing by ratios and proportions—the pattern by which the fundamental geometrical figure, the right triangle, scales up and collapses, out of which the metaphysics of the whole cosmos is constructed. The *proof* of the mean proportional was the construction of a rectangle equal in area to a square, and it was that understanding that made it possible for Thales and

10 The Metaphysics of the Pythagorean Theorem

the early Pythagoreans to grasp the famous hypotenuse theorem, for the internal structure of the right triangle is revealed by the areal equivalences of the figures constructed on its three sides. As we will see in chapter 3, it is a continued reflection on the triangle in the (semi-)circle, a theorem credited to both Thales and Pythagoras.

The main positive point I wish to emphasize about Babylonian mathematics is that more than a thousand years before the time of Thales and Pythagoras, we have evidence for geometrical diagrams that show the visualization of geometrical problems. Recently, this point had been made emphatically: "that [an analysis of both Egyptian and Babylonian texts shows that] geometric visualization was the basis for essentially all of their mathematics" because visualization is simply a natural and intuitive way of doing mathematics.[31] Below we have visualization of right-angled triangles and the "Pythagorean theorem": below left, we have a visualization of Old Babylonian problem text YBC 6967; and below, right, a visualization of the "Pythagorean theorem" in the Babylonian tradition, and finally, further below, the OB tablet YBC 7289.

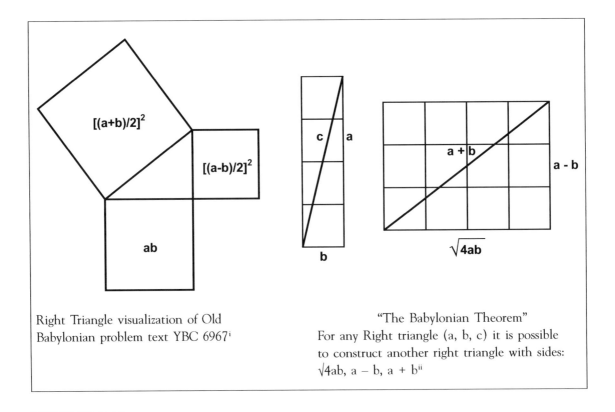

Right Triangle visualization of Old Babylonian problem text YBC 6967[i]

"The Babylonian Theorem"
For any Right triangle (a, b, c) it is possible to construct another right triangle with sides: $\sqrt{4ab}$, $a - b$, $a + b$[ii]

[i]After Rudman, p. 69.
[ii]After Rudman, p. 19.

Figure I.2.

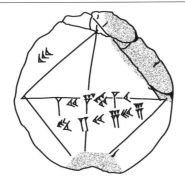 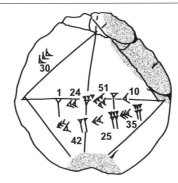

After YBC 7289
the 'Pythagorean Theorem'[i]

The diagonal displays an approximation of the square root of 2 in four sexagesimal figures, 1 24 51 10, which is good to about six decimal digits.
1 + 24/60 + 51/60² + 10/60³ = 1.41421296... The tablet also gives an example where one side of the square is 30, and the resulting diagonal is 42 25 35 or 42.4263888.

---

[i] As Janet Beery and Frank Sweeney describe this tablet at their website "The Best Known Babylonian Tablet" Home » MAA Press » Periodicals » Convergence » The Best Known Old Babylonian Tablet? "YBC 7289 is an Old Babylonian clay tablet (circa 1800–1600 BCE) from the Yale Babylonian Collection. A hand tablet, it appears to be a practice school exercise undertaken by a novice scribe. But, mathematically speaking, this second millennium BCE document is one of the most fascinating extant clay tablets because it contains not only a constructed illustration of a geometric square with intersecting diagonals, but also, in its text, a numerical estimate of $\sqrt{2}$ correct to three sexagesimal or six decimal places. The value is read from the uppermost horizontal inscription and demonstrates the greatest known computational accuracy obtained anywhere in the ancient world. It is believed that the tablet's author copied the results from an existing table of values and did not compute them himself. The contents of this tablet were first translated and transcribed by Otto Neugebauer and Abraham Sachs in their 1945 book, Mathematical Cuneiform Texts (New Haven, CT: American Oriental Society). More recently, this tablet was the subject of an article by David Fowler and Eleanor Robson [3], which provides insights into the probable methodology used to obtain such an accurate approximation for $\sqrt{2}$."

Figure I.3. Drawings after YBC 7289 in the Yale Babylonian Collection in New Haven, Connecticut. The diagonal displays an approximation of the square root of 2 in 4 sexagesimal figures, 1 24 51 10, which is good to about 6 decimal digits. 1 + 24/60 + 51/60² + 10/60³ = 1.41421296 . . . The tablet also gives an example where one side of the square is 30, and the resulting diagonal is 42 25 35 or 42.4263888.

Thus, we have a plentiful supply of evidence from Mesopotamia for geometrical diagrams long before the time of Thales, and even contemporaneous with him, and as we shall consider next, evidence for geometrical diagrams from Egypt, a location where we can certainly place him. The Babylonian evidence shows an awareness of the relationship between the side lengths of a right triangle. But it is my contention that the discovery of the hypotenuse theorem in the Greek tradition was a consequence of thinking through a *metaphysical* problem, of seeking to find the fundamental geometrical figure to which all rectilinear figures dissect—the right triangle—and we have no evidence that the right triangle played such a role in any metaphysical speculations from either Mesopotamia or Egypt.

D

Diagrams and Ancient Egyptian Mathematics: What Geometrical Knowledge Could Thales have Learned in Egypt?

Thales plausibly learned or confirmed at least three insights about geometry from his Egyptian hosts, and all of them involved diagrams: (1) formulas and recipes for calculating the area of rectangles and triangles, volumes, and the height of a pyramid (i.e., triangulation); (2) from the land surveyors, he came to imagine space as flat, filled by rectilinear figures, all of which were reducible ultimately to triangles to determine their area; (3) watching the tomb painters and sculptors, he recognized geometrical similarity: the cosmos could be imagined as flat surfaces and volumes articulated by squares, and each thing can be imagined as a scaled-up smaller version. The cosmos could be imagined—small world and big world—to share the same structure; the big world is a scaled-up version of the small world with which it is geometrically similar.

It is my purpose to set out a plausible case for what Thales could have learned by exploring some possibilities, but at all events I alert the reader to the speculative nature of my case. While the duration of time in which the surveyors worked, the architects built pyramids and temples, and the mathematics texts were produced spanned thousands of years, the surviving evidence from Egypt is exiguous; only a few papyri and rolls have so far been discovered. Clearly, there must have been much more evidence, but in its absence I can only explore some possibilities. Since I consider that some of the evidence suggests that geometrical problems were to be solved graphically, that is, visually, I explore the possibility that some of those problems might have been solved by using the red grid squares, which are plentifully evidenced in tomb painting and sculpture, but for which we have no surviving evidence in the mathematical texts.

Eudemus claims that Thales traveled to Egypt and introduced geometry into Greece. He never says that Thales learned all the principles of it there, nor does he claim that Thales learned the deductive method of proof from Egyptian sources. But Eudemus claims that Thales discovered many things about geometry, and in turn, investigated these ideas both in more general or abstract ways, and in empirical or practical ways:

Θαλῆς δὲ πρῶτον εἰς Αἴγυπτον ἐλθὼν μετήγαγεν εἰς τὴν Ἑλλάδα τὴν θεωρίαν ταύτην καὶ πολλὰ μὲν αὐτὸς εὗρεν, πολλῶν δὲ τὰς ἀρχὰς τοῖς μετ᾽αὐτὸν ὑφηγήσατο, τοῖς μὲν καθολικώτερον ἐπιβάλλων, τοῖς δὲ αἰσθητικώτερον.

Thales, who traveled to Egypt, was the first to introduce this science [i.e., geometry] into Greece. He discovered many things and taught the principles for many others to those who followed him, approaching some problems in a general way, and others more empirically.

First, Eudemus places Thales's travel to Egypt with his introduction of geometry into Greece. This suggests that he learned some geometrical knowledge in Egypt. The scholarly literature, even as recently as Zhmud's new study,[32] has downplayed such a connection. This is because Zhmud and others have come to regard "deductive proof" as central to Greek geometry, and to the role that Thales played in contributing to this direction; since no evidence survives in the Egyptian mathematical papyri that shows deductive strategies, the contribution of Egypt to Greek geometry has been reduced in apparent importance. Zhmud's objections to Eudemus's claim that geometry was learned and imported by Thales from Egypt are twofold: first, those like Thales had no way to communicate with the Egyptian priests (the Greeks never took an interest in learning foreign languages), and second, even if he could, there would be nothing for him to have learned about deductive reasoning since what we do have is only formulas and recipes for practical problem solving, without the explicit statement of any general rules to follow.[33]

I will argue the case in chapter 2 that Herodotus had it right when he claimed that the Greeks learned geometry from the Egyptians through land surveying, in addition to what they could have learned from priests and merchants about the contents of the mathematical papyri. Herodotus informs us that after the Milesians assisted Pharaoh Psammetichus (that is, Psamtik, c. 664–610) to regain his kingship and once again reunite upper and lower Egypt under a single Egyptian ruler, the grateful pharaoh allowed the Milesians to have a trading post in Naucratis, in the Nile delta; this location was very close to the pharaoh's capital in Sais.[34] Herodotus tells us that the pharaoh sent Egyptian children to live with the Greeks and so learn their language, to facilitate communication as their translators and interpreters.[35] Later, when Amasis ascended to the throne (c. 570), Herodotus tell us that these Greeks were resettled in the area of Memphis, that is, south of Giza, and served as his own bodyguard; Amasis needed to be protected from his own people.[36] Since the picture we get of the Greeks in Egypt shows them to be embraced warmly and gratefully by the pharaohs, when Thales came to Egypt it makes sense that he traveled with the continuing assistance and support of the Egyptian authorities, conversing with the Egyptian surveyors, priests, and merchants through these translators and interpreters.

In chapter 2 I will argue also that my sense of what happened in the originating stages of geometry in Greece is close to the one offered by van der Waerden in *Science Awakening*. In his estimation, Thales returned from Egypt and elsewhere with practical formulas and recipes and sought to prove them; this is how Thales, in his estimation, stands at the beginnings of geometry in Greece. What I add to van der Waerden's narrative is this: Thales returned to Miletus and shared with his compatriots what he learned from the surveyors, priests, and practical merchants who used the formulas and recipes contained in the Rhind Mathematical Papyrus (RMP) and Moscow Mathematical Papyrus (MMP) (both dating to the Middle Kingdom, c. 1850 BCE, that is, more than a thousand years before Thales came to Egypt). RMP problems 41–46 show how to find the volume of both cylindrical and rectangular based granaries; problems 48–55 show how to compute an assortment of areas of land in the shapes of triangles and rectangles; problems 56–60 concern finding the height or the *seked* (i.e., inclination of the face) of pyramids of a given square base. Thus the problems included formulas and recipes that showed how to divide seven loaves of bread among ten people (problem 4), how to calculate the volume of a circular granary that has a diameter of 9 and a height of 10 (problem 41), how every rectangle was connected inextricably to triangles that were its parts (problems 51 and 52), as problems for the land surveyors. His compatriots must surely have been as intrigued as they were skeptical. We can imagine members of his retinue asking: "Thales, how do you know that this formula is correct?" "How can you be sure?" And his replies were the first steps in the development of "proofs." I invite any teacher who has ever taught mathematics or logic to reflect on his or her own experiences; in the process of explaining countless times to many students how to figure out problems and exercises, the teacher comes to grasp ever more clearly the general principles and consequently deductive strategies—and so did Thales. The question of whether and to what extent Thales could have produced a formal proof, and thus the endless wrangling over whether and to what degree Thales invented, developed, and employed the *deductive method*, has bogged down the literature, fascinating though this question may be. The way to understand what happened in these earliest chapters is to ask, instead, what might have been found convincing by Greeks of the sixth century BCE, regardless of whether the lines of thought would meet the formal requirements that we later find in Euclid. From priests and practical people who divided food among the workmen and stored grain, Thales learned these formulas and recipes. Also in these mathematical papyri were formulas and recipes for determining the area of a triangle and a rectangle, calculating the height of a pyramid from the size of its base and the inclination (*seked*) of its face. It seems so plausible that Thales's measurement of a pyramid height by means of shadows was a response to learning how the Egyptians calculated pyramid height by their own recipes, and the ingenuity he brought to exploring afresh these long-held formulas.

The second part of Thales's instruction into geometry in Egypt came from the land surveyors; it is just this kind of knowledge that Herodotus alludes to.[37] Every year, after the Nile inundation, the surveyors carrying the royal cubit cord, supervised by a priest who wrote down and recorded the measurements, returned to each man who worked the land his allotted parcel; just this scene is captured in the tomb of Menna in the Valley of the Nobles. This means that each man had returned to him, by resurveying, a parcel of land of the same *area*. Taxes were assessed—the quantity of crops to be paid as tribute—in terms of the area of land allotted to each man.[38] It is my contention that, not infrequently, the land was so eviscerated by the surging, turgid flow of the Nile flood that parcels of land, customarily divided into rectangles (*arouras*), had to be returned in other shapes but with the same area. Through land surveying, the Egyptians came to understand that every rectilinear figure can be reduced to the sum of triangles. The evidence for this claim is suggested by the RMP problem 51 (and almost the same calculation of a triangular piece of land at MMP 4, 7, and 17) that offers instruction for the calculation of a *triangle of land*, and problem 52 for the calculation of a *truncated triangle of land*, a trapezium; here we have a window into just this reality of reckoning parcels of land that were not the usual rectangles, and the developing sense of areal equivalences between different shapes. It was here that Thales plausibly learned or confirmed equality of areas between geometrical figures of different shapes. When is a square equal in area to a rectangle? When is any polygon equal to a rectangle? The answer to all these questions of areal equivalence is driven by practical needs and shows that techniques of "land-measurement" (*geō + metria*) really do find application to the land itself; Thales was in a position to have learned or confirmed from the Egyptians *polygonal triangulation*—that every rectilinear figure can be divided into triangles, and by summing up the areas of the triangles into which every figure dissects, the area of every polygon can be reckoned. From the countless times the surveyors resurveyed the land, they developed an intuition for areal equivalences between shapes. I argue that the reduction of all figures to triangles is the quintessential metaphysical point that Thales learned or confirmed from Egyptian geometry, whether or not it was an explicit part of Egyptian teaching—the formulas relating triangles and rectangles show that it was implied.

The third thing that Thales plausibly learned or confirmed in Egypt is the idea of geometrical similarity. "Similarity" is the relation between two structures with the same shape but of different size; moreover, similarity was the principle that underlies and drives the microcosmic-macrocosmic argument, that *the little world and big world share the same structure.* Consequently, a vision burgeoned that the cosmos is a connected whole interpenetrated throughout by the *same* structure, and by focusing on the little world, we get deep and penetrating insights into the big world of the heavens and the cosmos itself that would otherwise be inaccessible. Whether the Egyptian tomb painters and sculptors grasped this principle and could articulate it, we cannot say with confidence, but their work displays it. And Thales might well have taken a lesson in geometrical similarity by watching them work.

We have abundant evidence left by ancient Egyptian tomb painters for the use of square grids, made with a string dipped into red ink, pulled taut and snapped against the wall,[39] and some ruled against a straightedge. The evidence is robust from as early as the Third Dynasty through Ptolemaic times. Below, left, we have the red gridlines still visible from the New Kingdom tomb of Ramose in the Valley of the Nobles; below, right, we have a figure on its original grid dating to the Saite period, contemporaneous with time of Thales's visit to Egypt.[40]

 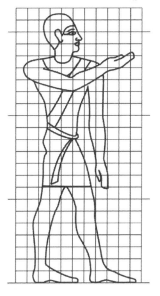

Figure I.4.

16   The Metaphysics of the Pythagorean Theorem

Robins also includes an example of Pharaoh Apries (c. 589–570 BCE) standing between two deities with the original grid. Thus, within the frame of time that Thales visited Egypt, the grid system for painting was still in use.[41]

There has been debate about just how the grids were used; they may have been used to make a small sketch on papyrus or limestone, for example, and then, a larger square grid might have been prepared on the tomb wall and the painter would transfer the details in each little square to the larger squares; that is, the tomb painting would be a scaled-up version of the smaller sketch with which it was geometrically similar. "Both primary and secondary figures on grids may frequently have been the work of apprentices attempting to follow the proportion style of the masters; the grid was not only a method for design but also one for teaching the art of proportion by means of workshop production."[42] This idea is represented in the diagram directly below, after the tomb of Menna; here we have depicted the surveyors with the measured cord and the scribe (with ink block in hand) to record the results:

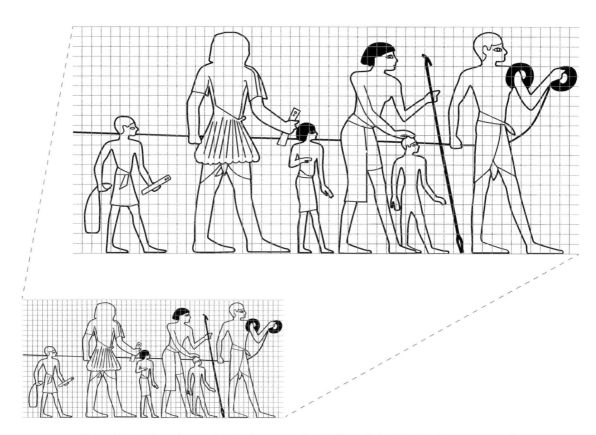

Figure I.5. After the tomb of Menna in the Valley of the Nobles (Luxor, Egypt).

We do have surviving examples of small sketches with red grids, such as an ostrakon (like that of New Kingdom Senenmut (below left),[43] or white board (below, right),[44] guiding the draftsman to produce human figures in appropriate proportions.

Figure I.6. Limestone shard of Senenmut with red grid lines (l), and white board with red grid lines (r). Artifacts owned by the British Museum.

Whether working on a small sketch or large tomb wall, both square grids would follow established rules of proportion, though the rules seem to have changed throughout the dynasties. Consider the illustration, below, where the rules established in the Older Canon are displayed: for standing figures 18 squares from base to the hairline, and proportional rules for seated figures.[45] In the New Kingdom, 19 grid squares to the neck and 22.5 to the top of the crown.[46]

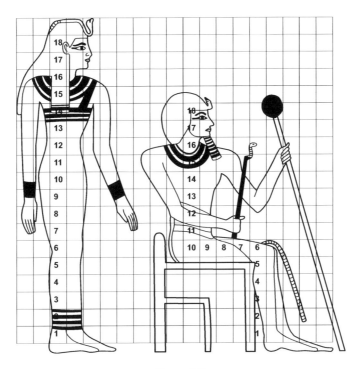

Figure I.7.

Some doubts have been expressed about whether working from a small sketch was a routine practice for the tomb painters, because no such copybooks have survived. The small sketches, such as the ones on an ostrakon and white board, might just as well have been a draftsman's practice piece or a model for an apprentice to copy.[47] But the important point underlying either interpretation is that space was imagined as flat and dissectible into squares, and the large and small world could be imagined as sharing the same structure. Any image could be scaled up, made larger, by increasing the size of the squares. The images would be identical because the proportions would remain the same, and the angular details of each square—smaller and larger—would correspond exactly. Insights into the big world could be accessible because the small world was *similar*; it shared the same structure.

What seems unambiguous is that the purpose of proportional rules for human figures was to render them in appropriate, recognizable forms. To do so, grids of different sizes were often prepared, but the proportional rules were followed.[48] Sometimes, a single large grid was placed over the tomb scene to accommodate the king, gods, vizier, or owner of a small tomb. Those figures were largest, and smaller figures, such as wives or workmen, were scaled down; they were painted in smaller size but retained comparable proportions to insure that they too were immediately recognizable. Sometimes, however, different-sized grids were used in the same scene or wall to represent the varying importance of the individuals depicted. Thus, the tomb painters imagined space as flat surfaces dissectible into squares, and the large world was projected as a scalable version of the small world. Since all squares divide into isosceles right triangles by their diagonals, a fortiori, Thales might well have come to a vision of the large world built up out of right triangles; the measurement of the pyramid height by its shadows focused on right triangles. Also, we see the same grid technique applied to sculpture. As Robins expressed it, "There can be no doubt that Egyptian sculptors obtained acceptable proportions for their figures by drawing squared grids on the original block before carving it."[49] When three-dimensional objects are also shown to be dissectible into squares, and thus triangles, a vision has been supplied of solid objects being folded up from flat surfaces. So, for a Thales measuring the height of a pyramid by its shadows at the time of day when every vertical object casts a shadow equal to its height, Thales had to imagine the pyramid and its measurement in terms of an isosceles right triangle, and the pyramid is easily imagined as four triangles on a flat surface folded up.

Let us think more about the geometrical diagrams Thales might have seen in Egypt. When Peet turns to examine the problem at MMP 6, the reckoning of a rectangle one of whose sides is three-quarters that of the other, he insists that the solution is to be found "graphically," although the papyrus does not contain such a diagram. He conjectures the following diagram:[50]

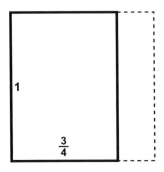

Figure I.8.

First, the figure is imagined as a square, that is an equilateral rectangle, and then problem 6 proceeds to compare the sides, reducing one side by one and one-third to find the desired rectangle. The main points to emphasize for our purposes are that the solution is imagined graphically, and that the solution is reached through construction of a diagram. The answer is found visually, and the uneven rectangle is reckoned by comparison with an even-sided rectangle, that is, a square.

Thus, we have mathematical problems that concern reckoning pyramid height, the very problem that Thales is credited with solving by measuring shadows and not by calculation. We should note that the pyramid problems in the RMP are, per force, problems involving triangulation. And to solve these problems—calculating the inclination of the pyramid's triangular face (i.e., *seked*) given the size of its base and height, or calculating the pyramid's height given the size of the base and its *seked*—the visual components are to set a triangle in comparison to its rectangle and square. In the RMP dealing with triangular areas (problem 51) the triangle is placed on its side, and its "mouth" is the side opposite the sharpest angle. It seems as if here, as in other area problems, the diagram is "schematic"; just as in the medieval manuscripts of Euclid and other mathematics texts, the diagram is not meant to be a literal, specific rendition of the figure in the problem, but only to suggest it more generally. Thus, historians of mathematics are forever debating the extent to which the figure matches the problem and whether or not this is a special case or just represents a range of cases, as here with triangles. Despite the earlier controversies, there is now general consensus that what had been a controversy concerning the formula for the area of a triangle has now been laid to rest: one-half base x height. Problems 49, 51, and 52 deal with the areas of rectangular and triangular pieces of land. Robins and Shute point out that the rough sketches have the triangles on their sides, not upright, as would be more familiar to us today.[51] Here is problem 51, below:

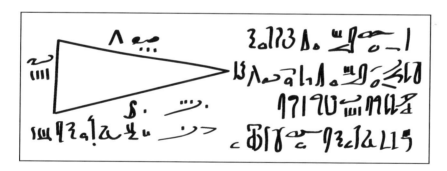

Figure I.9. After the Rhind Mathematical Papyrus (RMP) problem 51 in the British Museum.

Problem 51:[52]

> "Example of reckoning a triangle of land. If it is said to thee, A triangle of 10 khet in its height and 4 khet in its base. What is its acreage?
>
> The doing as it occurs: You are to take half of 4, namely 2, ***in order to give its rectangle***. You are to multiply 10 by 2. This is its acreage.
>
> | 1 | 400 | 1 | 1,000 |
> |---|-----|---|-------|
> | 1/2 | 200 | 2 | 2,000 |
>
> Its acreage is 2."

20   The Metaphysics of the Pythagorean Theorem

Here '2' must mean 2 thousands-of-land. (We should have expected 20 khet, but the answer '2' is in the unit of "thousands-of-land" which equals 20 khet.)[53] The striking phrase, as Peet points out, is "*This is its rectangle*" (*r rdj.t jfd.s*). Imhausen translates the sentence differently (into German): "*um zu veranlassen, dass es ein halbes Rechteck ist.*" "[Then you calculate the half of 4 as 2] so that it is half a rectangle."[54] Clagett translates "Take 1/2 of 4, namely, 2, in order to get [one side of] its [equivalent] rectangle."[55] But whichever translation we take, the main point is that the triangle's area is grasped in the context of its rectangle, with which it is inseparable. Peet argues that this problem, too, is meant to be solved graphically, and because it is not clear whether the diagram is schematic or a special case, he wonders how it might have been solved visually. Thus, he suggests a few possible diagrams of how it might have been imagined, but the main point for our purposes is to highlight that the triangle is grasped interdependently with its rectangle. The triangle was understood as a figure inseparable from the rectangle and the rectangle was understood inseparably from its square. And these interrelations of triangle, rectangle, and square show what Thales plausibly could have learned or confirmed in Egypt.

Peet's first conjecture about how to understand the problem visually is to try to sort out the technical terms in hieroglyphics to reach the view that this, below, is the likely diagram of the problem.[56]

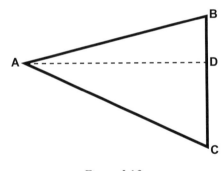

Figure I.10.

And then, since the problem explicitly identifies the triangle in the context of a rectangle—problem 51 includes the striking phrase: "This is its rectangle"—Peet speculates how the triangle might have been presented visually in the context of that rectangle, below.[57]

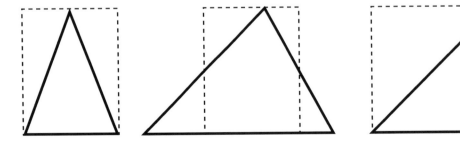

Figure I.11.

Acknowledging that all this is speculation, let us consider another way that it might have been presented diagrammatically, below.[58] Since the preparation of a square grid was familiar in tomb painting and sculpture, let us place the diagram on a grid to explore its visual solution. We can recall the famous story told by Diodorus of the two Samian sculptors of the sixth century BCE, each of whom made half a statue in different geographical locations—one in Egypt and the other in Samos—following the Egyptian grid system; when the two halves were finally brought together they lined up perfectly.[59] So, since in both tomb painting and sculpture, the grid system was employed, and its knowledge is confirmed by the Greeks from Samos contemporaneously with Thales, I begin with the square grid for this diagram. And since it should now be clear that the relations between triangle, rectangle, and square were inextricably interwoven, I follow through, as does Peet, with some efforts to represent these problems for the visual solutions that he recognizes were probably de rigueur. The whole square of 10 khet is equal to 100 setat, each made up of 100 cubits-of-land, and thus each strip 1 khet wide and 10 khets long is 10 setat of 100 cubits-of-land, and thus 1 strip is 1 thousand-of-land (1 = 1,000), and 2 strips is 2 thousands-of-land (2 = 2,000). Thus, the correct answer is the *triangle's rectangle*, that is, 2. And visually, the answer is immediately apparent.

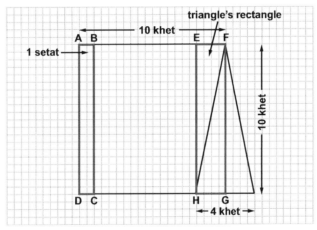

Rectangle ABCD = 1 thousand-of-land (= 1 setat)
Rectangle EFGH = 2 thousands-of-land (= 2 setat)

Figure I.12.

In problem 52, the challenge is to reckon the area of a truncated triangle of land, or trapezium:[60]

> "Example of reckoning a truncated triangle of land. If it is said to thee, A truncated triangle of land of 20 khet in its height, 6 khet in its base and 4 khet in its cut side. What is its acreage.
>
> You are to combine its base with the cut side: result 10. You are to take a half of 10, namely 5, **in order to give its rectangle**. You are to multiply 20 five times, result 10 (sic). This is its area.

The doing as it occurs:

1	1,000	1	2,000
1/2	500	2	4,000
		4	3,000
			Total: 10,000
			Making in land 20 (read 10)

This is its area in land."

Again, while Peet translates "You are to take a half of 10, namely 5, **in order to give its rectangle**" Clagett translates "Take 1/2 of 10, i.e., 5, in order to get [one side of] its [equivalent] rectangle."[61] Regardless of the translation it seems clear that the triangle is understood to be connected inextricably with its rectangle, of which it is half. The rectangle dissects into triangles; the triangle is the building block of the rectangle. Peet conjectured that this problem, too, was to be solved graphically.

I emphasize again that these problems show how *land* in triangular and truncated triangular shapes could be reckoned. These show plausibly the conceptual principles behind the practical work of surveyors, who sometimes, after the annual flood, had to return to the man who worked the land a plot of equal area but in a different shape. Guiding the calculation of areas of triangular and truncated triangular plots shows the kinds of flexibility needed since all rectangles reduce ultimately to the summation of triangles.

In the pyramid problems in the RMP, we are instructed how to find the inclination of the *seked* (that Peet calls the "batter") given the size of the base and its height, and how to find the height given its base and *seked*. In problem 56, the pyramid's base is 360 and its height is 250, and the answer is given in palms—the *seked* is 5 1/25. As Peet notes for problem 56, "The Egyptian was doubtless aware that the measurement he proposed to find, being a ratio and not a length, was independent of the unit of the original dimension."[62] Stated differently, the evidence shows that for more than a millennium before the time that Thales came to Egypt, Egyptian mathematics dealt familiarly with ratios and proportions. In both problems 56 and 57, it is assumed that 1 cubit = 7 palms, and that 1 palm = 4 fingers.

Problem 56:[63]

"Example of reckoning out a pyramid 360 in length of side and 250 in its vertical height. Let me know its batter (= *seked*).

You are to take half of 360: it becomes 180.
You are to reckon with 250 to find 180.
    Result 1/2 + 1/5 + 1/50 of a cubit
A cubit being 7 palms, you are to multiply by 7:

1	7
2	3 1/2
1/5	1 1/3 + 1/15
1/50	1/10 + 1/25

Its batter (= *seked*) is 5 1/25 palms."

In the visual presentation of the diagram, below, the pyramid's triangle is placed on a grid and its rectangle and square are shown.[64] Thus, we can see that the problem requires us to see just how many palms are in the square's base of 250. This is achieved by dividing 7 into 250, producing an answer of 35.71 units in 1 palm. Since we know that the baseline of the triangle's rectangle occupies 180 of the square's 250 units, and since we are required to discover how many of the 7 palms are occupied by 180 (to discover the *seked*): 180/35.71 = 5 palms and an excess of 1.45 units, since 5 x 35.71 = 178.55. Thus the stated answer of 5 1/25 palms can be seen graphically, that is visually.

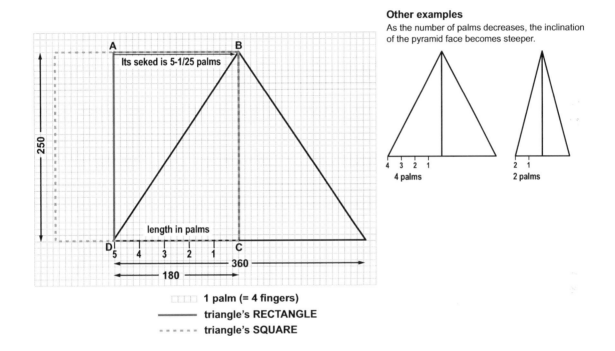

Figure I.13.

Finally, we come to problem 57 reckoning the height of a pyramid given the size of the base and *seked*:

> "A pyramid 140 in length of its side, and 5 palms and a finger in its batter (= *seked*).
> What is the vertical height thereof?
>
> You are to divide one cubit by its batter doubled, which amounts to
> 10 ½. You are to reckon with 10 1/2 to find 7, for this is one cubit.
>
> Reckon with 10 1/2 : 2/3 of 10 1/2 is 7.
> You are now to reckon with 140, for this is the length of the side:
>   Make two-thirds of 140, namely 93 1/3. This is the vertical height thereof."

24   The Metaphysics of the Pythagorean Theorem

To imagine the problem and its solution visually,[65] we again place the triangle face of the pyramid on the grid and identify its rectangle and square. In this case, the length of the square's base (not the rectangle) will be the same as the pyramid's height since, visually speaking, we can see that it is a side of the square. The baseline of the square is 93 1/3 cubits, and this must be the answer to the problem because the base line is also the height of the pyramid.

The base of the triangle is 140. The triangle must be isosceles—indeed, whenever the base is given and we are instructed to divide the base in half, the triangle *must* be isosceles—otherwise we would have been given two different measurements for the *seked*; the *seked* is 5 palms and 1 finger which is 5 1/4 palms since there are 4 fingers in each palm. Already from the previous problem we know that the inclination of the side of a triangle (= its *seked*) can be reckoned by comparing the base line of the triangle's rectangle with that of its square. So, here we have half the pyramid's base line as the short side of its rectangle, and this is 5 1/4 palms; the baseline of the square is 7 palms, and so the problem is to calculate this on the basis of the rectangle's baseline. Once we come to discover the square's baseline, a square can be constructed, and once we have done that we will know the pyramid's height that is the solution to the problem, thus:

```
5 1/4 palms   =   70 units (being half the triangle's base of 140)
2 1/4 palms   =   70 units
1/4 palm      =   70/21 = 3 1/2 units
1 palm        =   4 x 3 1/2 = 13 1/3 units
7 palms       =   93 1/3 units
```

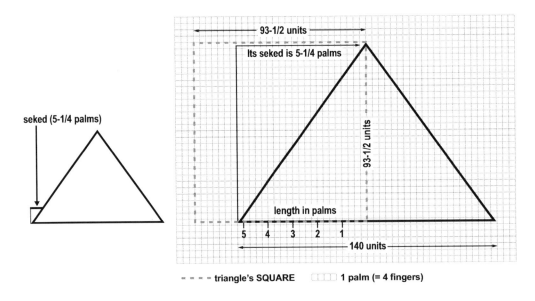

Figure I.14.

Thus, the Egyptian mathematical formulas practically applied in everyday life, polygon triangulation of land resurveying, geometrical similarity revealed by the grid technique of tomb painting and sculpting, and the inextricable connection of triangle, rectangle, and square, were the kinds of things that probably inspired, educated, challenged, or possibly just confirmed what Thales knew. The resolution of the problems visually requires us to imagine the routine production of diagrams. These are parts of Herodotus's report that the Greeks learned geometry from Egypt and that, according to Eudemus, Thales introduced geometry into Greece having traveled to Egypt. Minimally, there were some geometrical diagrams in Egypt for which we have evidence. There can be no doubt that there were many more, though about how they may have looked we can only conjecture. But certainly, here was one resource that plausibly inspired Thales's diagram-making.

E

Thales's Advance in Diagrams Beyond Egyptian Geometry

Eudemus claims not only that Thales traveled to Egypt, but that he achieved specific geometric results: our ancient reports suggest variously that Thales noticed (ἐπιστῆσαι) and stated (εἰπεῖν)[66] the base angles of an isosceles triangle are equal, demonstrated or proved (ἀποδεῖξαι) that a diameter bisects a circle,[67] and discovered (εὑρημένον) that if two straight lines cut each other the vertical opposite angles are equal.[68] Moreover, Proclus tells us that Thales discovered (εὗρεν) many others himself and taught the principles for many others to his successors (τὰς ἀρχὰς τοῖς μετ' αὐτὸν ὑφηγήσατο), approaching some empirically (αἰσθητικώτερον) and some more generally (καθολικώτερον).

On the authority of Eudemus, reported by Proclus, then, the origins of geometry for the Greeks begin with Thales. As Burkert summarized the report:

> Eudemus gives detailed reports about mathematical propositions, proofs, and constructions that Thales was supposed to have discovered, he distinguishes between tradition and deductions of his own, even recording an archaic locution used by Thales. This implies that there was a book, available to him or his authority, ascribed to Thales. The book in question must be that *On Solstices and the Equinox*. Whatever the situation may be with regard to authenticity, there obviously existed, in the sixth century, Ionic technical writings on problems of astronomy and the calendar, already with them, geometrical concepts—circles and angles—seem to take the place of Babylonian calculation. . . . With Thales the point is still a graphic or perceptible "showing" (δεικνύναι). But in the perspicuity itself there is a new element by contrast with the Babylonian "recipes." It is in the perceptible figure that mathematical propositions become clear in all their generality and necessity: Greek geometry begins to take form.[69]

I quote Burkert here at length because it is his work that galvanized a consensus to reject the connection of "Pythagoras" with the famous theorem that bears his name. Burkert's argument against Pythagoras and the theorem is that the reports that connect Pythagoras with the hypotenuse theorem, none of which he contends can be traced back securely to the fifth or fourth centuries, are too late to be reliable. But Burkert can connect geometrical knowledge

26　The Metaphysics of the Pythagorean Theorem

to Thales on the authority of Eudemus: Thales demonstrated (ἀποδεῖξαι), notices (ἐπιστῆσαι), states (εἰπεῖν), and discovered (εὑρημένον) things about geometry. Eudemus infers that Thales must have known that triangles are equal if they share two angles, respectively, and the side between them (= Euclid I.26) because it is a necessary presupposition of the method by which he measured the distance of a ship at sea, that Eudemus accepted as an accomplishment of Thales (Εὔδημος δὲ ἐν ταῖς γεωμετρικαῖς ἱστορίαις εἰς Θαλῆν τοῦτο ἀνάγει τὸ θεώρημα. Τὴν γὰρ τῶν ἐν θαλάττῃ πλοίων ἀπόστασιν δι' οὗ τρόπου φασὶν αὐτὸν δεικνύναι τούτῳ προσχρῆσθαί φησιν ἀναγκαῖον).[70]

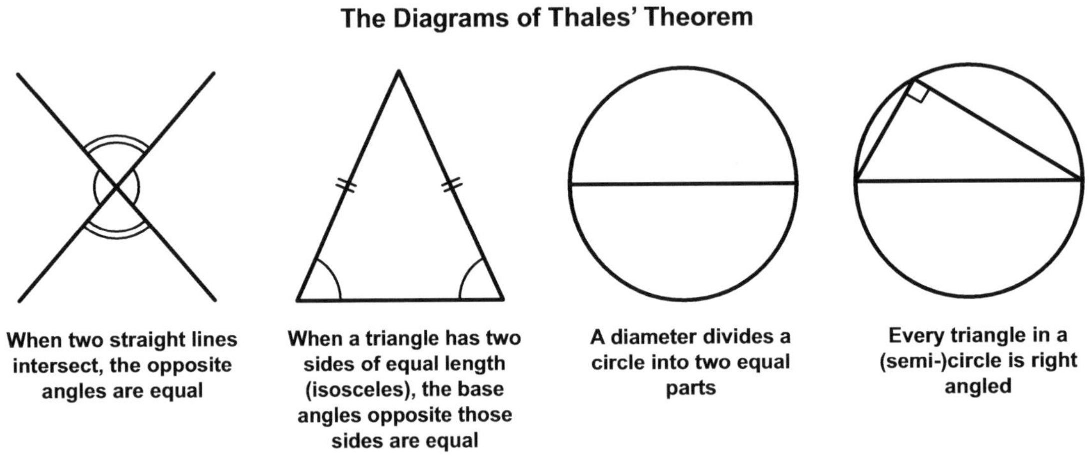

The Diagrams of Thales' Theorem

| When two straight lines intersect, the opposite angles are equal | When a triangle has two sides of equal length (isosceles), the base angles opposite those sides are equal | A diameter divides a circle into two equal parts | Every triangle in a (semi-)circle is right angled |

Figure I.15.

There is no doubt that the evidence is meager, but Eudemus had some written report that connected Thales to geometrical knowledge, and he filled in the blanks by his own understanding. Those who deny Thales knowledge of the principle of Euclid I.26 on the grounds that Eudemus merely infers it, and so clearly did not find such a claim in any ancient report to which he had access, have almost certainly got it wrong. When you begin to make practical diagrams and empirical tests of the achievements credited to Thales, and place them together, it seems quite clear that Eudemus had it just right, whether or not Thales stated his understanding as Euclid did later. Thus I contend that Thales understood equal and similar triangles and the relations between side lengths and their corresponding angles. My argument, presented in detail in chapter 2, is that this knowledge is both plausible and striking when one actually goes to Egypt and tries to measure the height of a pyramid by its shadows and stands near Miletus and tries to measure the distance of a ship at sea, as I have done with my students. My recommendation is not to simply talk about it but rather to go and do it; then we will be in a better position to think again about what Thales plausibly knew.

From the evidence preserved thanks to Eudemus, Thales was making diagrams—he must have, if we concede that proving anything at this early stage meant that equivalences and similarities were shown by making them *visible*. The meaning of a "proof" is literally expressed by the Greek term δείκνυμι—to "make something visible, to show or reveal." This early technique is referred to as an ἐφαρμόζειν proof—superposition—literally placing one thing on top of the other to show equivalences.⁷¹ This is just how the ancient tile standard works in the southwest corner of the Athenian Agora, which we return to in chapter 2; to test whether or not the tile you had been sold was of standard size, you brought your tile to the standard, literally picked it up and rotated it into place on top of that standard (ἐφαρμόζειν), and if the sides and angles lined up, here was proof of their equality:

Figure I.16.

Burkert accepts the view expounded earlier by von Fritz⁷² and Becker⁷³ and emphasized also by Szabó about ἐφαρμόζειν and δείκνυμι—the proof of something, at first, meant to make equivalences visible.⁷⁴ When Eudemus reported that Thales notices (ἐπιστῆσαι) and states (εἰπεῖν) that the angles at the base of an isosceles triangle are equal but uses the archaic expression "similar" (ὁμοίας—ἀρχαικώτερον δὲ τὰς [γωνίας] ὁμοίας προσειρηκέναι), it is clear that Eudemus had access to some written report, whether or not it was by Thales or merely attributed to him. Burkert is certain that this book had the title Περὶ τροπῆς καὶ ἰσημερίας; a book with such a title suggests the knowledge of the solstices and equinox, that is a solar calendar. And this seems right because had Thales measured the height of a pyramid at the time of day when the shadow length was equal to its height—as is reported by both Diogenes Laertius⁷⁵ and Pliny,⁷⁶ and as we discuss at length in chapter 2—he had to have made a solar calendar. I have made this measurement with my students a few times on the Giza plateau, and I can report that a measurement by this technique could only have been made successfully on very few days out of the whole year, and it would be as pointless as it would be preposterous to claim that it just so happened as a matter of luck or sheer coincidence that Thales showed up at just the right time on the right day. That Thales's younger contemporary, Anaximander, made a seasonal sundial—a sundial that indicated the solstices and equinoxes⁷⁷—fits seamlessly in the picture that both Thales and Anaximander had experience watching and measuring shadows cast by vertical objects (gnomons), had noticed and connected the various triangular and circular shapes that were displayed, and plausibly made diagrams to convince themselves and others of what they had discovered. This kind of "showing" by diagrams is, of course, both an exhibition of early stages of proving, and by connecting the steps, an early exhibition of deductive thinking. But focusing on "deduction" and the questions about how Thales may have learned it from Egypt or elsewhere leads us away from matters more relevant to these earliest accomplishments in geometry. What the reader should keep in mind is that, in these earliest stages, geometry revealed principles underlying our everyday experience of the world; it introduced a background of a stable kind of knowledge that, surprisingly, the world of divergent and ephemeral appearances did not, and yet this knowledge was relevant and applicable to that world. It was only because the first diagrams were *practical* that they could be applied to our world of experiences, and that the idea evolved that through recourse to geometry we could explain an intelligible structure that underlies and illuminates our world of experiences—that is, that *there was some underlying unity of all things*. Now it is hard to say whether Thales had the insight, the intuition, that there was an underlying unity underlying all differences in appearances, or whether it was through his geometrical investigations that he came to have the vision of an underlying unity. If Thales began with the metaphysical supposition that there was a single unity underlying all disparate appearances, he then sought and discovered in geometry a way to express how that unity alters without changing; geometry offered a way to describe and explain *transformational equivalences*—the structure of the alteration of appearances. Fiery, airy, watery, and earthy appearances—and all the multifarious combinations among them—are only altered forms of this basic unity; each is the same underlying unity whose structure is organized differently in space. This project was central to Thales's metaphysics.

Besides Eudemus, Aristotle's pupil, who is our foundational source of Thales's originating efforts in geometry, there is Aristotle himself, our foundational source about Thales's originating efforts in metaphysics; it is clear that Aristotle had access to some written report, probably by Hippias of Ellis, possibly Hippon of Samos, that informed him that Thales was the founder of philosophy because he identified a single underlying principle—ὕδωρ—an ἀρχή, out of which everything first comes and appears (ἐξ οὗ γίγνεται) and back into whose form everything returns upon dissolution (εἰς ὃ φθείρεται τελευταῖον);[78] there is no real change, only alteration of the primary stuff. Thus the substance remains but changes in its modifications: τῆς μὲν οὐσίας ὑπομενούσης, τοῖς δὲ πάθεσι μεταβαλλούσης.[79] To say that the substance remains but changes in its modifications is to deny real change itself—the multifarious appearances we experience are only *alterations* of this primary, self-subsisting unity. Thales called this substance ὕδωρ: while it remains the same, its appearance is altered. Nothing is either generated or destroyed since this sort of entity is always preserved: καὶ διὰ τοῦτο οὐδὲ γίγνεσθαι οὐθὲν οἴονται οὔτε ἀπολύσθαι, ὡς τῆς τοιαύτης φύσεως ἀεὶ σωζομένης.[80]

I take these claims by Eudemus and Aristotle to be fundamental to any narrative we construct about what Thales knew, speculative though such a narrative must be. If one grants this beginning, I invite my readers to consider this: It is impossible to believe that Thales announced to his compatriots that despite our experience, which is filled with differing appearances and apparent change, a single permanent unity underlies all these differing appearances—ὕδωρ—

that alters without changing—it is impossible to imagine Thales making this announcement without a member of his retinue, of his "school," standing up and saying: ὦ βέλτιστε, πῶς κάνε ("My good friend, How does this happen?"). And Thales's answer—the narrative connecting the metaphysical and geometrical themes to Thales—I am claiming has been lost but is echoed in Plato's *Timaeus* at 53Cff, which I discuss in chapter 3. This discussion, in a dialogue that certainly contains Pythagorean ideas, plausibly traces back to Pythagoras himself, but the original idea was proposed by Thales: the construction of the cosmos out of right triangles. Pythagoras learned this idea and project from Thales or a member of his school: Geometry explains the structure in terms of which this one unity alters without changing. Stated differently, we could say that geometry reveals the structure and rules for expressing transformational equivalences between appearances; all appearances are made from the same permanent unity, altered only in structure. To tell this story, Thales needed to discover the basic or fundamental geometric figure. He did; the figure is the right triangle. Let us consider the circumstantial case for this discovery and the evidence for it.

While credit for *proving* that there are two right angles in every triangle is attributed explicitly to the Pythagoreans by Proclus,[81] there has been conjecture about how it may have been known to Thales before them.[82] Allman, Gow, and Heath point out that Geminus stated that "the ancient geometers' observed the equality of two right angles in each species of triangle separately, first in equilateral, then isosceles, and lastly in scalene triangles, and that the geometers older than the Pythagoreans could be no other than Thales and members of the Ionic school."[83]

> ὥσπερ οὖν τῶν ἀρχαίων ἐπὶ ἑνὸς ἑκάστου εἴδους τριγώνου θεωρησάντων τὰς δύο ὀρθὰς πρότερον ἐν τῷ ἰσοπλεύρῳ καὶ πάλιν ἐν τῷ ἰσοσκελεῖ καὶ ὕστερον ἐν τῷ σκαληνῷ οἱ μεταγενέστεροι καθολικὸν θεώρημα ἀπέδειξαν τοιοῦτο: παντὸς τριγώνου αἱ ἐντὸς τρεῖς γωνίαι δυσὶν ὀρθαῖς ἴσαι εἰσίν.[84]
>
> In the same way, indeed, whereas the ancients (τῶν ἀρχαίων) theorized about each kind (εἴδους) of triangle separately that [it contained] two right angles, first in the equilateral [triangle] then in the isosceles [triangle] and later in the scalene [triangle], the moderns [οἱ μεταγενέστεροι, lit. the born-after] proved this general theorem: in all triangles the angles inside are equal to two right [angles].[85]

We shall review this matter in diagrammatic detail in chapter 2, but central to it is the deep and inescapable inference from this investigation that every triangle reduces ultimately to right triangles. From the Egyptians, Thales may well have first learned or confirmed that every polygon can be reduced to triangles to express its area—every polygon reduces to triangles. And from this exploration by the ancient geometers, Thales discovered that there were two right angles in every species of triangle and, coextensively, two right triangles in each triangle; this meant ultimately that every triangle reduces to a right triangle. This was one half of the metaphysical answer Thales was seeking, and it proved to be half of what the Pythagorean theorem meant to him.

30 The Metaphysics of the Pythagorean Theorem

First, there are six equilateral triangles that surround any point and fill up the space around it, and hence three equilateral triangles (we can see this immediately) above a diameter.

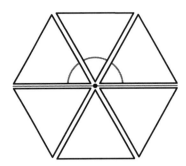

Figure I.17.

Now the sum of three of these angles around any point sums to a straight angle (= line), which is thus equal to two right angles. Since every equilateral triangle has three *equal* angles, then just as the three angles of the three equilateral triangles around the point sum to two right angles, so does the sum of the three equal angles in each equilateral triangle. Thus, there must be two right angles in each equilateral triangle.

Next, by dividing a square into two isosceles right triangles, the angle of each triangle summed to two right angles, since every square contained four right angles and the diagonal cuts the square in half.

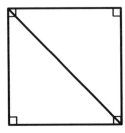

Figure I.18.

Once we are clear that every right triangle contains two right angles, and every triangle divides into two right triangles by dropping a perpendicular from the vertex to the opposite side, then not only does every triangle contain two right angles, but also every triangle divides into two right triangles ad infinitum, by continuing to drop a perpendicular from the right angle to the hypotenuse, shown directly below. So, whether we start with an equilateral triangle, an isosceles triangle, or a scalene triangle, then, by dropping a perpendicular from the vertex to its base, the summing of its angles to two right angles could be confirmed by completing its rectangle. Each rectangle contains four right angles and so by dividing each rectangle into two triangles, each triangle is revealed to have two right angles. Thus, had the investigations that there were two right angles in each species of triangle by Thales and his school—as Geminus reports—been carried out by this process (pictured below), it would have revealed at the same time that there are two right-angled triangles in every triangle, and each right-angled triangle contains two right angles. Here, then, we have the Greek version of the RMP 51 and 52 dictum: "There is the triangle's rectangle."

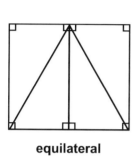
equilateral

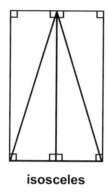
isosceles

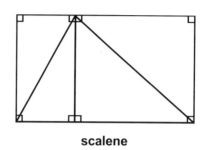
scalene

Figure I.19.

The grasping that there were two right angles in every triangle by showing, ultimately, that every triangle divides into two right triangles, was of great moment. This point was emphasized by Allman, Gow, and Heath. As Gow expressed it, "Hence he [Thales] could not have failed to see that the interior angles of a right-angled triangle were equal to two right angles, and *since any triangle may be divided into two right-angled triangles, the same proposition is true of every triangle*."[86] From these arguments, the most important point, often underemphasized if mentioned at all, is that *every triangle contains within it two right triangles*, not just that every triangle contains two right angles.

Every rectilinear figure reduces to triangles, and every triangle reduces to right triangles. The right triangle is the basic fundamental figure into which all other rectilinear figures dissect. This is arguably part of what was implied when Geminus claimed that *"the ancient geometers" observed the equality of two right angles in each species of triangle separately, first in equilateral, then isosceles, and lastly in scalene triangles*. The ancient geometers, before the time of the Pythagoreans, could only mean Thales and his retinue. But to see it, we must make the diagrams ourselves. It is my thesis that this process of exploring that there were two right angles in every triangle was central to the train of ideas that led Thales to discover the "Pythagorean theorem" by similarity—by ratios and proportions—and as we will explore in chapter 2, it followed decisively from his reflections on proving that the (tri)angle in the (semi-)circle was always right, which Diogenes Laertius reports from Pamphila.[87] Heath expressed the whole situation very well: "We have seen that Thales, if he really discovered that the angle in a semicircle is a right angle, was in a position, first, to show that in any right-angled triangle the sum of the three angles is equal to two right angles, and then, by drawing the perpendicular from a vertex of any triangle into two right-angled triangles, to prove that the sum of the three angles of any triangle whatever is equal to two right angles."[88] Thus, the point to be emphasized here

32   The Metaphysics of the Pythagorean Theorem

is that the plausible sequence of reasoning that allowed Thales to discover that every triangle contained two right angles was the very same sequence that allowed him to discover that every triangle reduces to the right-angled triangle, and thus that the right triangle was the fundamental geometrical figure.

The other part of the areal interpretation of the Pythagorean theorem that I am arguing Thales plausibly knew is that, while the right-angled triangle is the basic building block of all other figures because every figure reduces ultimately to it, when the right triangle expands and unfolds into larger and larger combinations and hence larger and larger appearances, or collapses into smaller and smaller right triangles, it does so in a pattern. The pattern became recognized as a continuous proportion. The idea of continuous proportions—or *mean proportions*—we examine at length in chapters 2 and 3. The proof of it is to show the areal equivalences drawn on its sides—to show when and that a square is equal to a rectangle. And this is central to the proof of the Pythagorean theorem, not at Euclid I.47 but rather at VI.31, the proof by ratios, proportions, and similar triangles. This was the line of thought by which Thales plausibly came to grasp the hypotenuse theorem.

F

The Earliest Geometrical Diagrams Were Practical:
The Archaic Evidence for Greek Geometrical Diagrams and Lettered Diagrams

I now turn to consider that practical diagrams were familiar to the Greek architects/engineers and to at least some of those working at monumental temple building sites long before the middle of the fifth century. Though the evidence that survives is meager, it provides an enormous window into the use of practical diagrams. When one sees that there is evidence in Greece already from the middle of the seventh century showing geometrical diagrams instructive for architectural purposes, and that lettered geometrical diagrams can reasonably be inferred as early as the middle of the sixth century, then, since Callimachus attributes to Thales the making of geometrical diagrams, and Eudemus testifies to them indirectly by attributing theorems to him, there can be no good grounds to deny that Thales was making geometrical diagrams. To see Thales making geometrical diagrams, we must set him, at least in the very beginning of his reflections on geometrical matters, in the context of those making practical diagrams. Moreover, having placed Thales in Egypt to measure the height of a pyramid, their long-established traditions could have shown him how diagrams were part of the discussion and solution of geometrical problems.

It had been thought, along the lines of Netz's approach, by those such as Heisel[89] that there were no surviving architectural drawings before Hellenistic times. Clearly, he did not yet know the drawings from the Artemision, published by Ulrich Schädler.[90] Schädler found geometrical sketches on tiles that had been burned in a fire, and so preserved, from the roof of an early temple of Artemis in Ephesus. The tiles date to the middle of the seventh century BCE, that is, before the time of Thales's flourishing. The drawings had been incised into wet clay.[91]

The concentric circles on the tile, below right, are a painted decoration of the edge of the roof. These tiles were made in an L shape (turn the L 90 degrees to the right) so that

from the roof to protect the wooden structures. Its outer face was fash- ves with a not quite semicircular lower edge.[92] These concentric circles

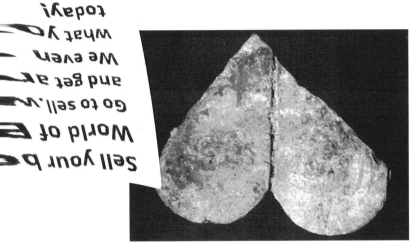

Figure I.20.

Schädler shows the details in his reconstruction illustration, below. The tiles are of normal size (as far as we can tell, since they are fragmented, but the thickness is the same). The drawing is also of 1:1 size of the later ornament and not scaled down, as can be seen from the superposition of drawing and ornament, below. Clearly, the drawing was not to be seen (why should it?). Schädler's claim is that it was a drawing of a working process, helpful for the workmen. What was to be seen was the ornamented border of the roof with the row of leaves and the concentric circles on them:[93]

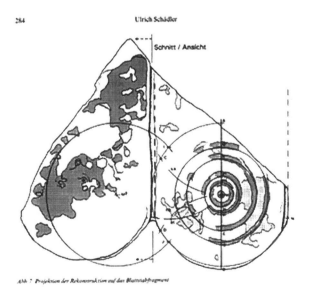

Figure I.21.

34   The Metaphysics of the Pythagorean Theorem

The geometrical sketch on the slab, below, is a drawing made to define (1) the curve of that lower edge of the roof tile and so the width of the "leaves," and (2) the diameters of the concentric circles, all mathematically quite exact, although the geometric method is just an approximation, but close enough to the mathematical proportions. The radius of the circle is 5.2 cm = 1/10 of the Ionic cubit. Thus, the circles in the drawing (that have been defined by the sketch) match the circles on the tile, and the point L defines the width of the "leaf," that is, the point where two leaves join and where the lower curved edge of the tile ends. Schädler's claim is that with the help of the sketch, the architect or workman defined the diameters/radii of the concentric circles, and the lower edge and width of the leaves, all on the basis (5.2 cm) of the Ionic cubit.

Figure I.22.

In his reconstruction drawing, Schädler clarifies the details on the tile, below:

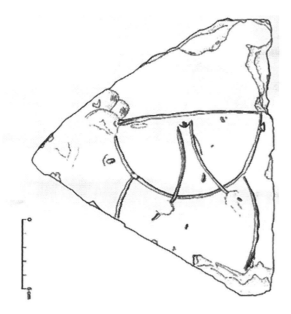

Figure I.23.

the sixth century, circa 550 BCE, that is, earlier than the time of Empedocles. [...] was driven from two sides separately as a time-saving strategy; Herodotus refers to it as "double-mouthed" (ἀμφίστομον).[95] I discuss the construction of the tunnel in relevant details in chapter 3, but here my discussion is directed specifically to the argument that Eupalinos had to have made a scaled-measured diagram, to achieve his successful results of the detour in the north tunnel—which is the excavator's claim: "There can be no doubt that an exact architectural survey of both tunnels had been established before the commencement of the correction. There can be no doubt either that the further digging was planned according to these plans so that the puncture could be successful with as little effort as possible and the highest certainty possible."[96] The case for such a plan is an inference from a study of the triangular detour in the north tunnel and the measure marks painted on the western wall of the tunnel itself. We have no surviving plan, but we do have the lettering on the western wall of the tunnel, and the inescapable picture that forms is one that points to a *lettered diagram*—a scaled drawing—that was in turn transferred to the tunnel wall. Here is the argument.

To drive the tunnel from two sides separately, since they meet directly under the smooth ridge of Mount Castro, the crest of which is enclosed within the city's fortified walls, Eupalinos had to have a clear idea from the start of the lengths of both tunnels, south and north. To do so, he staked out the hill and determined that both proposed entrances were fixed on the same level. This means, he set up stakes in straight lines—gnomons—running down from the summit of the hill, on both the north and south sides of the hill. Below, the white dots indicate where the stakes might have been placed down and along the hillside, from the top of the ridge of Mount Castro to the entrance of the south tunnel (= the lowest dot), all of which are within the city's fortified walls:

36   The Metaphysics of the Pythagorean Theorem

Figure I.24.

If we imagine that each white dot is a stake, Eupalinos had to measure carefully the horizontal distances between all the stakes on each side of the hill. By summing up all the horizontal distances, he had his hypothesis for the length of each tunnel half. In measures of our modern meter lengths, the south tunnel was approximately 420 m, while the north tunnel was 616 m. In any case, the tunnels were not of equal length, and the fact that they meet under the ridge is evidence that supports the hypothesis that the hill was staked out: this is why the tunnels meet directly under the ridge, and not midway between the two entrances. The simple geometry of this reasoning is presented below; the addition of the horizontal lengths on the right side, and the horizontal lengths on the left side, summed to the length of the tunnels.

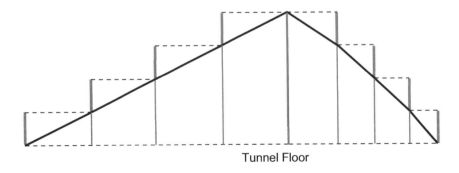

Figure I.25.

Eupalinos's task was to bring the straight line up and along the outside of the hill, into the hill itself as the tunnel line. As the digging proceeded, so long as he could turn back and still see the light from outside, he could be sure he was on track, or at least working in a straight line.

But in the north tunnel, after about 200 m, the rock began to crumble and Eupalinos feared a collapse. Kienast's analysis of the course of the tunnel shows that the detour took the form of an isosceles triangle. It is the planning for and controlling of this triangular detour that required the scaled-measured diagram. The strong evidence for the existence of a scaled-measured plan—and hence a lettered diagram—is the transposition and continuation of the fifth-than system of numbers painted in red on the western wall of the triangular detour. The letters on this tunnel will suggest at least an informal sketch with numbers or letters attached, which was the reference for painted letters on the wall as the digging progressed, to keep track of the progress; but the need for a very accurate diagram, and hence the plausible case for it, came when Eupalinos was forced to leave the straight line in the north tunnel.

In the south tunnel, the uninterrupted letters, painted in red on the western wall, show us the marking system of lengths. The letters, from the Milesian system of counting numbers—I, K, Λ, M, N, Ξ, O, Π, Q (10 through 90)—starting with I = 10, K = 20, Λ, = 30 and so on, run in tens to a hundred, P. The measure marks begin from the exact place at which the digging started and continue without interruption through the south tunnel. This was what Kienast refers to as System 1—the original marking system for tunnel length, painted on the western wall. Since the measure marks are at intervals of 20.60 m and are in units of ten, Eupalinos selected as his module 2.06 m, and divided the whole tunnel length into some 50 parts, for a sum total of approximately 1,036 meters. And since this length of 2.06 m does not correspond to any other basic unit of foot or ell—Samian, Pheidonic, Milesian[97]—Kienast suggested that it seemed reasonable to claim that Eupalinos invented his own tunnel measure, his own tunnel module.[98] In the context of early Greek philosophy, here was another example of the "one over many," just like the other contemporaneous standardizations of weights and measures, and of course, coinage. The architects of the Ionic stone temples in the sixth century—the Samian Heraion, Dipteros I, c. 575, the archaic Artemision c. 560, and the Didymaion of Apollo c. 560—also needed a module to plan their gigantic buildings, so that the architectural elements would display the proportions they imagined and thus the finished appearance they intended, and accordingly each was reckoned as a multiple or submultiple of this module. Vitruvius tell us that their module was column-diameter.[99] When Anaximander expresses the size of the cosmos in earthly diameters that he described as analogous in shape and size to a 3 x 1 column-drum—9 + 1 Earth diameters to the wheel of stars (9/10), +9 to the moon wheel (18/19), + 9 to the sun wheel (27/28)—he is not only making use contemporaneously of a modular technique, but he has adopted exactly the architects' module—column diameter—to measure the cosmos. It is in this sense that Anaximander came to imagine the cosmos by architectural techniques, because he came to grasp the cosmos as built architecture, as architecture built in stages like the great temples.[100] The fact that Anaximander wrote the first philosophical book in prose at precisely the same time that the Samian and Ephesian architects wrote prose treatises is no mere coincidence. It suggests the overlap and interaction of these two communities of interest.[101]

Thus, in the south tunnel the Milesian letters painted in red on the western wall allowed Eupalinos to keep track of his progress. The original plan is preserved, then, through the spacing of the letters on the wall, and suggests that there was an original diagram consisting of two straight lines—north tunnel and south tunnel—that met at distances directly under the ridge, and each tunnel length was expressed in units of 10 of Eupalinos's tunnel modules of 2.06 m, following the Milesian system of letters: beginning with the Greek letter iota (I = 10) and following the Greek alphabet to the letter rho (P = 100), after which the series begins again and continues to sigma (Σ = 200), and then beginning once more and ending in T (T = 300). The letters painted on the western wall, then, correspond to the same letters on a diagram; thus, the existence of such a diagram is an inference from the lettering on the wall. This is not a matter of keeping track of a few marks; the tunnel is divided into some 50 parts of 20.60 m each, and consequently required dozens of letters painted on the western wall to follow the progress of tunnel length.[102] This whole matter of how Eupalinos worked is both complicated and clarified by and through problems he encountered in the north end. When Kienast began to study the measure marks in the north end, he first discovered that they had been *shifted inward* on the triangular detour. When and why?

We now understand that as Eupalinos supervised the digging in the north end, he kept track of the progress in carrying out his plan by measure marks, just as he did in the south tunnel. He knew the total length of the north tunnel by counting up the horizontal distances between the stakes, as expressed in the system of Milesian numeration, which allowed him to follow the progress as the digging proceeded. In the north tunnel only one letter from the original series is still visible on the western wall, the letter Λ. All the other letters of the original marking system are now hidden behind strengthening walls constructed *after* the completion of the tunnel to insure that the walls remained stable. But the Λ is at exactly the right place to correspond to the original marking system, which starts, just as in the south end, exactly where the digging began. It was by means of these lettered marks that Eupalinos checked the progress in the tunnel excavation. Then, after about 200 m, the rock began to crumble and Eupalinos feared collapse. He was now faced with the problem of figuring out how to make a detour and still arrive at the originally planned meeting place directly under the ridge. When Kienast examined the triangular detour that comes next, he discovered that the letters are no longer in the places they would be expected. Instead, these measure marks appear "shifted inwards" toward the meeting point; they are in the same uniform distances, but they had shifted. How shall we account for this?

Kienast's theory is that after some 200 m of digging in the north tunnel, Eupalinos encountered crumbling stone and, fearing a collapse, had to abandon the straight line. Since the natural contour of the stratigraphy folded to the northeast, Eupalinos dug away from the crumbling stone toward the west. Kienast's theory is that, before the detour was begun, Eupalinos carefully inspected the terrain on top of the hill under which the detour would be needed, and noticing the conditions of the stone and stratigraphy, imagined rather exactly how long the tunnel would need to detour to the west before the tunnel line could be turned back to reach the meeting point. It is at this moment that Eupalinos made a scaled-measured diagram, or added it to the original diagram of the tunnel line. The tunnel shows that Eupalinos planned the tunnel detour in the form of an isosceles triangle. He would dig westward at a specific angle—roughly 22 degrees—and when the stone again became safe, he would turn back at exactly the same angle.

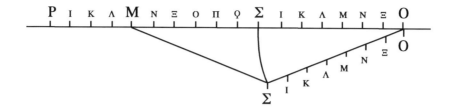

Figure I.26.

Now, here is where we have the evidence for Eupalinos's use of a scaled-measured diagram. He knew the original straight line, and hence the distance left to dig from M to rejoin the originally planned straight line at O. But, having left the straight line at M' he placed his compass point on O, on the originally planned straight line, and swung an arc that began from Σ—almost the midway point between M and O—and intersected the vertex of his proposed triangular detour and marked it Σ on his scaled-measured diagram *and* on the tunnel wall itself at the vertex of the triangular detour when he reached this point. Kienast's argument is that the painting of Σ at the vertex *proves* that Σ was intentionally part of the design plan. Now, since Σ on both the original diagram and the triangular detour in the tunnel itself were both radii of the same circle (i.e., arc), the length left to dig from Σ to I, K, Λ, M, N, Ξ, O (O is on the originally planned straight line) he already knew by the original plan. That sequence was followed on the second leg of the triangle from the vertex at Σ to O. All those letters are in the original intervals of length and painted on the second leg of the triangular detour.

The unresolved problem with this proposal is that the detour caused the total length of the tunnel to increase, and Eupalinos needed to know—as he made his scaled-measured diagram—exactly how much extra digging was required to complete this detour, and so, he began to *count backward*, and *mark backward on the diagram*, the intervals (i.e., same lengths) from the vertex of the detour to the original straight line, now on the first leg of the triangular detour. The result was to shift the measure marks inward, toward the meeting point. This Kienast explains as System 2 or the *new marking system*, shifting inward as a result of the triangular vertex at Σ (Ϙ, Π, O, Ξ, N, M . . . Λ, K, I . . . and so on).

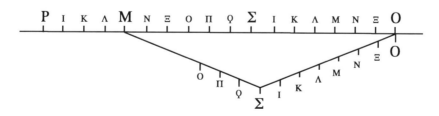

Figure I.27.

Counting back to reach 'M,' the distance lengthened by the detour would have become apparent, without any complicated arithmetical computation, below. The placement of M on the first triangle leg, below, and on the tunnel wall itself, is not where the M should have appeared, as it does on the original straight line. This distance between the 'M' on the first triangle leg and its original positioning is exactly the prolongation of the tunnel as a result of the detour; the additional distance from the "shifted M" to the 'M' on the original straight line is a distance of 17.59 m.

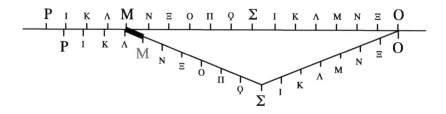

Figure I.28.

Thus, the planned detour took the shape of a triangle; by digging westward, and then back eastward at the same angle, he was imagining the detour as an isosceles triangle. Both angles turned out to be roughly 22 degrees, and if the base angles of an isosceles triangle are equal then the sides opposite them must also be equal in length—this is one of the theorems attributed to Thales.

Kienast's thesis, then, is this: Since there is the letter Σ painted in red on the west wall of this detour at the vertex of this triangle, along with other Milesian letters in the series, Eupalinos *must* have made a scaled-measured diagram with the original straight line marked with Milesian letters for increments of 10 modules each, along with the proposed detour. Then with the aid of a compass whose point was set on 'O' where Eupalinos planned to rejoin the original straight line, he drew an arc from the Σ on the straight line to the vertex of the triangle-detour. Since both segments OΣ on the diagram—the original segment on the straight line and the segment on the second side of the triangle detour—are radii of the same circle (of which this is an arc), the length of both segments must be the same. Then, as work proceeded to rejoin the originally planned straight line at 'O' in the second side of the triangle detour, the same distances were transposed to the west wall of the second leg of the triangle to follow the progress Ι, Κ, Λ, Μ, Ν, Ξ, Ο. By the scaled-measured diagram—*before* the digging of the detour started—the first leg of the triangle could be marked off in lettered distances—in reverse—Σ, Ϙ, Π, Ο, Ξ, Ν, Μ, and the lengthened distance caused by the detour would become immediately apparent—17.59 m without the added complexity of arithmetical computation. Thus, by means of the scaled-measured plan—the lettered diagram—he knew *before* he started the detour a very close approximation of the extra digging required to complete the tunnel successfully.

And finally, there is written on one of the strengthening walls—which means it was added *after* the tunnel was finished—the word PARADEGMA; painted on both sides, left and right, is a vertical measure mark:

|ΓΑΡΑ Δ Ε ΓΜ Α|

Figure I.29.

The remarkable thing is that the distance from the left vertical measure mark to the right is almost exactly 17.59 m,[103] and for this reason Kienast, following the suggestion of Käppel, embraced the view that Eupalinos celebrated his calculation of the prolongation of tunnel.[104] Since *PARADEGMA* is painted on a strengthening wall, it was clearly painted *after* the work was completed and so could not have served a purpose in the process of digging the tunnel.[105]

G

Summary

In summary, then, we have considered how Netz presents a narrative about the shaping of deductive reasoning that, per force, includes a view about the origins of geometry. His narrative focuses on the lettered diagram as a way to mark definitively and distinctively its starting point. Netz also insists that we focus on a diagram that is not a representation of something else but rather is the thing itself; earlier architectural diagrams, even if there were any, are ruled out of court as "geometry" per se. Of course, Netz is welcome to tell his story however he wants and for his own purposes, but if it includes accounting for the origins of geometry, his own modern disciplinary biases preclude him from seeing the beginnings. Netz argues that geometrical diagrams of Anaxagoras and Hippocrates were "perhaps the very first diagrams," but they were not. He emphasizes the importance of lettered diagrams, the earliest of which he dates to Aristotle, but we have evidence for lettered diagrams some *two centuries earlier*. He claims that "the mathematical diagram did not evolve as a modification of other practical diagrams,"[106] but the case I have set out suggests to the contrary that they did. Whatever happened by the time of Anaxagoras and Hippocrates and transformed their enterprise and the very ontological status of their diagram-objects, these results came after more than a century of diagram-making by architects/engineers, and the workmen who had to produce the elements that were specified by their designs. The evidence that survives may not be robust, but to anyone who understands it—since none of it was meant to be saved—it proves that already by the middle of the seventh century, and perhaps earlier still, the architects or workmen used geometrical methods to define measurements and proportions of certain details of their buildings, that measurements and proportions of the temples were based on numbers, mathematically exact, not just casually selected or random. The diagrams guided the success of their mathematically exact intentions. The evidence means that geometrical diagrams may very well have been a customary technique in sixth-century architecture and engineering, and familiarity with them may well have been widespread through the communities that engaged in these kinds of activities.

To begin to grasp Thales's diagram-making, we must place him first within the context of practical diagram-making, and possible influences through Mesopotamia and certainly from

Egypt regarding the determination of areas of land, the Egyptian evidence suggesting that rectilinear figures could one and all be reduced to triangles. The evidence also includes broadly contemporaneous Greek applied geometry of architectural techniques at the building sites of monumental stone temples. I have already argued at length regarding the compelling reasons to place Anaximander, his younger contemporary, there as an attentive observer.[107] Whether Anaximander participated more actively in architectural projects we cannot say, but the kinds of practical achievements with which he is credited—making a seasonal sundial, a map of the inhabited Earth, a map or model of the cosmos with proportional distances to the heavenly wheels of sun, moon, and stars—suggest both interest and competence in matters of applied geometry. Burkert is certain that the book to which Eudemus had access was *On Solstices and Equinox*, whether or not it was written by Thales, in part because he acknowledges the existence of Ionic technical writings in the sixth century on problems of astronomy and the calendar. But what Burkert failed to connect adequately was the existence of architectural *prose* treatises in the middle of the sixth century by Theodorus on the temple of Hera in Samos and by Chersiphron and Metagenes on the temple of Artemis in Ephesus, reported to us by Vitruvius, that almost certainly included the rules of proportion for the architectural elements, and may well have contained diagrams; these I have discussed elsewhere.[108] Netz insightfully aligns diagram-making and prose writing in exploring and explaining the rise of geometrical knowledge, but their alignment took place *earlier*; those ingredients are initially woven together by the middle of the sixth century in the enterprises of the architects building monumental temples and writing their prose books. Anaximander, who we know to be the first philosophical author to write in prose, because a sentence of it is preserved by Simplicius, is also credited with writing an *Outline on Geometry* (ὅλως γεωμετρίας ὑποτύπωσιν ἔδειξιν).[109] In the illuminated light of this broader context of architectural and philosophical prose treatises, the plausibility that Anaximander produced some work on geometry—whether in the one book that we know to have existed, or in some other book now lost—is enhanced. Had Anaximander written something on geometry, it's hard to imagine it without some diagrams. Netz's explicit doubts about diagrams in Anaximander's work may not be well founded.[110] Papyrus would have been comparatively inexpensive for Milesians, whose trading colony had long been established in Egypt, and the inclusion of diagrams would not have been prohibited by cost, though later copies of such a papyrus roll might have had to be shortened and diagrams omitted.

The relevance of monumental temple architecture and architectural prose writings to the development of geometrical thought has been vastly underappreciated; those such as Netz are hardly in a position to see it, because the practical and applied evidence is dismissed out of hand as banausic. But the term "banausic" that Netz uses suggests that what was being done was not carried out at a refined or elevated level. Netz's labeling suggests that such efforts were "mundane." But a reflection on the Eupalinion suggests, quite to the contrary, that the extraordinary achievement in precision and accuracy of the tunnel as a result of working from a lettered diagram in scale was hardly mundane. When Burkert wrote *Weisheit und Wissenschaft: Studien zu Pythagoras, Philolaus und Platon* (1962), Coulton's important 1977 study *Ancient Greek Architects at Work* was not yet written; by the time Netz was reflecting on ancient architecture, Coulton's 1977 study, on which he relies so heavily, was no longer current in relevant respects. Kienast published *Die Wasserleitung des Eupalinos auf Samos* in 1995, and Schädler's publication in *Griechische Geometrie im Artemision von Ephesos* came out in 2004. The geometry in ancient architecture and engineering has largely escaped the appreciative notice of some of the historians of mathematics.

The essence of *practical diagram-making* is the realization that geometrical relations provide a nonchanging platform on which our world of imprecise appearances can be placed; it is through the making of practical diagrams that a deep and penetrating insight into a structure underlying nature emerges and is delineated. True, it was not the awareness on which Netz focuses, of a mathematical object that exists in a timeless realm itself, and merits investigation as a thing itself without connection or application to anything else. But at the beginning, geometrical thought and its diagrams displayed an awareness of a timeless realm that nevertheless had an illuminating connection to our ephemeral world, distinct but not fully disengaged from it. It was only because a relation was envisaged between geometrical diagrams and the world to which it was applied in so illuminating a way that the metaphysical inquiries of those such as Thales and his compatriots embraced geometry. The awareness of the applications of timeless geometrical structures to our time-bound world opened the door to the usefulness of geometry to resolve the metaphysical question: How does an underlying unity appear divergently by altering without changing? If only Thales could discover the fundamental geometric structure into which all other figures dissect, then by repackaging and recombining them, he could account for the multiplicity of appearances by alteration without admitting change.

Chapter 1

The Pythagorean Theorem

Euclid I.47 and VI.31

What are the chances that Pythagoras, or even Thales before him, *knew* the famous mathematical proposition that has come down to us under the name "Pythagorean theorem?" The question is not whether he *proved* it, or even if he stated it in the form of a theorem. Could either of these men who lived in the sixth century BCE have *understood* it? And if so, in what senses? Since, until recently, scholarship has tended to discredit "Pythagoras" not only with regard to the theorem but to any contributions to mathematics, could a circumstantial case be constructed that points to such understanding, contrary to the prevailing views? To even approach the question, we must be very clear about what we are looking for. In these earliest chapters of the Greeks in geometry, perhaps the mathematical proposition was grasped differently than in the forms in which it has come down to us in our modern education; indeed, it may have been grasped differently in Euclid himself.

So, let us begin by exploring the Pythagorean theorem. How did Euclid present it at the end of the fourth century BCE? How might it have been understood two-and-a-half centuries earlier by the philosophers of the sixth century BCE? Or to put the matter differently, *what do you know when you know the Pythagorean theorem?*

First, let us be clear that the theorem we are exploring is the one that claims a law-like relation among the sides of a right triangle, a law-like relation between line lengths and figures drawn on the sides of a right triangle. It has been my experience that most educated people today respond to this question about the Pythagorean theorem by producing a formula—$a^2 + b^2 = c^2$—and offer as an explanation that when asked to calculate the length of a side of a right triangle given the lengths of two other sides, they apply the formula to derive the answer. But this formula is algebra; the Greeks of the time of Euclid, and certainly before, did not have algebra, nor does the surviving evidence show that they thought in algebraic terms.[1] So this poses a false start from the outset. If we begin with Euclid, and then hope to look back into the earlier investigations, we must focus on the relation between line lengths and the areas of figures constructed from them. Can Pythagoras, or Thales, be connected with an understanding that brings together the lengths of the lines with the areas of figures constructed on them, and if so, how and why?

46   The Metaphysics of the Pythagorean Theorem

To begin this exploration, I shall set out (A) the most famous presentation of the theorem from Euclid's Book I, Proposition 47, and then explain the claims that must be connected to effect the proof. I next set out the mathematical intuitions[2] that anyone would be assumed to know to connect what is connected in *this* proof. Next I shall set out (B) the largely neglected, second version of the proof, sometimes called the "enlargement" of the Pythagorean theorem as it is presented in Euclid VI.31, and follow this by explaining the claims that must be connected to effect this proof; I then set out the mathematical intuitions that anyone would be assumed to have to know to connect what is connected in the proof. Next (C), I explain the related concepts of "ratio," "proportions," and "mean proportional" as they apply to this discussion. Afterward (D), I try to clarify further the idea of the mean proportional (or geometric mean) by contrasting it with an arithmetical mean. And finally (E), I provide an overview and summary connecting to the idea of the *metaphysics* of the hypotenuse theorem that anticipate the arguments of chapters 2 and 3, which offer a broader and wider picture of how Pythagoras, and Thales before him, plausibly knew the theorem. Taking this approach puts us in a position to know precisely what we are looking for, had Thales or Pythagoras grasped these relations between line lengths and the areas of figures constructed on them.

A

Euclid: The Pythagorean Theorem I.47

First, (i) I will present the proof of the theorem as it appears in Euclid I.47, then (ii) I will explain the strategy of the proof in a reflective way, and third (iii) I will set out the *ideas* that must be connected—the *intuitions*—whose connections are the proof. The importance of this third approach is central to my project because without being misled by the ambiguous question "Could a sixth century Greek *prove* this theorem?"—appealing to Euclid as a paradigm of "proof"—we are asking instead "Could a sixth century Greek have grasped the relevant ideas and connected them?" And if so, what is the nature of the evidence for them, and moreover, what does this tell us about *reasoning*, formal and informal, and why the Greeks might have been investigating these things?

(i) The Pythagorean theorem of Euclid I.47 (following Heath):

> Ἐν τοῖς ὀρθογωνίοις τριγώνοις τὸ ἀπὸ τῆς τὴν ὀρθὴν γωνίαν ὑποτεινούσης πλευρᾶς τετράγωνον ἴσον ἐστὶ τοῖς ἀπὸ τῶν τὴν ὀρθὴν γωνίαν περιεχουσῶν πλευρῶν τετραγώνοις.

> "In right-angled triangles the square on the side subtending the right angle is equal to the squares on the sides containing the right angle."

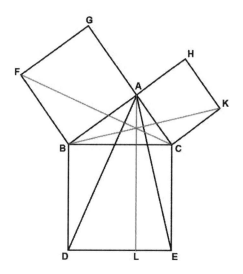

Figure 1.1.

Let ABC be a right-angled triangle having the angle BAC right;

I say that the square on BC is equal to the squares on BA, AC.

For let there be described on BC the square BDEC, and on BA, AC the squares GB, HC; **[I.46]** through A let AL be drawn parallel to either BD or CE, and let AD, FC be joined.

Then, since each of the angles BAC, BAG is right, it follows that with a straight line BA, and at the point A on it, the two straight lines AC, AG not lying on the same side make the adjacent angles equal to two right angles;

Therefore CA is in a straight line with AG. **[I.14]**

For the same reason BA is also in a straight line with AH. And, since the angle DBC is equal to the angle FBA, for each right, let the angle ABC be added;

Therefore, the whole angle DBA is equal to the whole angle FBC. **[C.N. 2]**

And since DB is equal to BC, and FB to BA, the two sides AB, BD are equal to the two sides FB, BC respectively;

and the angle ABD is equal to the right angle FBC;

Therefore, the base AD is equal to the base FC, and the triangle ABD is equal to the triangle FBC. **[I.4]**

Now the parallelogram BL is double the triangle ABD, for they have the same base BD and are in the same parallels BD, AL. **[I.41]**

And the square GB is double of the triangle FBC, for they again have the same base FB, and are in the same parallels FB, GC. **[I.41]**

[But the doubles of equals are equal to one another.]

Therefore, the parallelogram BL is also equal to the square GB.

Similarly, if AE, BK be joined, the parallelogram CL can also be proved equal to the square HC; therefore, the whole square BDEC is equal to two squares GB, HC. **[C.N. 2]**

And the square BDEC is described on BC, and the squares GB, HC on BA, AC.

Therefore the square on the side BC is equal to the squares on the sides BA, AC.

Therefore etc.

QED [OED]

48   The Metaphysics of the Pythagorean Theorem

(ii) Reflections on the strategies of Euclid I.47:

The strategy of the proof is to create, first, squares on each side of a right triangle. Next, draw a perpendicular line from the right angle of the triangle to the base of the square on the hypotenuse BC. The result is that the square divides into two rectangles, each of which will be shown to be equal to the square drawn on each of the remaining two sides, respectively, of the original triangle ABC.

To show this, then, a line is drawn from the right angle, vertex A, to point L parallel to both sides of the square drawn on the hypotenuse, BD and CE. This creates rectangle BL, which is then divided by drawing a line from point A to D, the left corner of the square, creating at the same time triangle ABD. The next step is to show that rectangle BL is twice the area of triangle ABD because both share the same base BD between the same two parallel lines. Even now we must keep in mind that the strategy of the proof is to show that the square on the hypotenuse BC, now divided into two rectangles, is equal to the sum of the squares on the legs AB and AC.

Next, the argument shows that triangle ABD is equal to triangle FBC, because sides AB and BD are together equal to sides FB and BC, the angle ABD is equal to the angle FBC, and base AD is equal to base FC. This is side-angle-side (SAS) equality, which Euclid proves in theorem I.4. And because triangle FBC and square GB share the same base FB between the same parallels FB, GC, the square is double the area of triangle FBC. And then, since triangles FBC and ABD are equal, the rectangle BL that is also double the triangle ABD is also equal to the square GB, because things equal to the same thing are also equal to each other. At this point, the proof has shown that the square on the longer side of the right triangle is equal to the larger rectangle into which the square on the hypotenuse is divided.

Then, AE and BK are joined, creating triangles AEC and BKC, and they are shown to be equal to each other, ceteris paribus. And because both rectangle CL and triangle AEC share the same base CE, and are between the same parallel lines, the rectangle is double the area of triangle AEC. And since the triangle AEC is equal to the triangle BKC, the square HC must be double the area of triangle BKC, since both share the same base KC between the same parallel lines, and thus the rectangle CL is equal to the square HC, since things equal to the same thing are also equal to each other.

Finally, the proof ties together these two parts by showing that since the rectangle BL is equal to the square GB, and the rectangle CL is equal to the square HC, and the two rectangles BL and CL placed together comprise the largest square BDEC, thus the squares on each of the two sides, taken together, are equal to the square on the hypotenuse.

What follows, immediately below, then, is the sequence of this proof diagrams:

These two triangles are of equal area because they share two sides equal and the angle between those sides (SAS equality, I.4):

The Pythagorean Theorem 49

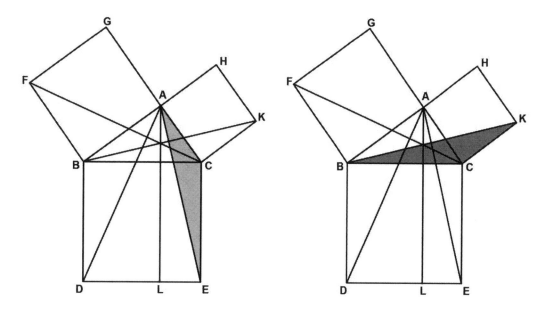

Figure 1.2.

The rectangle has twice the area of the triangle since they share the same base and are within the same parallel lines AL and CE (I.41).

AL and CE are parallel and share the same base CE

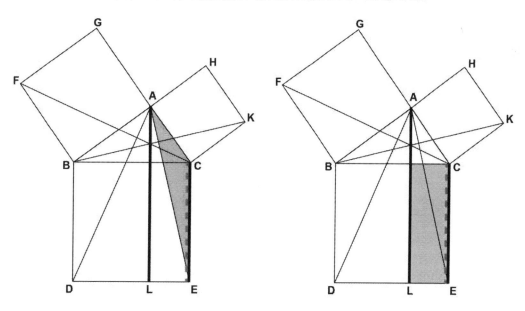

DASHED RED LINE [┊] is the base shared in common

BOLD BLACK LINES [|] are the parallel lines between which the figures are constructed

Figure 1.3.

50   The Metaphysics of the Pythagorean Theorem

The square has twice the area of the triangle, since they too share the same base and are within the same parallel lines KC and HB:

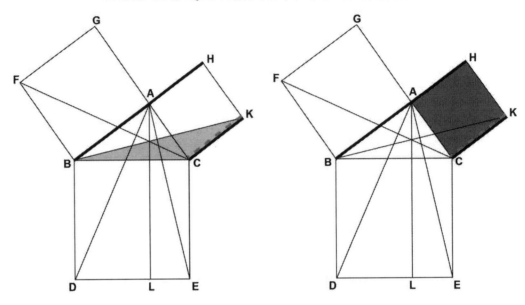

DASHED RED LINE [⁝] is the base shared in common

BOLD BLACK LINES [|] are the parallel lines between which the figures are constructed

Figure 1.4.

Thus, the square on the shortest side of the right triangle has the same area of the smaller rectangle into which the square on the hypotenuse has been divided because each is double the area of the triangle that shares the same base, between the same parallel lines, and each of those triangles is equal to one another. The rectangle and square are equal because they are double triangles equal to one another.

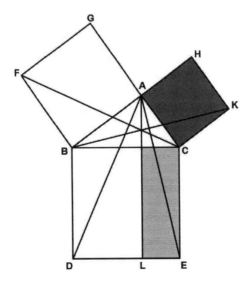

Figure 1.5.

At this point I wish to modify what I have just written. I have expressed this point of equality of figures as equalities of area, though this is not Euclid's way of expressing the matter. He customarily speaks of *figures* being equal, not areas. Looking back to the sequence of propositions beginning with I.35, this is a point that needs to be considered. When he says that parallelograms are equal or triangles are equal, it is the figure that is equal and not some number attached to it. By appeal to Common-Notion 4, things that coincide with one another are equal to one another. From there, however, the remaining Common-Notions allow Euclid to conclude that figures that do not coincide are also equal. This point is worth emphasizing for we are looking at figures, square AK and rectangle CL, that are equal despite their very different shape.

Now the same strategy proceeds to show that the square on the longer side of the right triangle is equal in area to the larger rectangle into which the square on the hypotenuse is divided.

First, these two triangles are shown to be equal in area because they too share two sides in common and the angle between them (SAS equality, I.4):

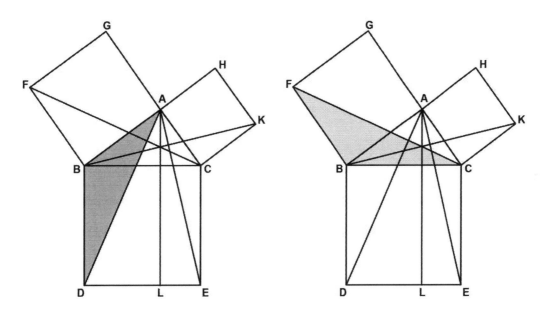

Figure 1.6.

52 The Metaphysics of the Pythagorean Theorem

Since both the triangle and the rectangle share the same base and are between the same parallel lines BC and AL, the rectangle has twice the area of the triangle:

BD and AL are parallel and share the same base BD

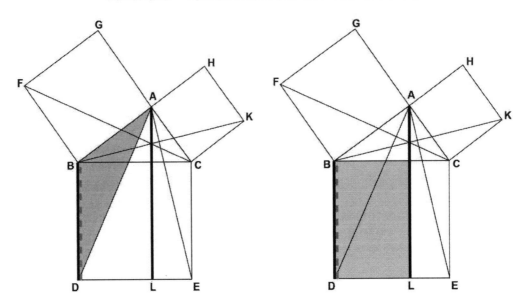

DASHED RED LINE [⦂] is the base shared in common

BOLD BLACK LINES [|] are the parallel lines between which the figures are constructed

Figure 1.7.

And since both this triangle and this square share the same base and are between the same parallel lines FB and GA, the square has twice the area of the triangle:

FB and GC are parallel and share the same base FB

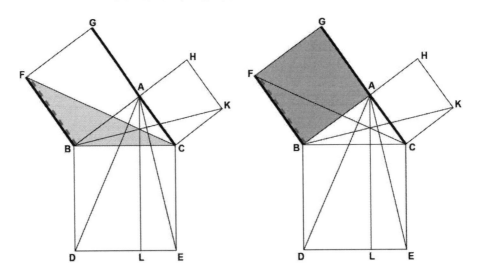

DASHED RED LINE [⦂] is the base shared in common

BOLD BLACK LINES [|] are the parallel lines between which the figures are constructed

Figure 1.8.

And so the square on the longer side of the right triangle has the same area as the larger rectangle into which the square on the hypotenuse is divided, because each is double the area of the triangle that shares the same base between the same parallel lines, and each of those triangles is equal to one another:

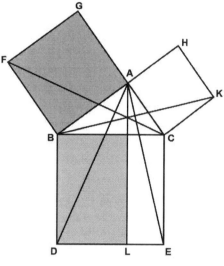

Figure 1.9.

And thus, the areas of the squares on each side of the right triangle are equal to the two rectangles, respectively, into which the square on the hypotenuse is divided:

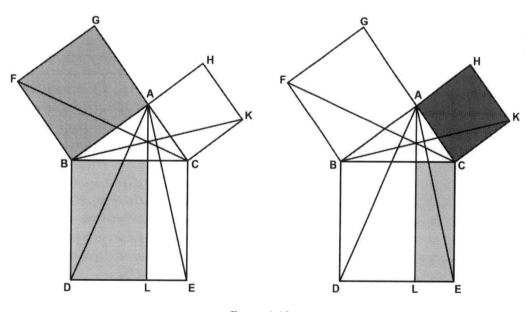

Figure 1.10.

And thus, the square on the hypotenuse is equal to the sum of the squares on the two sides:

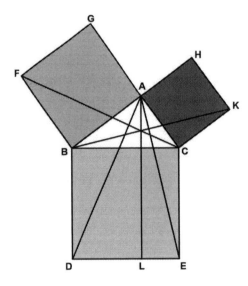

Figure 1.11.

(iii) The geometrical intuitions: the sequence of ideas that are connected in the proof:

What I have presented, immediately above, then, is the argument sequence, of the formal proof for Euclid's I.47, the so-called "Pythagorean theorem," following Heath's version of Heiberg's text. But let us now review the proof trying to get clear about the basic ideas that have to be connected to produce the sequence of thoughts that comprise *this* proof. Stated in another way, we remind ourselves that each step in a formal proof is justified by an appeal to an abstract rule—this is key to understanding the *deductive method*. These rules are appealed to in each step of Euclid's formal proofs. So we now reflect on what Euclid thinks we have to know to follow the sequence of I.47. Could it be that our old friends of the sixth century BCE grasped the ideas and their connections, whether or not they could produce such a "proof"—that is, technically speaking, this irrefragable chain of connected thoughts justified by explicitly stated rules of derivation? This is the question to be kept in mind as we continue. So, what ideas would have to be grasped?

It is a fair generalization to say about Euclid's *Elements* that the central ideas that run throughout, but especially in Books I–VI, are equality and similarity. If we may be allowed to place the themes in more modern terms, we might describe the contents of Books I–VI as being concerned with *congruence, areal equivalence,* and *similarity*—though there is no term in ancient Greek that corresponds with what we call "congruence." The theorem at I.47 relies on equalities and areal equivalences; the theorem at VI.31 relies on extending these ideas to include similarity, and the ratios and proportional relations that this entails, explored in Book V. The matter of *equality* among triangles is taken up in Book I propositions 4, 8, and 26. What these propositions show is that *if certain things are equal in two triangles, other things will be equal as well*. And the overall, general strategy of Book I shows that the common notions are axioms for equality and include a fundamental test, namely, that things that can be superimposed are equal; I.4 begins an exploration of equal figures that are equal via superimposition, that is, they are equal and identical in shape; I.35 shows that figures can be equal that cannot be superimposed, that is, figures can be equal but not identical in shape; I.47 shows that it is possible to have two figures of a given shape, a square, that can be equal to a third figure of the same shape.

I.4: (SAS—Side-Angle-Side) Two triangles are equal if they share two side lengths in common and the angle between them:

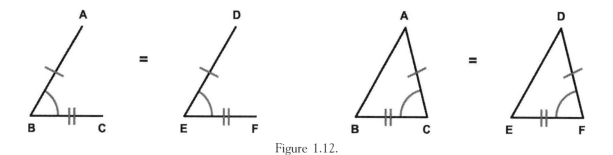

Figure 1.12.

I.8: (SSS—Side-Side-Side) Two triangles are equal (i.e., congruent) if they share the lengths of all three sides:

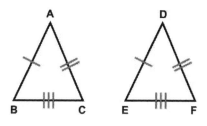

Figure 1.13.

I.26: (ASA—Angle-Side-Angle) Two triangles are equal (i.e., congruent) if they share two angles in common and the side length between them:

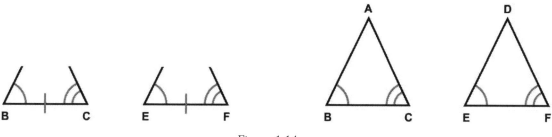

Figure 1.14.

The third theorem of equality, I.26, is explicitly credited to Thales by Proclus on the authority of Eudemus.[3] Two triangles are shown to be equal—congruent—if two sides and the angle contained by them are equal (SAS), if all three sides are equal in length (SSS), and if two angles and the side shared by them either adjoining or subtending are equal (ASA). The last of these is credited to Thales because Eudemus *inferred* it was needed for the measurement of the distance of a ship at sea. We shall investigate such a measurement in chapter 2. An understanding of it was needed to grasp the measurement, by one approach, while an understanding of similarity was needed by all the other approaches. But let us get clear about the general matter; one does not come to grasp angle-side-angle equality without having recognized side-side-side and side-angle-side equality. The equality and similarity of triangles are principles abundantly clear already by the measurement of pyramid height, as we shall also explore in the next chapter, despite the endless controversies in the scholarly literature of whether Thales or someone in the sixth century could have grasped the idea of equality or similarity.

In addition, almost certainly, the earliest attempts to prove equality—equality between triangles, for example—were by superposition; they were ἐφαρμόζειν proofs. This means that one triangle was quite literally placed on top of another, showing that the lengths *coincided* (ἐφαρμόζειν) and that the angles met at the same places and to the same degree. In the language of I.4, with the triangle ABC fitted on DEZ with point A placed on point D while straight line AB is fitted on DE, point B fits on E because AB is equal to DE. To show these kinds of equality, to quite literally make them visible (i.e., "to prove"—δείκνυμι), triangles, squares, rectangles, and all rectilinear figures were capable of being *rotated* to fit. Euclid does not like this, and tries to avoid it whenever possible—because all physical and material demonstrations are subject to imperfections—but certainly this is how "proof" began for the Greeks in the sixth century BCE and before.[4] "Proof" was a way of making something *visible*,[5] something for all to see and confirm. We must keep this in mind as we continue; the archaic Greek minds of persons such as Thales and his compatriots could envision equality by *rotating* one object—or a figure in a diagram—one over another to see equality of fit. To make this point clear, I refer to the famous "tile standard" in the southwest corner of the Athenian Agora. If one had purchased, say, a roof tile and had any doubts about whether it was standard fare, he had only to take the tile to the standard, rotate it about so it would line up with it, edge to edge and angle to angle, and test it for fit. If by visible inspection, the tile "fit," there was the *proof* of its equality. Clearly, "proof" consisted in making the equality "visible."

Figure 1.15.

Now, when a parallelogram, say a rectangle, is claimed to be double a triangle—and this means double the *area* (χωρίον)—such a demonstration cannot be effected as easily by the ἐφαρμόζειν technique. Let us keep in mind that for Euclid a χωρίον is not a number but the space contained within a figure, an area set off by a fence, as it were. Now before we follow Euclid further, let's keep in mind the question of whether and how areal equivalence could be shown by "superposition."

Let us recall how equality between figures is demonstrated in Euclid Book I, since the proof of the Pythagorean theorem at I.47 requires that two triangles be shown to be equal; then the argument is divided into two parts. First, one of the triangles is shown to be half the area of a rectangle (i.e., a parallelogram) since they are constructed between the same parallel lines and on the same base, and the other is shown to be half the area of a square since they too are both constructed between the same parallel lines and on the same base, and since one of the triangles is half the area of the rectangle, and the other equal triangle is half the area of the square, the square on one side of the right triangle is equal to one of two rectangles into which the square on the hypotenuse is divided. Next, the same strategy is used to show that the other rectangle into which the square on the hypotenuse is divided is equal to the area of the square on the other side of the right triangle, ceteris paribus, because they are double triangles equal to one another. Clearly, one of the deep underlying intuitions is to grasp that two figures can be equal and unequal at the same time, though in different respects: in one respect, they can both have the same area, and yet be different in shape. Another key intuition is to grasp *how* it is that the triangles are half the area of the parallelograms *when constructed between the same parallel lines and on the same base*, and that every parallelogram is of equal area when constructed within the same parallel lines on that same base. For to grasp this idea we have to challenge our intuition that *perimeter is a criterion of area*, which it is not. To grasp this idea, we have to imagine space in such a way that rectilinear figures are imagined within parallel lines—space is conceived fundamentally as a flat surface structured by parallel straight lines, and figures unfold in these articulated spaces. When we imagine in this fashion, we discover that every triangle constructed on the same base, between two parallel lines, *regardless of its perimeter*, will have the same area. Moreover, every parallelogram constructed on the same base as the triangles will have double the area of every triangle, also regardless of its perimeter. To explore this mathematical intuition, we must think through the idea of parallel lines in Euclid's Book I.

To explore this intuition, we should clearly understand that the Pythagorean theorem of I.47 is *equivalent* to the parallel postulate—postulate 5—and this helps us to see that the sequence of proofs from I.27 through I.47 all rely on grasping the character of parallel lines and parallel figures drawn within them. There is no explicit claim that Euclid understood it this way. However, Proclus divides his commentary on the propositions in Book I according to whether the proposition depends on parallel lines or not. That is, the dependence of the Pythagorean theorem on the propositions depending on parallel lines—particularly I.35–41—was understood. By exploring these interwoven interconnections, we can see more clearly the geometrical structure of space for the ancient Greeks.

To identify them both as *equivalent* means that *they mutually imply each other*. To grasp the idea of the parallel postulate, then, is to grasp what follows from the assumption that "If a straight line falls on two straight lines and makes the interior angles on one side less than two right angles, the two straight lines if produced indefinitely meet on *that* side on which are the angles less than two right angles."

58   The Metaphysics of the Pythagorean Theorem

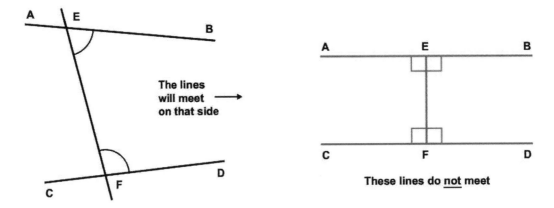

Figure 1.16.

Playfair's axiom is another description of Euclid's fifth postulate; it expresses the same idea by claiming that if there is some line AB and a point C not on AB, there is *exactly one and only one line* that can be drawn through point C parallel to line AB. But it is important to point out that Proclus proposes just this understanding of fifth postulate in his note to Euclid I.29 and I.31.[6]

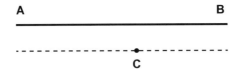

Figure 1.17.

The parallel postulate implies that:

1. In any triangle, the three angles are equal to two right angles

2. In any triangle, each exterior angle equals the sum of the two remote interior angles

3. If two parallel lines are cut by a straight line, the alternate interior angles are equal, and the corresponding angles are equal.

Thus, let us list the parallel theorems of Book I, propositions 27–33:

I.27: If a straight line falling on two straight lines makes the alternate angles equal to one another, the straight lines will be parallel to each other.

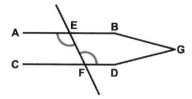

Alternate angles are equal and lines are proven to be parallel

Figure 1.18.

I.28: If a straight line falling on two straight lines makes the exterior angle equal to the interior and opposite angle on the same side, or makes the interior angles on the same side equal to two right angles, the straight lines will be parallel to one other.

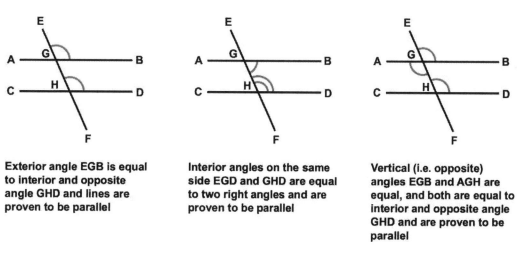

Exterior angle EGB is equal to interior and opposite angle GHD and lines are proven to be parallel

Interior angles on the same side EGD and GHD are equal to two right angles and are proven to be parallel

Vertical (i.e. opposite) angles EGB and AGH are equal, and both are equal to interior and opposite angle GHD and are proven to be parallel

Figure 1.19.

I.29: A straight line falling on a parallel straight line makes the alternate angles equal to one another, makes the exterior angle equal to the interior and opposite angle, and makes the interior angles on the same side equal to two right angles.

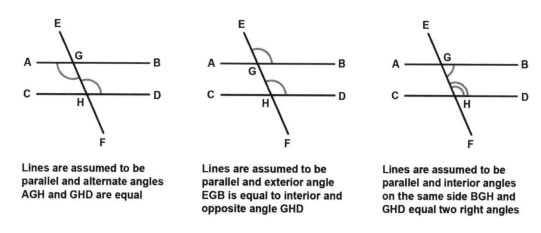

Lines are assumed to be parallel and alternate angles AGH and GHD are equal

Lines are assumed to be parallel and exterior angle EGB is equal to interior and opposite angle GHD

Lines are assumed to be parallel and interior angles on the same side BGH and GHD equal two right angles

Figure 1.20.

I.30: Straight lines parallel to the same straight line are also parallel to one another.

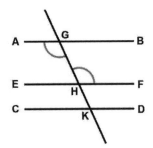 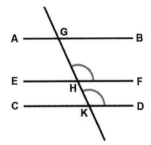 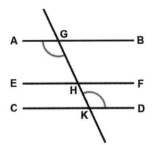

Lines AB and CD are assumed to be parallel to EF, AB is proven to be parallel to CD, and alternate angles AGH and GHF are equal

Lines AB and CD are assumed to be parallel to EF, AB is proven to be parallel to CD, and exterior angle GHF is equal to interior and opposite angle GKD

Lines AB and CD are assumed to be parallel to EF, AB is proven to be parallel to CD, and angle AGK is also equal to alternate angle GKD

Figure 1.21.

I.31: Through a given point to draw a straight line parallel to a given straight line.

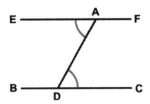

If a straight line from A to BC makes the alternate angles EAD and ADC equal, then EF and BC must be parallel.

Figure 1.22.

I.32: In any triangle, if one of the sides be produced, the exterior angle is equal to the two interior and opposite angles, and the three interior angles of the triangle are equal to two right angles.

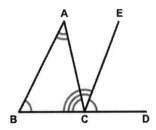

Figure 1.23.

This is the proof, sometimes called "Thales's theorem," that the angles in any triangle sum to two right angles. This is because Thales is also credited with the proof—or some understanding—that every triangle in a (semi-)circle is right, by Diogenes Laertius on the authority of Pamphile, a proof that presupposes the theorem that the base angles of isosceles triangles are equal, and this has been understood to require the prior understanding that the angles of every triangle sum to two right angles. The "proof" of I.32 is credited to the Pythagoreans by Proclus, as we shall discuss in the next chapters, but this does not preclude the likelihood that Thales understood its interconnections; as we considered in the Introduction, working out Geminus's claim that the "ancient" (= Thales and his school) investigated that there were two right angles in each species of triangles—equilateral, isosceles, scalene—means that one can plausibly argue that the principles here would have been understood.

I.33: The straight lines joining equal and parallel straight lines (at the extremities which are) in the same directions (respectively) are themselves also equal and parallel.

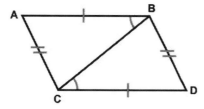

Figure 1.24.

The central idea, for our purposes, is to display the fundamental propositions or theorems that lead to the proof of I.47. They include the parallel propositions and then the areal relations and equivalences between triangles and parallelograms constructed on the same base between the same parallel lines. In setting out theorems I.27–33 we begin to see how Euclid I–VI imagined the world of geometrical objects to be rectilinear on flat surfaces, and that space is imagined as a vast template of *invisible lines*, parallel and not, in the context of which rectilinear figures appear. In that world, parallelograms on the same base are one and all equal to each other, triangles on the same base are also equal to each other, and since every parallelogram can be divided by its diagonal into two equal triangles, every parallelogram has twice the area of every triangle drawn on the same base. To these concerns we, and Euclid, now turn.

Thus, parallelograms on the same base and in the same parallel lines, as in the illustration below, are equal to each other (35, 36) (NB, at I.34, "parallelogrammic *areas*" are first introduced; before this proposition, figures are merely said to be equal, but not equal in area. Again, let us be clear that for Euclid, "area" is not a numerical measure but rather a space contained by the parallelogram. Thus χωρίον is the space contained itself.)

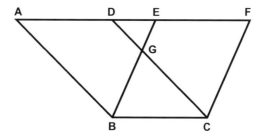

Figure 1.25.

62   The Metaphysics of the Pythagorean Theorem

The proof follows the sequence that, first, ABCD and EBCF are both parallelograms on the same base, BC. This means that AD is parallel to BC, EF is also parallel to BC, and since the opposite sides of a parallelogram are equal, AD is equal to EF.

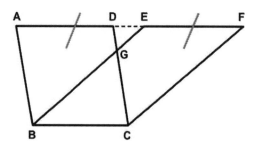

Figure 1.26.

Thus, if DE is added to each, lines AE and DF will be equal. On the other hand, AB is equal to DC, being the opposite sides of parallelogram ABCD. From this it follows from I.4 (SAS equality) that triangle AEB is equal to FDC since sides EA and AB are equal respectively to sides FD and DC, while angle EAB is equal to FDC (AB and DC being parallel).

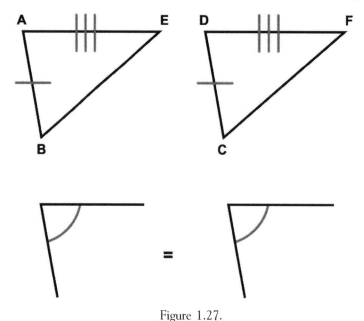

Figure 1.27.

Now, the argument continues, if triangle DGE is subtracted from each, it makes the trapezium ABGD equal to trapezium EGCF, because when equals are subtracted from equals the remainders are equal.

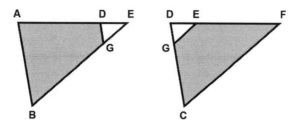

Figure 1.28.

And now, when one adds triangle GBC to both trapezia, the whole parallelogram that results, ABCD, is equal to the whole parallelogram EBCF. Thus, parallelograms on the same bases between the same parallel lines are equal to one another.

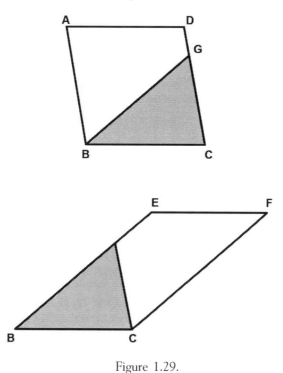

Figure 1.29.

In the next step toward I.47, below, triangles on the same base and between the same parallel lines are equal in area to one another, Euclid I.37:

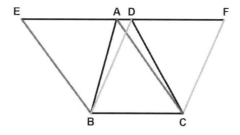

Figure 1.30.

64   The Metaphysics of the Pythagorean Theorem

As the diagram at Euclid I.37 makes visible, *triangles on the same bases and in the same parallels have equal area*. Euclid has already established that parallelograms have opposite sides equal, and that parallelograms on the same base and in the same parallel lines are equal in area to each other. Now he extends these arguments to triangles. He places two triangles, BAC and BDC, on the same base, BC, and then creates a parallelogram out of each, EBCA and DBCF, and each of them is equal to the other, since they too are on the same base and between the same parallel lines.

Since the triangles bisect the equal parallelograms—AB bisects EBCA, and FB bisects DBCF—the triangles must be equal because each is half the area of the equal parallelograms.

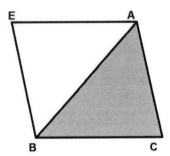 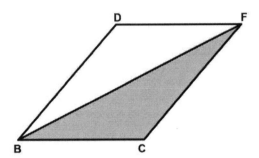

Figure 1.31.

Thus the proof that triangles on the same bases and in the same parallel lines are equal to each other follows from the proof for areal equivalence between parallelograms. *Note that every parallelogram can be imagined as being built out of two triangles, created by a diameter—a diagonal—that bisects it, and thus every parallelogram is imagined as being dissectible into triangles. The deep intuition is that all parallelograms reduce ultimately to triangles, and that the relation between triangles can be illuminated by projecting their areas into parallelograms composed of them.*

And then, finally for grasping the sequence in I.47, we have I.41, where Euclid proves that every parallelogram that shares the same base as any and every triangle, within the same parallel lines, must always have double the area of any and every triangle.

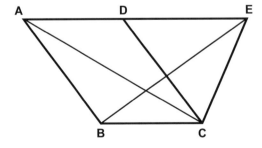

Figure 1.32.

Euclid's diagram I.41: If a parallelogram has the same base with a triangle and both are within the same parallels, the parallelogram is double the area of the triangle. The proof is to construct a parallelogram ABCD and triangle EBC, both on the same base BC, and between two parallels AE and BC. First, let us focus on the point that both of the triangles ABC and EBC, below, have equal area because both are constructed on the same base between the same parallel lines.

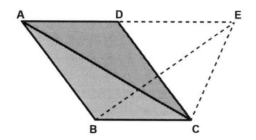

Figure 1.33.

Then, since AC is the diagonal of the parallelogram ABCD, triangles ABC and EBC are equal, since both are on the same base BC and between the same parallel lines. And since AC is the diameter of the parallelogram ABCD and bisects it, the triangle ABC is half the area of the parallelogram, and since triangle ABC is equal to triangle EBC, then the parallelogram is twice the area of any triangle sharing the same base, between the same two parallel lines.

So, in summation, let us be clear that the mathematical intuitions required for proving the Pythagorean theorem of I.47 include equality of triangles, and the areal relations among triangles and parallelograms; between two parallel lines, and sharing the same base, parallelograms—squares and rectangles—have twice the area of triangles; and *perimeter is not a criterion of area*. It seems commonsensical that as the perimeter of a figure increases, so does its area, and this is usually true. But figures constructed between parallel lines and on the same base *all* have the *same* area, regardless of their perimeters. In the diagram below, both triangles ABC and DBC have the same area, but DBC has a *much greater* perimeter.

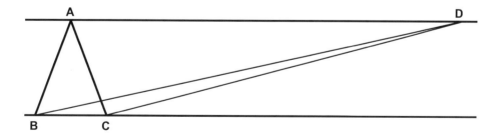

Figure 1.34.

Let us be clear about this diagram and its meaning. Triangle ABC has the same area as triangle DBC. Had the parallel lines AD and BC been extended in the direction of D so that D was placed hundreds or even thousands of miles away, triangle DBC *would still have the same area* as triangle ABC. It is in this sense that *perimeter is not a criterion of area*. Thus, the perimeter of triangle ABC might be, say, eight units (adding up the lengths of the sides), while the perimeter of triangle DBC might be hundreds of units (!) but nevertheless have the same area. In his commentary on Euclid, Proclus observes this point with care:

> Such a misconception is held by geographers who infer the size of a city from the length of its walls. And the participants in a division of land have sometimes misled their partners in the distribution by misusing the longer boundary line; having acquired a lot with a longer periphery, they later exchanged it for lands with a shorter boundary and so, while getting more than their fellow colonists, have gained a reputation for superior honesty. . . . This has been said to show that we cannot at all infer equality of areas from the equality of the perimeters; and so we should not be amazed to learn that triangles on the same base may have their other sides lengthened indefinitely in the same parallels, while yet the equality of their areas remains unchanged.[7]

Thus one of the mathematical intuitions that underlie the proof of I.47 is this principle that, contrary to a certain kind of common sense, perimeter is not a criterion of area, though often it appears to be. In the two triangles above on base BC, ABC has the same area as DBC, even if base BC were imagined to be somewhere in the United States with point A directly above, while point D was imagined, within the same parallel lines, to be in South America! The mathematical intuition reflects the specific conditions of parallel lines and figures constructed between them on the same bases, and on flat surfaces.

B

The "Enlargement" of the Pythagorean Theorem: Euclid VI.31

(i) The Pythagorean theorem of Euclid VI.31 (following Heath):

> Ἐν τοῖς ὀρθογωνίοις τριγώνοις τὸ ἀπὸ τῆς τὴν ὀρθὴν γωνίαν ὑποτεινούσης πλευρᾶς εἶδος ἴσον ἐστὶ τοῖς ἀπὸ τῶν τὴν ὀρθὴν γωνίαν περιεχουσῶν πλευρῶν εἴδεσι τοῖς ὁμοίοις τε καὶ ὁμοίως ἀναγραφομένοις.
>
> *In right-angled triangles the figure on the side subtending the right angle is equal to the similar and similarly described figures on the sides containing the right angle.*

Let ABC be a right-angled triangle having the angle BAC right;

The Pythagorean Theorem

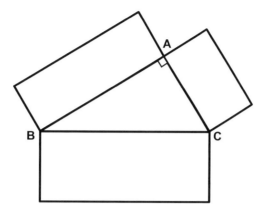

Figure 1.35.

I say that the figure on BC is equal to the similar and similarly described figures on BA, AC.

Let AD be drawn perpendicular.

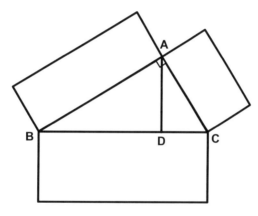

Figure 1.36.

Then since, in the right-angled triangle ABC, AD has been drawn from the right angle at A perpendicular to the base BC, the triangles ABD and ADC adjoining the perpendicular are similar both to the whole ABC and to one another. **[VI.8]**

And since ABC is similar to ABD, therefore, as CB is to BA, so is AB to BD. **[VI Def. 1]**

And since three straight lines are proportional, as the first is to the third, so is the figure on the first to the similar and similarly described figure on the second. **[VI.19, porism]**

Therefore, as CB is to BD, so is the figure on CB to the similar and similarly described figure on BA.

For the same reason also, as BC is to CD, so the figure on BC is to that on AC;

so that, in addition, as BC is to BD, DC, so the figure on BC to the similar and similarly described figures on BA, AC. **[V.24]**

But BC is equal to BD, DC;

Therefore the figure on BC is also equal to the similar and similarly described figures on BA, AC.

Therefore etc.

QED [OED]

In this proof, the figure is the εἶδος. In Book VI, sometimes the generic phrase used is ὅμοια σχήματα, "similar figures," but the term τετράγωνον ("square") does not appear here. The proposition proved is that for *any figure*, similar and similarly drawn (εἴδεσι τοῖς ὁμοίοις τε καὶ ὁμοίως ἀναγραφομένοις),[8] the figure on the hypotenuse has an area equal to the sum of the areas of the figures drawn on the other two sides.

At first look, the diagram, and proof, at VI.31 seems surprising to many, since they had learned the "Pythagorean theorem" as the *square* on the hypotenuse was equal to the sum of the *squares* on the two sides, and the appearance of rectangles does not quite fit into that earlier understanding. The diagram presented here is the work of Heiberg's edition of 1881 and reproduced by Heath and the many who followed this tradition, but it is not in the oldest manuscript tradition, though this point must be modified slightly. The rectangles in the diagram of VI 31 are no mere invention of Heiberg, for Commandino's edition already has rectangles on each side of the triangle in his Italian translation of 1575.[9]

The oldest manuscript traditions for VI.31, however, do *not* contain *any* figures at all constructed on the sides of the right triangle, though caution may be in order since diagrams in the medieval manuscript tradition tend to be schematic. Thus, conic sections tend to be represented as circles or parts of circles. Nevertheless, Codex B and Codex b both have what appears to be an isosceles right triangle (right angle not identified but clearly at 'A'.[10] By common sense, it is obvious that dividing an isosceles right triangle in half yields two parts that equally make up the whole, and each part is similar by being exactly half the area of the whole, and of course in the same shape. The continuous division of isosceles right triangles, by the perpendicular from the right angle to the hypotenuse, continues to reduce each to two similar triangles, continuing the process of halving, and thus the continuation into equal right triangles ad infinitum. Areal equivalences are most immediately apparent in the case of *isosceles* right triangles. This was a point emphasized also by von Fritz in his important essay that deals with Thales, Pythagoras, and the discovery of the irrational.[11] The diagram below is repeated in both Codex B and Codex b (below left). I have inserted the right-angle box in the smaller diagram (above right) for the sake of clarity.

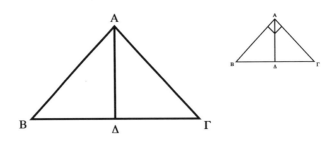

Figure 1.37.

(ii) Reflections on the strategies of Euclid VI.31:

The proof works by the following series of steps. First, a right triangle is constructed, and then a perpendicular, AD, is drawn from the right angle to the side that subtends it, namely, the hypotenuse. This perpendicular divides the right triangle into two smaller right-angled triangles, each of which is similar to the other, and to the whole, largest triangle; this is proved in VI.8. This means that all three triangles have their angles respectively equal, and the sides about the equal angles respectively proportional. As the largest triangle's hypotenuse length, BC, is to its shortest side, AC, so the smallest triangle's hypotenuse length, AC, is to its shortest side, DC (geometrical ratios require that "like" be compared with "like"). Moreover, the word for proportion is ἀναλογία, and indeed, the proportion sets out an analogy. Now, since the shortest side of the largest triangle, AC, is also the longest side (i.e., the hypotenuse) of the smallest triangle, so we have the respectively equal sides in the same ratio, and note that, consequently, AC appears twice:

$$BC : AC \quad :: \quad AC : DC$$

Thus, we have three straight lines that are proportional—BC, AC, DC—so as the ratio of the length of the first, BC, is to the length of the third, DC, so the area of the figure on the first, BC, is in the same ratio to the area of the similar and similarly described figure on the second, AC (porism to VI.19). And this means that BC has to DC a *duplicate* ratio of BC to AC by definition (Book V. def. 9). And so if any three magnitudes M, N, and L are in proportion, that is M : N :: N : L, then we can call the ratio M:L the duplicate ratio of M : N (or N : L). Thus, as the *lengths* of the longest and shortest sides of triangle ABC stand in duplicate ratio (first to the third), so do the *areas* constructed on the longest and middle lengths (first to the second). The areas in the same ratio are duplicated (i.e. squared).

Then, by the same steps of reasoning, BC : BD is the same as the ratio of the figure on BC to the figure on BA. Thus, BC : BD is the duplicate ratio of BC to BA. Thus, the ratio of the lengths of the sides—first to the third—is the same as the areas constructed on those sides respectively—first to the second. As the lengths of the sides increase, the area of figures constructed on them increases in the duplicate ratio of the corresponding sides. Stated anachronistically and in more modern terms, the *areas* scale in the duplicate ratio of the linear scaling, that is, the increase of line length.

And then the argument appeals to theorem V.24 in the theory of proportions. The theorem V.24 is: If a first magnitude has to a second the same ratio as a third has to a fourth, and also a fifth has to the second the same ratio as a sixth to the fourth, then the sum of the first and fifth has to the second the same ratio as the sum of the third and sixth has to the fourth. In effect, the principle of V.24 is that if A : B :: C : D and E : B :: F : D, then (A + E) : B :: (C + F) : D.[12]

(iii) The geometrical Intuitions—the sequence of ideas that are connected in the proof:

(a) First, there are general issues about *proportion* that are presupposed, and this requires some reflection on the idea of *ratios*, (b) next, there are ideas about *similarity* presupposed, and (c) there are ideas that connect lengths with areas, bringing together ratios, proportion, and similarity, and this discussion will lead to the principle of "duplicate ratio," and to the understanding that all polygons are reducible ultimately to triangles. To understand duplicate ratio we must first be clear about arithmetical vs. geometrical means, and then show how lengths and areas are interwoven with it.

C

Ratio, Proportion, and the Mean Proportional (μέση ἀνάλογον)

To understand the proof of the *enlargement* of the Pythagorean theorem in Book VI, by similar triangles, we must clarify the key, relevant terms of the theory of proportions articulated in Book V. Euclid tells us that a "ratio" (λόγος) is a sort of relation in respect of size between two magnitudes of the same kind (Def V.3), and a "proportion" (ἀναλογία) is a special relation between two ratios. Accordingly, Euclid states, "Let magnitudes which have the same ratio be called proportional" (Def. V.6). Thus, a ratio provides a comparison of the relative sizes of two magnitudes, and since all the theorems of Book VI depend on ratios, the theory of proportions, and hence ratios (Book V), precedes the discussion of similar figures (Book VI). Now I follow James Madden in his extremely clear presentation of ratio and proportion in Euclid:[13]

> We think of a ratio as a number obtained from other numbers by division. A proportion, for us, is a statement of equality between two "ratio-numbers." When we write a proportion such as $a/b=c/d$, the letters refer to numbers, the slashes are operations on numbers and the expressions on either side of the equals sign are numbers (or at least become numbers when the numerical values of the letters are fixed).
>
> This was not the thought pattern of the ancient Greeks. When Euclid states that the ratio of A to B is the same as the ratio of C to D, the letters A, B, C and D do not refer to numbers at all, but to segments or polygonal regions or some such magnitudes. The ratio itself, according to Definition V.3, is just "a sort of relation in respect of size" between magnitudes. Like the definition of "point," this tells us little; the real meaning is found in the use of the term. It is in the rules for use that we find the amazing conceptual depth of the theory.
>
> The definition that determines how ratios are used is Euclid V.5. This tells us how to decide if two ratios are the same. The key idea is as follows. If we wish to compare two magnitudes, the first thing about them that we observe is their relative size. They may be the same size, or one may be smaller than the other. If one is smaller, we acquire more information by finding out *how many copies of the smaller we can fit inside the larger*. We can get even more information if we look at various multiples of the larger, and for each multiple, determine *how many copies of the smaller fit inside*. So, *a ratio is implicitly a comparison of all the potential multiples of one magnitude to all the potential multiples of the other*. (Italics added for emphasis.)

These notions are justified by the very nature of magnitude and ratio as set out in the very first definitions of Book V.

When I turn to consider the metaphysics of this proposition at the end of the chapter, I will emphasize that a ratio, then, tells us how many small worlds fit into the big world—how many microcosms fit into the macrocosm. The whole cosmos is conceived aggregately out of small worlds. Here is the connection to "similarity."

The first theorem of Book VI, proposition 1, states that triangles of the same height are proportional to their bases, and thus two triangles that have the same height but one is on a base twice that of the other, the triangle on that larger base is *twice the triangle* on the smaller base. Stated differently, this means that one triangle is to another as one base is to another, and so *twice the triangle base* means twice the *area*.

A crucial consideration for this point of departure is that it connects magnitudes of different kinds: areas and lengths.[14]

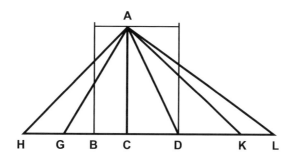

Figure 1.38.

The proof of VI.1 is built on the same foundation as I.47, namely, I.35, 38, 41. The strategy is to create equimultiples of the areas of segments in order to apply definition 5 of Book V. Parallelograms on the same base and between the same parallel lines have equal area. Triangles on the same base between the same two parallel lines have equal area, and parallelograms sharing the same base within those same parallel lines have twice the area of the triangle on the same base. Book VI then introduces *proportions*, and this enables the focus to explore figures *not* on equal bases.[15] Book VI, then, begins by showing that if some base is larger or smaller, the figure itself is in *excess* or *defect* of the other. While the height remains the same, as the base increases, so does the comparative area—the ratio of the magnitude of one to the other. In VI.1 Euclid shows that the areas increase exactly in proportion to the increases of the bases; in VI.19, however, we are speaking about similar figures all of whose sides are increased according to a certain ratio, and in that case, the increase of the areas is in the *duplicate ratio* by which the sides increase. At the same time we should observe that "similarity" is the foundation of the microcosmic-macrocosmic argument form. The little and the big figures that are *similar* share the same structure, the same shape, but as these figures increase in magnitude, their areas increase, and they increase in the duplicate ratio (i.e., the squared ratio) of their sides.[16] Before continuing to follow the sequence, I turn to clarify the key idea of the "mean proportional" or "geometrical mean."

D

Arithmetic and Geometric Means

Let us be clear about the difference between an "arithmetic mean" and a "mean proportional" (μέση ἀνάλογον) or "geometrical mean." An arithmetic mean is an *average*, or midpoint, between two magnitudes or numbers. If we sought the arithmetic mean between two magnitudes, one 18 units and the other 2 units, the arithmetic mean would be 10. We could describe it as the midway or halfway point between the two extremes. We could reckon it by adding both together and dividing by 2. Thus, the arithmetic mean is the midpoint between two extremes, or in comparing magnitudes, a length that is the average between two other lengths that are extremes to its mean.

While the arithmetical mean has a common difference with respect to the extremes, the geometric mean has a common ratio. Thus, the difference between the extremes is double the difference between an extreme and the mean, in the case of an arithmetic mean, while the ratio between the extremes is the duplicate ratio between an extreme and a mean in the case of a geometric mean. Now, if I may restate this relation in terms that the mathematician will not appreciate but that non-mathematicians may find helpful: the mean proportional or geometric mean is a multiplicative, not arithmetical, function. Thus, while the arithmetical mean of 2 and 18 is 10, the geometrical mean is 6. This means that there is a ratio or factor 3, multiplied by the first number, that when multiplied again by that same factor results in 18—the third number 18 is in the *duplicate ratio* of the factor that separates and connects the first two numbers, 2 and 6. Thus, the mean proportional between 2 and 18 is 6: 2 multiplied by 3 is 6, and 6 multiplied by 3 is 18. In this case, the factor is 3. The idea behind the mean proportional that I shall emphasize is that this relation expresses a principle of *organic growth*—it is a continuous proportion. However, the duplicate ratio is a statement of fact, not a process, but what it signifies is the fact that as line lengths increase in continuous proportions—grow— certain relations hold between the line segments and the areas of similar figures drawn on them.

The reader should note that the numerical interpretation of the mean proportional works out well for 2 and 18, but in many other cases we will not find whole-number solutions (that is, integers). In the case of 9 and 18, for example, the factor turns out to be 9 times $\sqrt{2}$. In Euclid, Books I–VI, no number is given or reckoned, so how shall we understand the idea of a "mean proportional" in terms of lengths and areas? Let us consider in more detail the idea of "blowing up" or "scaling up"—that is *enlarging*—a figure. If we do so imaginatively, every dimension "blows up" uniformly. Another way to express this is to say that each angle or dimension "scales." For those who have used some type of overhead projector—a slide projector or PowerPoint projector, for example—the image projected onto a screen is much larger than the slide or computer screen, but all the relations of magnitudes and angles between the parts in the image that are enlarged have the *same ratios* between their parts. Thus, if the image of a geometrical figure, say a triangle, is projected on a screen and we seek to make the image larger as we zoom in (or smaller as we zoom out), the relation between the sides remains the same—remains in the same ratio—and so the image is blown up proportionately. If we had a way to measure the percentage of enlargement, that percentage would be the *factor* by which the whole image scales up. We could call that factor 'k', and the point is that when this factor is repeated to produce a larger triangle, and yet an even larger triangle—or a smaller and smaller one—this factor is the multiplicative function of the mean proportional. Clearly, an understanding of this principle was known to every artist working in archaic times—for instance, the vase painter—when the same scene was replicated on larger and larger vases, and it was the same procedure employed in Egyptian tomb painting, by means of proportional grids, as we already considered in images and discussion earlier.[17] And certainly the production of larger and larger vases required the potter to understand what procedures needed to be controlled to make a vase, or its painted images, twice the height or width of a smaller example. Each image or material is scaled up. The successful examples prove that the ancient artisans knew how to produce these results.

In every right-angled triangle, the perpendicular drawn from the right angle to the hypotenuse divides the hypotenuse into two parts: the length of this perpendicular is the mean proportional between the two lengths into which the hypotenuse is divided. But let us be clear—it is not the arithmetical average length between the two segments of the hypotenuse, but rather, the μέση ἀνάλογον, the mean *proportional* between the two extremes in terms of a common factor. That common factor is the percentage of enlargement, or reduction, that connects the segments in proportion, the factor in terms of which the figures scale.

Now let us continue in the sequence leading to the enlargement of the hypotenuse proof, VI.8: The perpendicular from the right angle of a triangle to its hypotenuse divides the hypotenuse into two right triangles similar to each other and to the larger, whole triangle. Each of the triangles has equal angles and the sides about those equal angles are respectively proportional.

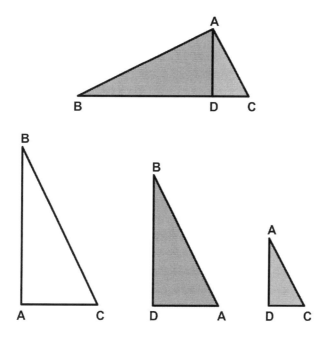

Figure 1.39.

74   The Metaphysics of the Pythagorean Theorem

The reader will notice that, in the above diagram, the three triangles have been rotated to display that they are all similar—the shape is the same—and thus the angles in the figure are the same, but the lengths have changed. Euclid, as we have already considered, tries to avoid rotation and physical comparisons because of their inherent imperfections, but this is almost certainly how Thales, Pythagoras, and Greeks of the sixth century BCE would have shown similarity, and compared shapes, angles, lengths, and areas. Note also that the porism or corollary of VI.8 is that *the perpendicular from the right angle to the hypotenuse is the mean proportional* (μέση ἀνάλογον), *or geometric mean, between the two lengths into which the hypotenuse is divided by it*. The point is that once you know that ABD is similar to CAD, you know that BD : AD :: AD : DC, for the ratios of corresponding sides must be the same. So *by the very definition of* a "mean proportional," AD is a mean proportional between BD and DC.[18] The mean, then, is the middle proportional length between two lengths that stand to them as extremes—one longer and one shorter—that connects the two in a "continuous" proportion.

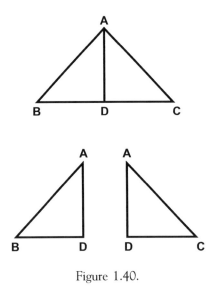

Figure 1.40.

Our next step is to construct three lengths that are proportional since this is a building block en route to VI.31 by VI.13, 17, and 19. This is shown in VI.11. The idea is to begin with any two line segments; from them, a third proportional is constructed. Consider the diagram below:

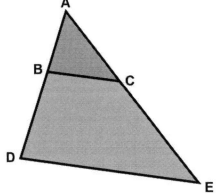

Figure 1.41.

We begin with line segments AB and AC. We construct BD in a straight line with AB, and equal to AC. Then we connect points B and C, and extend AC in a straight line to point E so that when we connect points D and E they will be parallel to BC. Segment CE is the third proportional to AB, AC, because AB is to BD as AC is to CE by VI.2). But BD is equal to AC, so that AB is to AC as AC is to CE. This is the definition of a third proportional.

Next, VI.13 shows the construction for that porism that the perpendicular from the right angle to the hypotenuse is the "mean proportion" between the two magnitudes into which the hypotenuse is now divided. This is the construction of the *mean proportional*. As I will argue later, this was the plausible route by which both Thales and Pythagoras came to grasp the theorem of VI.31; it is the principle by which the right triangle expands and unfolds, or contracts and collapses—*organic growth*. The route was through Euclid III.31—but as I will explore in the next chapter not by the proof that appears in Euclid, which is credited to both Thales and Pythagoras, that every triangle inscribed in a (semi)circle is right-angled. It also presupposed an understanding of the mean proportional between two extremes. To guarantee a right angle, every triangle inscribed in a (semi-)circle is right.

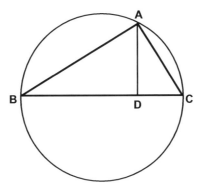

Figure 1.42.

The next crucial step to our goal of understanding the intuitions presupposed in VI.31 is Euclid VI.16 and 17. These are conclusions that follow from the corollary of VI.8, that the perpendicular drawn from the right angle to the hypotenuse is the mean proportional. The theorem of VI.16 is: *If four straight lines be proportional, the rectangle contained by the extremes is equal to the rectangle contained by the means; and, if the rectangle contained by the extremes be equal to the rectangle contained by the means, the four straight lines will be proportional.*

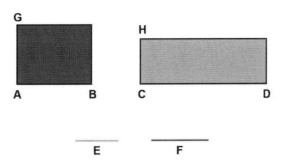

Figure 1.43.

76   The Metaphysics of the Pythagorean Theorem

The proof strategy is to begin with four straight lines AB, CD, E, and F that are proportional so that AB is to CD as E is to F, and to show that the rectangle made by AB, F is equal in area to the rectangle made by CD, E.

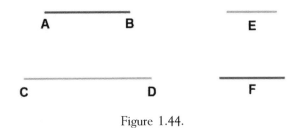

Figure 1.44.

Thus, when two line lengths are proportional so that the longer CD is to the shorter AB as the longer F is to the shorter E, then the rectangle made by the longer CD with the shorter E is equal in area to the rectangle made by the shorter AB with the longer F.

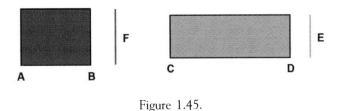

Figure 1.45.

The principle of showing that rectangles whose sides are proportional are of equal area, then, is that "equiangular parallelograms in which the sides about the equal angles are reciprocally proportional are equal."

And the next step, connecting the areas of rectangles and squares, and hence the mean proportional—or geometrical mean—is shown in VI.17: If three straight lines be proportional, the rectangle contained by the extremes is equal to the square on the mean; and, if the rectangle contained by the extremes be equal to the square on the mean, the three straight lines will be proportional.

From VI.16 we know that if *four* straight lines are proportional, the rectangle made from the longest and shortest is equal in area to the rectangle made by the two other segments (or means); *this is the principle of showing that the rectangle made by the extremes is equal to the rectangle made by the means*. Now given *three* straight lines *in continuous proportion*, the same principle is applied, but this time the rectangle made from the extremes—the longest and shortest—is equal to the rectangle made by the means. Since the rectangle made by the "means" has two *equal* sides, the rectangle is equilateral, and thus a square. It is my conjecture that this is how Thales came to know that the square on the perpendicular from the right-angled triangle was equal in area to the rectangle made from the two extremes into which the hypotenuse was divided by it—this was his discovery of the μέση ἀνάλογον. But it is also plausible to suppose that Thales's intuitions about the geometric mean were reflected later in what Euclid preserves through I.45, namely, the construction of a parallelogram equal to a given rectilinear figure in a given rectilinear angle.

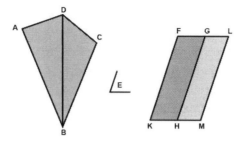

Figure 1.46.

However, the route that led Thales to this insight may be, I will argue in the next chapter, that the grasping of a continued proportion of three straight lines is part of understanding the cosmos as unfolding or growing (or collapsing) in a regular and orderly way; the discovery of this pattern, this order, had as its visual proof the areal equivalences created by the line lengths—the rectangle and square. Connectedly, this is how Thales came to discover the areal relation among the sides of a right triangle when he searched further to discover the relation among the sides of the two similar right triangles into which the perpendicular partitioned the hypotenuse. The right triangle contains within itself these dissectible relations.

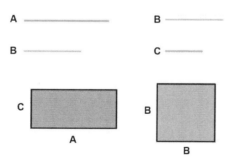

Figure 1.47.

The idea of the geometrical mean in a continuous proportion—a relation among three proportional line lengths, and consequently areas of figures constructed on them—is a principle of organic growth. It is my thesis, to be developed in the next chapter, that the discovery of the mean proportional—and hence continuous proportion—provided an answer to a question that looms behind Thales's doctrine that everything is ὕδωρ, how the cosmos grows, how the right triangle expands into different shapes, and hence different appearances.

In Euclid, the idea that we know mathematically as "squaring"—which is a term of geometrical algebra, and not part of Euclid's technical parlance, nor the parlance of the early Greek mathematicians or philosophers—is presented as "in duplicate ratio" (ἐν διπλασίονι λόγῳ); the expression is employed to describe a geometrical fact about the relation between side lengths and the areas of figures drawn on them, not a process or operation to be performed. Let us try to clarify this ancient Greek idea. A duplicate ratio is not a geometric version of an operation, algebraic or otherwise. It refers to a ratio obtained from a certain proportion: it refers to a state of affairs rather than a procedure. Thus, if you have a proportion with three terms, A, B, and C so that B is the middle term (that is, A : B :: B : C), then A : C is the duplicate ratio of A : B. The chief geometrical fact is that if two rectilinear figures, for example triangles, are similar, then the ratio of their areas is the duplicate ratio of their corresponding sides ("is the duplicate ratio" not "is produced by duplicating the ratio"!). Indeed, when we consider the proof of VI.19, below, we begin with similar triangles with corresponding sides BC and EF; next we bring forth a third proportional BG, that is, a line such that BC : EF :: EF : BG; then we show that the ratio of the areas (via VI.15) is the same as BC : BG. For Euclid, then, the duplicate ratio simply *is* so; it is not *produced* by an operation on the sides like squaring.[19]

To express this idea, Euclid now follows by showing that when a mean proportional is constructed, magnitudes—that is *lengths*—that stand in the relation of duplicate ratio—the first to the third—the figure constructed on those bases has magnitudes—that is *areas*—that also stand in duplicate ratios—the first to the second—to the sides on which they are similarly drawn. This means, then, that given three proportional lines, when the length of the first stands to the third in duplicate ratio, the areas of similar figures constructed on the first and the second stand in the same duplicate ratio. Thus the area of the similar and similarly drawn figure on the first length and the one constructed on its mean proportional or second length (separating it from the third and smallest length), stand in the same ratio as the first and third line lengths. In the context of Thales's cosmic imagination, the cosmos expands in continuous proportions, and the areas of figures constructed on those sides increase in the duplicate ratio of the corresponding side lengths—this is integral to grasping the meaning of the μέση ἀνάλογον.

The next step to VI.31 is shown in the proposition of Euclid VI.19, below, that "Similar triangles are to one another in the duplicate ratio of their corresponding sides." The proof begins with identifying triangles ABC and DEF as similar, and thus the angle at B is equal to the angle at E, AB is to BC as DE is to EF, and BC corresponds to EF. Given this, Euclid claims that triangle ABC has to triangle DEF a ratio duplicate of that which BC has to EF. His proof now is to construct a third proportional, BG, by drawing a line from vertex A to G, on the side opposite BC (cf. the construction of a third proportional given in VI.11), and so BC is to EF as EF is to BG. And so BC is to BG in the duplicate ratio of triangle ABC is to ABG. But triangle ABG is equal to triangle DEF, so triangle ABC also stands in the duplicate ratio that BC has to EF.

The Pythagorean Theorem 79

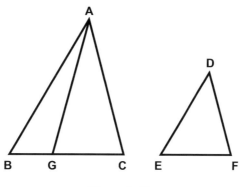

Figure 1.48.

There is one more crucial step in grasping the mathematical intuitions that underlie the enlargement of the Pythagorean theorem at VI.31, but it is not actually used in that proof. But I claim that this theorem has remained unappreciated in grasping the original discovery of the hypotenuse theorem for the Greeks of the sixth century; that is the proof at VI.20 that *similar polygons are divided into similar triangles, and into triangles equal in multitude and in the same ratio as the wholes, and the polygon has to the polygon a ratio duplicate of that which the corresponding side has to the corresponding side.*

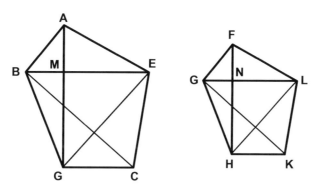

Figure 1.49.

What this theorem shows—the deep mathematical intuition—is that all rectilinear figures can be dissected ultimately into triangles, and similar figures dissect into similar triangles, and, following VI.19, the corresponding sides of similar triangles are in the duplicate ratio, and thus so are the corresponding sides of all polygons compounded of them.

All polygons—all rectilinear figures—reduce to triangles, and every triangle reduces ultimately to right triangles—by dropping a perpendicular from vertex to opposite side, and then the enlargement of the Pythagorean theorem at VI.31 shows that all right triangles dissect ad infinitum into right triangles by dropping a perpendicular from the right angle to the hypotenuse opposite, as shown in VI.8. In the diagrams below, the largest triangle (which could be *any* triangle where the perpendicular falls within the triangle) is divided into two

80   The Metaphysics of the Pythagorean Theorem

right triangles by dropping a perpendicular from any of its vertices to the side opposite; then, in the next four sequences from left to right, the operation continues indefinitely by dropping a perpendicular from the right angle to the hypotenuse, thus dividing that side into two right angles, and thus two similar right triangles. *The deep mathematical intuition is that it is right triangles all the way down.* The right triangle is the fundamental geometrical figure into which all rectilinear figures dissect.

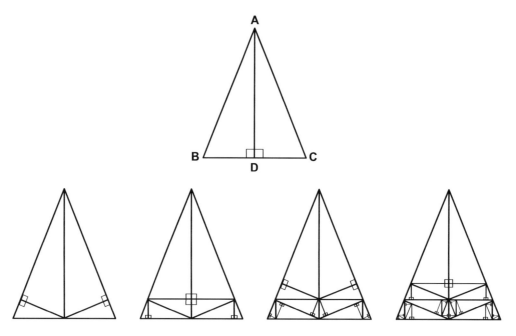

Figure 1.50.

The principle of adding ratios in V.24 is then applied in VI.31: V.24 says that if A : B :: C : D and E : B :: F : D then (A + E) : B :: (C + F) : D. This principle may strike us as adding fractions with a common denominator. If one were to compare it to what we do with fractions, it would be multiplication, but in Greek mathematics it is somewhat mysterious. It is almost by definition that the ratio compounded of U : V and V : W is U : W.[20] Here, V.24 is employed to show that as a line length is composed of two parts, and thus is the sum of those two parts, so the figures projected on the line lengths with which they are associated stand in the same added relation as the side lengths. Thus, the figures on the line lengths associated with the segments sum to the figure on the line length that is the sum of those parts.

Thus we now reach VI.31, the *enlargement* of the Pythagorean theorem. In the three similar triangles, BAC, BDA, and ADC, the corresponding sides opposite to the right angle in each case are BC, BA, and AC. *The triangles therefore are in the duplicate ratios of these corresponding sides.* In modern terms, we would say this means that the areas of those triangles stand in a ratio that is the square of the ratio in which the sides stand. As Giventhal summed up this matter, "We know that all pairs of corresponding sides stand in the same ratio, because that's part of what is *meant* by being similar. Suppose all the sides stood in the ratio 2 : 5, then the areas would stand in the ratio 4 : 25. Recall that the whole concept of 'duplicate ratio' was essentially the ancient's way of understanding the more modern concept of squaring,"[21] but let it be emphasized that "duplicate ratio" is not an operation—unlike squaring—but a statement of a fact about line lengths and the similar figures drawn on them.

It is my thesis that for Thales and Pythagoras, the mathematical intuitions that underlie VI.31 are the revelations that all rectilinear figures reduce to triangles and ultimately to right triangles—this is the microcosmic-macrocosmic principle. Inside every right triangle are two right triangles . . . dissectible forever. When the right triangles scale up or collapse—when the cosmos expands or contracts—it does so in a pattern of geometric mean proportions, that is, in the duplicate ratios of their corresponding sides. The proof of VI.31 appears with the diagram of a right triangle surrounded by **rectangles,** or perhaps they are **parallelograms**, but in any case they are not squares. Thus, here I have reached what I regard to be the conclusion of my investigation in this chapter, to figure out what someone knows if they know the "Pythagorean theorem." This is what we were searching for, and in the next two chapters we will explore whether a plausible case can be constructed that Pythagoras, and Thales before him, knew this and by what mathematical intuitions they discovered that the right triangle was the fundamental geometric figure.

E

Overview and Summary: The Metaphysics of the Pythagorean Theorem

The familiar refrain from mathematicians and historians of mathematics is that Euclid's *Elements* is free from metaphysical presuppositions, and in one sense this is certainly true because there is no explicit discussion of such matters. But stating it this way, in my estimation, obfuscates the metaphysical background, and metaphysical projects, that plausibly gave rise to the thinking that is expressed and systematized in the *Elements*. Euclid was the last in a line of Greek geometers who organized and presented *Elements*, and he was so successful that all the earlier efforts disappeared. My thesis is that it was in the context of metaphysical inquiries that geometrical techniques were first embraced; it was metaphysical inquiries that appealed to geometry as a way to discover the ultimate underlying geometrical figure, just as the earliest philosophers sought to identify the ultimate underlying principle or unity of nature.

If this is so, it should come as no surprise that even much later with Euclid's compilation of proofs, an original metaphysical problem that motivated geometrical investigations might still be sensed. Proclus himself declared that there was a metaphysical underpinning:[22] the whole purpose of Euclid was to get to the construction of regular solids of Book XIII.[23] This is why Euclid's *Elements* is set up this way, according to Proclus. The construction of the regular solids is a continuation of the original project that emerged when it was discovered that the right triangle was the fundamental geometrical figure. The regular Platonic solids—the four elements, and hence the whole cosmos, discussed in the *Timaeus*—are all built explicitly out of right triangles.

Euclid enters the picture some two-and-a-half centuries *after* the time of Thales and Pythagoras, and despite how slowly things may have developed at first into a systematization of geometry, for more than a century prior to Euclid's compendium (c. 300 BCE) of geometric efforts, formal proofs were a focus. Netz has already conjectured, and it is not implausible, that probably by the last quarter of the fifth century, the vast majority of propositions that make up Euclid's *Elements* were already known.[24] Gow even conjectured that perhaps Thales had known the contents of what later became Books I–VI;[25] if so, he knew the "Pythagorean theorem"—*both* proofs. Proofs became a focus epistemologically; they provided answers to questions such as "How do you know that?" "How can you be sure?" What are the interconnections, and presuppositions, of the claims that are being made? What are the implications of each proven theorem for the ones that follow? But at the beginning, with Thales, there could not have been a system, but rather there were some speculations that were furthered by reflecting on things geometrically. By making countless diagrams, probably in wet ground, scratched into the dirt, sand, or potter's clay, and reflecting on what relations could be discovered in those figures, they worked out the implications so far as they might have unfolded. There were no geometric texts circulating in Greece before the middle of the sixth century—the first *Elements* of which we know is the work of Hippocrates of Chios in c. 440 BCE—nor do we have any indications of a tradition engaged in discussing such matters. A possible exception is the architects in eastern Greece (Ionia), who, having only had previous experience building small structures, now began contemporaneously building monumental temples out of hard stone and writing about them in their prose treatises.[26] It seems entirely plausible that as Thales explained geometrical relations over and over again to his puzzled and quizzical compatriots, he began to grasp the general principles that stood behind them. It is an experience familiar to teachers everywhere. This might well have been the beginning of deductive strategies and proofs.

In bringing this chapter to a close, let us gather the mathematical intuitions presupposed to connect the sequence of thoughts in both proofs, and then set the stage for the following chapters on Thales and Pythagoras, which show the metaphysical speculations that plausibly led them onto these connections, now that we know what we are looking for. It is my contention that the mathematical intuitions that underlie Euclid's Book VI—especially similar triangles—were in nascent form grasped already by Thales, Pythagoras, and the Pythagoreans of the sixth and early fifth century BCE.

In review, then, we can see what anyone needs to grasp had they followed the trains of thoughts provided in the formal proofs that have come down to us from Euclid. To grasp I.47, that in right triangles the area of a square on the hypotenuse is equal to the sum of the areas of the squares on the two sides, one must (i) understand the idea of equality of triangles, (ii) understand proofs by *shearing*—that between parallel lines, parallelograms on the same base are equal in area, and so are triangles on the same base, and furthermore that parallelograms have twice the area of triangles sharing the same base, because every parallelogram, rectangle or square, is divisible into two equal triangles by its diagonal—and these relations require us to grasp, surprisingly, that perimeter is not a criterion of area. These ideas are pivotal to grasping I.47, and can be shown to be *equivalent* to Euclid's fifth postulate—to grasp the one is to grasp the other, whether or not one is clear that one has done so.[27] My point in emphasizing the interrelation of I.47 and the parallel postulate is to draw the readers' attention to the idea that the Pythagorean theorem of I.47 is something discoverable in the process of working out the interrelations of a line falling on two parallel lines and the figures that can be constructed on them. When one articulates space in accordance with Euclid's parallel theorems, one will see that the Pythagorean theorem is an inescapable consequence. And the entire conception of geometry—Books I–VI—presupposes a grasping of space as fundamentally flat. The Pythagorean theorem reveals a feature of right-angled triangles, and placed between parallel lines, the areal equivalences of the figures drawn on its sides follows a law or rule.

The fifth postulate states that when a straight line falls on two straight lines and makes the interior angles on one side less than two right angles, the two straight lines if produced indefinitely will meet on that side on which are the angles less than two right angles. Expressed as Playfair's axiom, the postulate can be rephrased: Given a straight line AB and a point C not on that line, there is exactly one and only one line drawn through point C that is parallel to the line AB. And so one is forced to imagine the internal relations that follow apodictically from the intersection of a straight line on two parallel lines. And moreover, all of geometrical space is conceived as expressible in rectilinear figures on a flat field defined by parallel lines, and *every one of those figures can be dissected ultimately to triangles*. The consequences of the parallel postulate are that in any triangle the three angles sum to two right angles, each exterior angle equals the sum of the two remote interior angles, the intersection of a straight line on two parallel lines makes the alternate angles equal and the corresponding angles equal, and from these it is implied that the square on the hypotenuse is equal to the sum of the squares on the two sides.[28]

And let us keep in mind that while this theorem is restricted to *right* triangles, *every triangle into which all rectilinear figures can be reduced is reducible ultimately to right triangles, and thus what is true of right triangles is true of all triangles, and a fortiori true of all rectilinear figures into which they dissect*. Heath, and Gow before him, emphasized just this point. Finally, let us observe that this proof displays areal equivalence; it has nothing to do with the assignment of numeric or metric values such as 3, 4, 5 for the lengths of the sides, an example of the famous "Pythagorean triples." No numbers at all are supplied or involved with the proof, nor anywhere in the first six books of Euclid are numbers even mentioned. Proclus says that I.47 was Euclid's discovery, and we have no reason to suppose it was not. In any case, however, it was not the route by which Thales and Pythagoras made their discoveries; it is my thesis that the route was through and along to VI.31—this was the earlier discovery for the sixth-century Greeks.

Let me add one more qualification in my argument about the meaning I attach to "discovery." When I argue that Thales and Pythagoras plausibly *discovered* the hypotenuse theorem by a certain train of thoughts, I do not mean to argue that the theorem was not known long before them in Mesopotamia and elsewhere, as Neugebauer argued almost a century ago when he translated the Cuneiform tablets.[29] And for all we know, perhaps Thales was aware of this mathematical curiosity from some travel to Babylon, or noticed it in the use of knotted cords by Egyptian builders or even patterned in tiles, or had it conveyed to him by someone on the caravan route from China and India that ended in Miletus and Ephesus. My claim is that their "discovery" came in the context of metaphysical inquiries, and there is no evidence of the theorem being discussed by any other culture in such a context, though in the process of working out this problem for the Greeks, geometrical figures took on a life all their own, and a discipline grew up separately to treat them.

The proof of VI.31 is by a completely different line of reasoning than I.47; it depends on an understanding of ratios and proportions, and similarity, and rests on how the ratios of the lengths of the sides of a right triangle produce areal equivalences in the same ratios. The length of every line conjures a figure unfolded from that magnitude. The least number of lines

84   The Metaphysics of the Pythagorean Theorem

to enclose a space is three—a triangle. Each magnitude conjures a triangle (and a square, and all other polygons) as extrapolations of its length. And every rectilinear figure, every rectilinear magnitude, reduces ultimately to a right triangle. And from there, it is *right triangles all the way down*. Let us be clear how these ideas are connected. First, we focus on decreasing microcosms and then expand into increasing macrocosms.

Let us consider the deep intuition about Euclidean space—and a fortiori space for the Greeks of the Archaic period as well—in the following way. If one reflects on it, one sees that to state the Pythagorean theorem at all—in the case of VI.31—we must be able to refer to figures that are similar but not equal.[30] Even in I.47, where squares are depicted, whether or not the squares on the legs are equal, the square on the hypotenuse will be greater than the others, that is, similar but not equal. For this reason the Pythagorean theorem is referring to a very deep aspect of Euclidean geometry.[31] In this sense, VI.31 directs itself to the root of the problem, to explain *how* figures can be similar without being equal.[32] The metaphysical problem for Thales was to explain how appearances could be similar without being equal, to show how everything had a single underlying unity—ὕδωρ—and thus divergent appearances were nonetheless similar, even though in an important way they were not equal to it. By my account, young Pythagoras learned this project from Thales or members of his school and developed and refined it. To account for differences in appearances without accepting change, only alteration, the solution was the application of areas—the rules of *transformational equivalences* between appearances—to show how all triangles could be expressed as rectilinear figures, equal in area to other rectilinear figures or similar to them. We shall explore the application of areas theorems in chapter 3.

Every triangle can be divided into two right triangles by dropping a perpendicular from the vertex to the side opposite, below left:

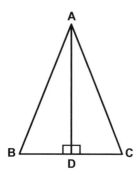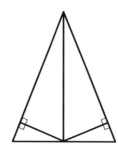

Figure 1.51.

And every right triangle can be reduced continuously into two smaller and similar right triangles by drawing a perpendicular from the right angle to the hypotenuse, above, middle and right; thus the division makes each of the two triangles similar to each other and to the whole, larger triangle, ad infinitum. The perpendicular from the right angle to the hypotenuse is the mean proportional, or geometric mean, in length between the two parts into which the hypotenuse is thus partitioned. And what this means is that the ratio of the longest length of the hypotenuse (BD) to the shortest length (DC) is such that the square—or indeed *any* figure—drawn on the longest side is in the same ratio as the square (or *similar* figure) drawn on the perpendicular. Stated in yet another way, BD : AD :: AD : DC and thus BD : AD : DC stand in a continuous proportion. The rectangle made by the first and third (BD, DC) is equal in area to the square on the second (AD). The ratio of the lengths of the first (BD) to the third (DC) is the same ratio as the figure on the first (BD) is to the figure on the second (AD). The visualization that these line lengths stand in continuous proportions is the areal equivalences of the figures constructed on the sides, below.

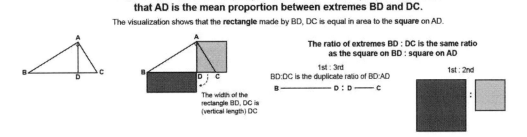

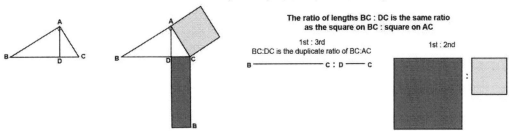

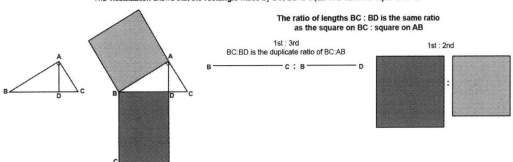

Figure 1.52.

Moreover, there are two other mean proportionals in the proof: the ratio of the length of the hypotenuse of the largest triangle is to its shortest side as the length of the hypotenuse of the smallest triangle is to its shortest side. It just so happens that the shortest side of the largest triangle is also the hypotenuse of the shortest. This "shared side" is the mean proportional connecting the largest and smallest triangles, and thus the three lengths together form a continuous proportion. Moreover, the ratio of the length of the hypotenuse of the largest triangle is to its longest leg as the length of the hypotenuse of the medium-sized triangle is to its longest leg. And it just so happens that the longer of the two sides of the largest triangle is also the hypotenuse of the medium-sized triangle—its mean proportional. Thus, the mean proportional—or geometric mean—between the longer and shorter sides, or extremes, is not the "average" length between them but rather the "scaled-up" length between them, by the blow-up factor. The shortest side blows up by a factor that reaches the length of the mean proportional, and that mean proportional blows up or scales up in turn *by the same factor* to reach the length of the longest magnitude. *The compounding of the "extremes"—the two sides, longest and shortest—produces a rectangle whose area is equal to a square the side of which is the mean proportional.* In the geometric algebra that commentators refer to but that I am insisting is perfectly misleading to understand what the ancient Greeks were doing and how they did it, is: $a \times c = b \times b$, or $ac = b^2$.

From these considerations we are finally able to express clearly the answer to the question "What do *squares* have do with the Pythagorean theorem?" Surprisingly, and in an important way, *VI.31 shows that squares have nothing special to do with it!* Any similar figures similarly drawn (i.e., proportionately scaled) on the three sides—not only squares—stand in the "Pythagorean relation": the areas of the figures on the two sides sum to the area of the figure on the hypotenuse. And so, equilateral triangles, semicircles, polygons of all dimensions, even irregular figures, similar and similarly drawn, *all* stand in the same areal relation. Consider these figures, below:[33]

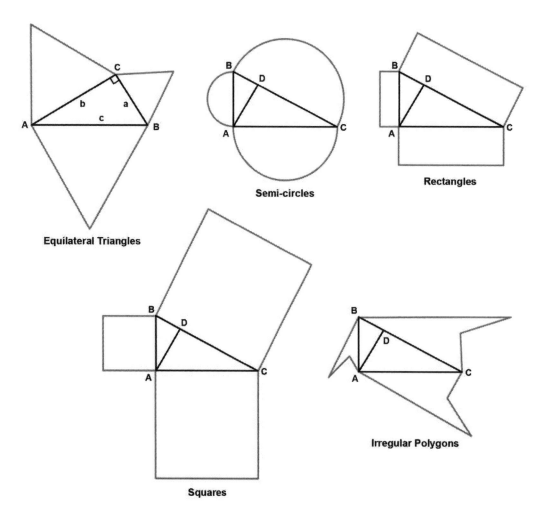

Figure 1.53.

So why did so many of us learn the Pythagorean theorem as focused on squares? What do squares have to do with it? What has misled many scholars from thinking through what the hypotenuse theorem could have meant to our old friends of the sixth century BCE is our education that instructed us algebraically, and focused (if it focused at all) on I.47, which turns out to be a special case of VI.31. We have no evidence that the Greeks, even through the time of Euclid, thought of the relation among the sides of the right triangle as $a^2 + b^2 = c^2$. We have no evidence that the Greeks expressed geometric problems algebraically—whether or not these theorems were amenable to algebraic expression—and to think through the hypotenuse theorem as algebra misses *how* the Greeks came to see the matter and its insights. Moreover, it's not squares but rather "squaring" that is central and relevant, but this must be understood geometrically, not algebraically, if we are to grasp this idea the way the ancient Greeks did. What holds the areal relations together is that the similar and similarly drawn figures *all scale with the square of the linear scaling*—this fact the Greeks referred to as "duplicate ratio." This quintessentially important point deserves to be reemphasized: the area of similar figures scales as the square of the linear scaling. But, in terms of the Greeks of the archaic and even classical periods, this means that as the length of the sides increases, the areas of the figures constructed on those lengths *increase in the duplicate ratio of the corresponding sides*.

So think of it this way: take any figure similar and similarly drawn on the sides of a right triangle, tile them with squares and "expand" them proportionately (= scale them).[34] If you take a square of side length 'a', and then if we rescale a figure made up (approximately) by squares, then because the area of each square scales by the square of the scaling, so does the total area. The smaller the squares we started with to tile the figure, the more accurate the approximation of the area. In modern mathematical terms, calculus is all about limits, and integrals are exactly the mathematical way of defining area by taking a limit of the sum of areas of small squares, and tiling the figure as the size of the squares decreases to zero. Of course, the Greeks did not have calculus any more than they had algebra, but could it be that they thought of area in this way, or at least had such an intuition about it? Perhaps Thales saw such examples of tiling in Egypt or elsewhere? But this kind of thinking operates on the principle of the microcosmic-macrocosmic argument, that the structure of the little world and the big world is similar; the tiniest appearance, and the most gigantic, all share the same structure. An area was expressible as a square, and a square reduces to (isosceles) right triangles by a diagonal. The narrative that constitutes my thesis is that, beginning with Thales, space was imagined as reducible to triangles, and as discussed in the Introduction, this was fundamental to how the Egyptian land surveyors understood area (Rhind Mathematical Papyrus, problems 51 and 52). Much later, this project was developed with the calculus, namely, tessellation.[35] Could it be that the ψῆφοι, or diagrams made of "pebbles" by the Pythagoreans, were place-keepers for basic areal units, such as triangles or squares?

Whatever that case may be, if you know that the area of a square is proportional to the square of the side length—in the duplicate ratio—then the same holds for a figure tiled by squares, so it holds for any figure since we can approximate arbitrarily well by tiling. In principle you could do the same if you started with knowing that the area of some other type of figure scaled according to the square of the linear scaling. This certainly would work for right triangles in general, equilateral triangles, or rectangles, all of which could be used to tile a figure well; you might have more difficulty tiling a figure with circles, for example. But if you started with the formula for area of any particular figure, which need not be a square, that would tell you that *the figure's area scales as the square of the linear scaling*, and the argument (from integral calculus) just given about tiling would then imply that the same is true for any figure. The scaling law for areas of similar triangles is formulated in VI.19 this way: similar triangles are to one another in the duplicate ratio of the corresponding sides. The corollary ("porism") to this proposition extends this property to arbitrary figures, not just squares as shown in I.47.

From a modern point of view, then, the fact that $a^2 + b^2 = c^2$ for every right triangle is more or less equivalent to saying that the sum of the angles is 180 degrees, or two right angles (NB, both of which do *not* hold on a curved surface but only on a flat one; on a sphere, the angles can add to more than two right angles, and $a^2 + b^2 > c^2$; for example, on the Earth, consider a triangle made up of the North Pole and two points on the equator 90 degrees apart; this has all right angles and equal sides, and the sum of the angles is 270 degrees, and $a^2 + b^2 > c^2$ clearly since all sides have equal length. On a negatively curved surface such as a horse saddle, the opposite is true, and so $a^2 + b^2 < c^2$). What all this means is that the standard form of the *Pythagorean theorem is based on an assumption that we are working on a flat surface*. Let me clarify this a bit: The Pythagorean theorem is a statement that *if* we are working on a flat surface and the triangle is right *then* $a^2 + b^2 = c^2$. On a sphere the Pythagorean theorem still holds, but has a different (generalized) form. This form becomes the standard $a^2 + b^2 = c^2$ in the limit when the curvature tends to zero. Hence it is not the theorem that breaks on the sphere but the equality $a^2 + b^2 = c^2$.[36] By my thesis, the supposition that space is conceived as flat surfaces is how sixth-century Greeks started as they danced first with geometry; this was their deep mathematical intuition, that the cosmos was conceived fundamentally to be built out of flat surfaces, and layered up and/or folded up into volumes. This is a very important point to emphasize because, as Couprie has argued persuasively and at length about the astronomical imagination of Anaximander and Anaxagoras, they conceived of a flat Earth, and much of the difficulty of grasping their astronomical and cosmological insights follow from thinking through a view that is very foreign to a modern audience—a *flat* Earth.[37]

There is one more diagram crucial to fill in this picture that supplies the metaphysical background for this whole enterprise because it, along with "duplicate ratio," is quintessential to grasp a principle that underlies VI.31, though it is not explicitly referred to in that theorem. In my estimation, it harkens back to a much older time, the time of Thales and Pythagoras, and to the Egyptians before them. This is the proposition at VI.20: *Similar polygons are divided into similar triangles, and into triangles equal in multitude and in the same ratio as the wholes, and the polygon has to the polygon a ratio duplicate of that which the corresponding side has to the corresponding side, below.*

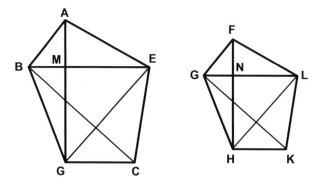

Figure 1.55.

The proposition and proof of VI.20 reflects the grandest scheme of concerns that gets incorporated in the treatment of similar figures, namely, that all polygons are reducible to triangles, and the areal relations stand in the same duplicate ratios of triangles to which all can be reduced ultimately.

Chapter 2

Thales and Geometry

Egypt, Miletus, and Beyond

No doubt, in the infancy of geometry, all sorts of diagrams would be drawn, and lines in them, by way of experiment, in order to see whether any property could be detected by mere inspection.

—Thomas Heath[1]

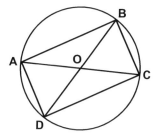

We may imagine Thales drawing what we call a rectangle, a figure with four right angles (which it would be found, could be drawn in practice), and then drawing the two diagonals AC, DB in the annexed figure. The equality of the   pairs of opposite sides would leap to the eye, and could be verified by measurement, Thales might then argue thus. Since in the triangle ADC, BCD, the two sides AD, DC are equal to the two sides BC, CD respectively, and the included angles (being right angles) are equal, the triangles are equal in all respects. Therefore the angle ACD (i.e., OCD) is equal to the angle BDC (or ODC) whence it follows (by the converse of Euclid I.5 known to Thales—*the base angles of an isosceles triangle are equal, and if the base angles are equal, the sides subtended by the equal angles are also equal*) that OC = OD. Hence OA, OD, OC (and OB) are all equal, and a circle described with O as center and OA as radius passes through B, C, D, also. Now AOC, being a straight line is a diameter of the circle, and therefore ADC is a semicircle. The angle ADC is an angle in the semicircle, and by hypothesis it is a right angle. The construction amounts to circumscribing a circle about the right-angled triangle ABC which seems to answer well-enough Pamphile's phrase about [Thales] "describing on a circle a triangle which is right-angled."[2]

—Heath, *Manuel*, 1931, pp. 87–88

A

Thales: Geometry in the Big Picture

Who Thales was, what exactly he did and how, his importance for the development of philosophy, and for our purposes here his contributions to the earliest stages of geometry in Greece, has been the source of endless controversy. On geometrical matters, some scholars have argued that Thales probably made no contributions at all and the details in the doxographies are so late as to make them more or less worthless testimony. Other scholars have argued that probably all the achievements in geometry with which Thales has been credited are broadly appropriate and fitting. And then, of course, there are the scholars who hold positions somewhere between these two extremes.

D. R. Dicks is among the most negative of the naysayers, and he followed in the spirit of the historian of mathematics, Otto Neugebauer, who regarded all the evidence as too late to be persuasive.[3]

> The picture of Thales as the founder of Greek mathematics and astronomy, the transmitter of ancient Egyptian and Babylonian science to Greece, and the first man to subject the empirical knowledge of the Orient to the rigorous, Euclidean type of mathematical reason cannot be defended. . . . For this picture there is no good evidence whatsoever, and what we know of Egyptian and Babylonian mathematics makes it highly improbable, since both are concerned with empirical methods of determining areas and volumes by purely arithmetical processes, and neither shows any trace of the existence of a corpus of generalized geometrical knowledge with "proofs" in the Euclidean style.[4]

B. L. van der Waerden, Neugebauer's student, has one of the most optimistic readings of crediting Thales with the theorems. For him, the crux of the matter was to determine precisely what it meant to accept the report of Proclus, on the authority of Eudemus, that Thales stands at the *beginning* of ancient mathematics. If it meant that Thales made basic mathematical discoveries, then the claim is patently false, but if it meant that he systematized and logically organized the ancient mathematics he discovered from Egypt and Babylon, then this is precisely how he stands at the beginning. The evidence from Egypt and Mesopotamia shows that more than a millennium before Thales's time, we have extensive evidence of just these mathematical discoveries, and so Thales did not invent the theorems empirically; he adopted these earlier traditions. Since we are now aware of these mathematical traditions "[it] knocks out every reason for refusing Thales credit for the proofs and for the strictly logical structure which Eudemus evidently attributes to him."[5] So, the central claim for van der Waerden is that Thales stands at the beginning of *logically organizing and systematizing the knowledge* of older civilizations. As he elaborates, van der Waerden's point of view is this:

> A closer look at the propositions ascribed to Thales, shows that, instead of belonging to the early discoveries of mathematics, they form part of the systematic, logical exposition of mathematics. At the start, in the first excitement

of discovery, one is occupied with questions such as these: how do I calculate the area of a quadrangle, of a circle, the volume of a pyramid, the length of a chord; how do I divide a trapezoid into two equal parts by means of a line parallel to the bases? These are indeed the questions with which Egyptian and Babylonian texts are concerned. It is only later that the question arises: How do I prove all of this?[6]

Thus, van der Waerden's assessment is that "proofs" of such statements as the base angles of an isosceles triangle are equal, vertical angles are equal, a diameter divides a circle into two parts, and so on, are just the kinds of theorizing that belong to *the early systematic exposition of the logical structure of geometry*—in this sense, Thales stands at the beginning. Greek mathematics was constructed out of the remnants of Egyptian and Babylonian civilizations—from long-dead wisdom—and Thales embraced and erected on them a new edifice where thinking does not tolerate obscurities or doubt the correctness of the conclusions acquired. While the older traditions provided examples, like useful school texts, there was no evidence for an interest in the theoretical assumptions that underlie them. Therefore, for van der Waerden, Thales's position at the beginning of Greek geometry initiates a change in character of what geometric knowledge meant, a kind of knowledge that offered certainty and was produced by a deductive methodology that would state ever more explicitly the rules by which it reasoned.

Walter Burkert offers an assessment somewhere in between but closer to van der Waerden. Burkert begins his reflections on Thales with the passages from Aristophanes *Clouds*. We have a scene that opens in front of Socrates's Thinking-Shop, flanked by two statues, Astronomy and Geometry. One of his students described how Socrates is scattering ashes over a table and bending a cooking spit into shape to serve as a pair of compasses (διαβήτης), and under cover steals a coat. "Why do we go on admiring Thales?" inveighs Strepsiades, a would-be student and father of horse-loving, debt-ridden son Pheidippides. Burkert declares "This shows that for the Athenian public Thales and geometry belong together."[7] This is the same conclusion he reaches in mentioning the passage in Aristophanes's *Birds* where Meton enters Cloudkookooland and announces "I want to survey [γεωμετρῆσαι] the air for you! He will apply his ruler so that the circle may become a square, which will look like a star." All that Pithetaerus can say is "This man's a Thales!" (ἄνθρωπος Θαλῆς).[8] Thus, for the Greeks almost a century and a half after the death of Thales, his name still conjures a connection to geometry. From the reports about Thales by Eudemus, reported by Proclus, Burkert concludes that there was a book ascribed to Thales that was the basis of these reported achievements in geometry—he suspects *On the Solstices and Equinoxes*, whether or not it is authentic. Burkert subscribes to Becker's analysis that all the theorems attributed to Thales can be derived simply from considerations of symmetry,[9] and von Fritz shows that the proofs can one and all be demonstrated by superposition (ἐφαρμόζειν), consciously avoided in later geometry.[10] So, while Burkert can entertain credit for Thales's innovations in geometry, they very much mark the beginnings because the method of demonstration, although deductive, is fundamentally empirical—placing one triangle on top of another to demonstrate equality, or that angles are equal, or that the diameter divides the circle into two equal parts. "With Thales the point is still a graphic or perceptible 'showing' (δεικνύναι)[11] . . . It is in the perceptible figure that mathematical propositions become clear in their generality and necessity: Greek geometry begins to take form."¹² Thus, the theorems attributed to Thales plausibly belong to him, but their subject matter is driven by empirical considerations. Burkert leaves his reader with the impression that Thales may very well be connected with ideas in these theorems, but he is cautious about the degree to which they have met the rigor of formal proof.

Is there some way to bring new light into these endless and divisive debates? I propose to try to do so by exploring the many diagrams that display the theorems with which Thales is credited and the ones that show the experiments in measuring the height of a pyramid and the distance of a ship at sea for which we also have tertiary reports. I invite the readers to join in this approach to discover a new point of view with which to revisit these controversies. A key point to keep in mind is that *if* these theorems are really connected with Thales's achievements, they must represent only a handful of the countless diagrams I suggest that Thales made, discussed, and reflected on. If readers will follow me in making the diagrams, they will see they lead to an ongoing inquiry into the structure of a certain kind of space. Thales is credited with starting a "school," and it is plausibly in this school that so much of the sketching and discussion took place. "Proof" at its earliest stages in the time of Thales meant making something *visible*, showing forth and revealing, and diagrams were part of what came to be a program.¹³ The Greeks did not have an old tradition of mathematical wisdom, as did the Egyptians and Mesopotamians. And it seems entirely plausible that, returning from Egypt and elsewhere, Thales brought back some of this ancient wisdom and shared it with his compatriots. But many of Thales's compatriots, I surmise, did not travel so widely, nor had they thought at length about these matters and geometrical diagrams, if at all. His compatriots would take nothing for granted, and it seems plausible that they asked for proof of the claims, some way to show why a reasonable person might believe the kinds of geometrical wisdom that Thales sought to share. "Proofs" were a way of making sense of things, a way of persuading oneself and others; they brought together cognitive organization with bodily awareness. The geometrical diagram was part of the new process of "proof"; it offered a way to show or make visible something that Thales claimed, and moreover the diagrams offered a way of uniting body and mind. Even if we trust in only a weaker case than that of van der Waerden, and find it merely plausible that Thales is connected to the theorems, then we should still keep in mind that these few diagrams are tokens of the panoply of sketches Thales made in the sand, and perhaps in clay and on animal skins, or perhaps on the rolls of papyrus he might have brought home from the Milesian trading colony of Naucratis in the Nile delta, reflecting on themes geometrically. Already at the beginning of the third century BCE, Callimachus in his *Iambi* records of Thales that someone "found the old man in the shrine of Apollo at Didyma scratching on the ground with a staff, and drawing a figure (γράφοντα τὸ σχῆμα)."¹⁴ With this view we can appreciate better the comment made by Heath with which this chapter began: "No doubt, in the infancy of geometry, all sorts of diagrams would be drawn, and lines in them, by way of experiment, in order to see whether any property could be detected by mere inspection." This is the picture of Thales, I propose, that places Thales at the origins of geometry for the Greeks.

For the introduction of "proof," then, I imagine a slightly different emphasis in my answer about its origins, though one that is in concert with van der Waerden. Van der Waerden's thesis is that the basic theorems or propositions of geometry had already been discovered in Egypt and Mesopotamia, and it is from these sources that Thales became aware of many of them, and that he began the process of organizing this knowledge. I accept this part of the thesis. However, what I would like to add is this: when Thales returned to Greece from his travels to Egypt and elsewhere with this knowledge, and presented it to his compatriots, he found himself among quizzical and skeptical associates. To convince them of the soundness and reliability of this geometrical knowledge, his replies to their doubts and disbelief are the sources for his *proving*.[15] What motivated his "proving" was the skepticism he encountered among his Greek compatriots. "Well, that is very well that you propose that the base angles of an isosceles triangle are equal," one of his compatriots might have said, interrupting Thales as he spoke, "but best of men, how do you know this, how can you be so sure?" Thales's replies were the beginnings of "proof."

If one inspects a range of the important and influential studies on early Greek philosophy in general and on Thales in particular, while almost all of them mention and discuss the doxographical reports of Thales's discoveries or contributions to geometry, there is usually not a single geometrical diagram or sketch in these studies—nothing, or almost nothing.[16] In the seminal works on early Greek philosophy by the classical scholar Zeller in the last quarter of the nineteenth century,[17] and then Burnet in the first quarter of the twentieth century,[18] and then again in Kirk and Raven (and even later with Schofield) in the last half of the twentieth century through several editions containing an essay of some twenty-five pages on Thales, not a single geometrical diagram appears.[19] In the encyclopedic six-volume work by W. K. C. Guthrie, the first volume of which was published in 1962, there is not a single geometrical sketch in discussions about Thales.[20] In more recent works by Barnes,[21] again, there is not a single geometrical sketch. And in important studies by R. E. Allen,[22] by C. C. W. Taylor containing an essay by Malcolm Schofield on Milesian beginnings,[23] and by Roochnik,[24] yet again, there is not a single geometrical diagram or sketch. In Richard McKirahan's study there is a single diagram offering to illustrate Thales's geometrical conception of measuring the distance of a ship at sea—just one.[25] And in a recent full-length book on Thales by O'Grady—*Thales of Miletus*—more than 300 pages long, yet again there is not a single geometrical sketch, though the topic of geometrical achievements is explored in no less than two chapters.[26] In these studies by classical scholars and those writing in the spirit of analytic philosophy, there have been no efforts to explore *visually* the subjects on which Thales is reputed to have reflected. And lest one think that this predilection is limited to classical and analytic approaches, it should be pointed out that not only is there no sketch or geometrical diagram in Hegel's volume I of his *History of Philosophy*, nor in any of the works published by Heidegger on ancient Greek philosophy, but even in Husserl's important study *On the Origins of Geometry*, and Derrida's commentary on it, the reader will not encounter even a single geometrical diagram.[27] How can one adequately explore the topic of the origins of ancient Greek geometry, or the plausibility of Thales's place in it, without geometrical diagrams?

The answer to this question—Why have scholars not explored *visually* the testimonies that Thales made geometrical discoveries or proofs?—and the astonishing lack of sketches, might be presented the following way. In the recent studies by Barnes and McKirahan, and the studies many years earlier by Burnet, almost the same argument is delineated. Proclus, on the authority of Eudemus, credits Thales with measuring the distance of a ship at sea. Then, Eudemus *infers*—ascribes to him—that Thales must have known the theorem that we have come to know from Euclid as I.26, that triangles are congruent when they share two equal angles and one side equal between them. Thus, our scholars *infer* that Eudemus had no evidence in front of him to credit Thales with the theorem, but only argues a priori since he regarded it as a presupposition of such a measurement: *Thales must have known it*. And then, our chorus of scholars suggests that it might very well have been an empirical assessment. Thus, there is no need for geometrical diagrams because the evidence for them is entirely unsecure. This is where the secondary literature stands.

The procedure I have adopted turns these approaches on their heads. First, let us keep in mind that I already argued in the Introduction for the existence of geometrical diagrams in Greek Ionia—scaled-measured diagrams, no less—dating to the middle of the sixth century, and other mathematically precise diagrams in the middle of the seventh century. Thus, first start with the reports of Thales measuring the height of the pyramid and measuring the distance of the ship at sea. Start making geometrical diagrams; try alternate diagrams; place those diagrams next to one another and see if these lend any clarity to what someone would have to know, or probably know, to have made those measurements. As Heath said, right on the mark, "No doubt, in the infancy of geometry, all sorts of diagrams would be drawn, and lines in them, by way of experiment, in order to see whether any property could be detected by mere inspection."[28] But in the efforts to explore these beginnings, and Thales's place in them, scholars have routinely avoided making and exploring the geometrical diagrams. My hypothesis was that to begin to understand Thales and his geometrical speculations, we have to understand that he must have made countless diagrams—with a stick in sand, or in the potter's wet clay—and discussed them with his retinue, his school, and with anyone who cared to listen. Until scholars spend a lot of time sketching the geometrical diagrams, they might never understand what Thales was doing with geometry because it represents a kind of thinking in a *visual* mindset. Ensconced in that mindset, the ideal of "proof" as δεικνύναι will become apparent in all its richness.

The problem of contention that has stood in the way of making progress in our understanding of Thales is the endless debates about whether he could have "proved" this or that theorem. Let us be clear that a formal proof offers a series of steps of reasoning each of which is justified by an abstract rule. If we should come to doubt any step, we appeal to the rule to overcome the doubt, and so reach a conclusion beyond a reasonable doubt. We have no evidence in the sixth century for what later appears in Euclid as the Ὅροι (Definitions), Αἰτήμαται (Postulates), or Κοιναὶ ἔννοιαι (Common-Notions). So let us not get distracted from the endless wrangling about "formal proof." Von Fritz recommended, wisely, that we must recognize that whatever Thales and other sixth-century compatriots were doing in geometry, their claims had to be persuasive by the standards of the time.[29] Our approach is to figure out the kinds of things that anyone would have to know to follow the lines of thought that we shall explore for Thales. Can he be connected with the famous hypotenuse theorem? My thesis is that he can, by the general route preserved in Euclid VI.31 (not I.47), because it is a consequence of similar triangles, by ratios and proportions, and I contend that Thales understood similar triangles. We shall see where and how it leads, given the evidence attributable to Thales.

Instead of mentioning the measurements and then focusing on dismissing the theorems that later authors claim Thales discovered, proved, or simply understood, we should have better success if we begin by going out with our students to the Giza plateau and the coast near Miletus and try the measurements ourselves, as I have done; as we recreate these experiments and make diagrams of the measurements themselves, we can see what one discovers by and through the process. This is how we shall now proceed. When we do so, the geometrical diagrams come to take on a life by themselves. Then we begin to think—to *imagine*—as did Thales.

B

What Geometry Could Thales Have Learned in Egypt?

B.1 Thales's Measurement of the Height of a Pyramid:

Technique 1: When the Shadow Length Was Equal to Its Height

We have two doxographical reports that Thales measured the height of a pyramid by its shadow at a time of day when the shadow length was equal to its height:

> Hieronymous says that he [sc. Thales] even succeeded in measuring the pyramids by observation of the length of their shadow at the moment when our shadows are equal to our own height.
> —Attributed to Hieronymous, fourth century BCE, by Diogenes Laertius, second century CE[30]

> —Thales discovered how to obtain the height of pyramids and all other similar objects, namely, by measuring the shadow of the object at the time when a body and its shadow are equal in length.
> —Pliny[31]

First, we shall investigate this technique, and then the other technique, reported by Plutarch, when the shadow was unequal but proportional to the pyramid's height.

More than a thousand years before Thales could have arrived in Egypt, the Egyptians could calculate the height of a pyramid. Problem 57 in the RMP,[32] a document produced by a scribe Ahmose and datable to the Middle Kingdom and not implausibly to the Old Kingdom itself[33] shows us that given the base of a pyramid and its *seked*,[34] or inclination of the pyramid face, the height can be reckoned. The point to be absorbed is that when Thales arrived on the Giza plateau, the Egyptian priests *already knew the answer* to the question: What is the height of the pyramid? This is a point rarely mentioned in the literature on Thales's alleged measurement. And the answer they knew was in royal cubit measures. Had Thales measured the shadow *in* Egypt of either the pyramid of Khufu or Khaefre, the shadow would have been more than one hundred feet long. It is difficult to believe that he carried with him from Miletus a cord of such length. He almost certainly did so by means of the royal cubit, knotted cord—provided by his hosts with the assistance of the ἁρπεδονάπται or "rope fasteners," a Greek term applied to the surveyors—and so his answer was plausibly in precisely the same measure that the Egyptian priests knew for a millennium and probably longer.[35] What plausibly made the measurement so memorable was that, in the same cubit measures, it must have been startlingly close to what the priests already knew. What Thales did was to show that the height could be calculated in a very different way from the formula used in the RMP and MMP, and this might well have led him to reflect on the difference in their approaches, and to consider what these different approaches may have meant.[36] Comparing the results and techniques might also have opened a door for Thales to learn geometrical wisdom from his Egyptian hosts.

98 | The Metaphysics of the Pythagorean Theorem

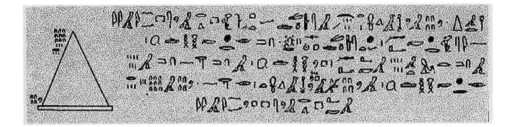

Figure 2.2. After the Rhind Mathematical Papyrus (RMP) in the British Museum.

In the RMP problems 57, above, and 58 deal with architectural structures, and pyramids in particular. The object is to show how there is a formal relation between the size of the base, the height of the pyramid, and the inclination of the pyramid's face, that is, the *seked*. The ancient Egyptians are generally interpreted to not have the concept of an "angle" as the Greeks developed it. Instead, the *seked*, or unit of slope, is defined as the lateral displacement in palms for a drop of 7 palms (= 1 royal cubit). Thus, begin at the corner of the base of the pyramid and measure up 1 cubit (vertically), and from there measure the distance across until the pyramid face is reached (horizontally). The displacement from the cubit, measured vertically, to the inclined face measured horizontally, is the *seked*, below:

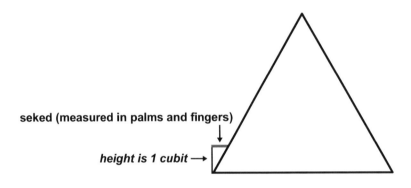

Figure 2.3.

The *seked*, then, is taken to be half the width of the base divided by the height, both of which are expressed in cubits, and then multiplied by 7. In problem 57 the illustration shows that the pyramid is only roughly sketched (i.e., schematic) because the inclination is clearly steeper than the 53° 10' that the middle pyramid on the Giza plateau displays, that of Kha-ef-re (= Chefren); the *seked* is given as 5 + 1/4 palms, which is equivalent to the slope of the middle pyramid. The side of the base is given as 140 cubits, and the problem is to reckon the pyramid's height. The instructions are to take one cubit and divide it by the *seked* doubled, which amounts to 10 1/2. Then we reckon with 10 1/2 to find 7, for this is one cubit. Then 2/3 of 10 1/2 is 7. Then, take 2/3 of 140, which is the length of the side, to make 93 1/3. This is the height of the pyramid.[37]

Let me add another consideration here, an historical one, that I will come back to shortly—the testimony by Plutarch, who records Thales's measurement of the height of a pyramid in the form of a dialogue with Niloxenus addressing Thales. He recounts that *"Pharoah*

Amasis was particularly pleased with your [sc. Thales's] measurement of the pyramid."[38] *If* we are to accept this testimony, late as it is, a few important points are worth considering, and almost never mentioned in the secondary literature. First, Amasis ascended to the throne in c. 570 BCE, and so, if Plutarch has the right pharaoh, Thales's measurement came rather late in his life—he must have been not younger than fifty-five years old, if he was born circa 625. Second, the very idea that a Pharaoh marveled at Thales's measurement makes no sense unless he had some information about the pyramid height from his priests, who followed the calculations along the lines of the RMP, and these could only have been in royal cubits. Thus, Plutarch's claim suggests a bigger picture in which Thales made use of the royal cubit cord—the one used by surveyors—producing a result in the same units of royal cubits that must have been remarkably close to the answer derivable from the tradition of the RMP. Otherwise, Amasis was in no position to marvel at anything Thales did by setting up a stick and reasoning proportionately. To have particularly pleased the Pharaoh the answer must have been very close to what was known already, and in cubit measures.

By contrast with the directions in the RMP, Thales or anyone else who attempts to measure the height of an object at the time of day when the shadow it casts is equal to its height discovers important things when they try to make the measurement on the Giza plateau. I've made the measurement a few times with my students:

Figure 2.4.

100　The Metaphysics of the Pythagorean Theorem

The measurement above was made on 23 May 2013 at 8:33 a.m.,[39] a special time when the shadow was equal to the pyramid height because at just this time on this day the height of every vertical object was equal to its own shadow, and the sun was due east. In the foreground, I have set up a gnomon and was able to confirm that the shadow was equal to its own height, and as in Plutarch's testimony I also set up a gnomon at the extremity of the shadow cast by the pyramid.

Thus, Thales, or anyone who makes the measurement, has to imagine a right angle with vertical and horizontal line segments of equal length. When the angle of the sun's light is factored into this picture, it becomes obvious that one is forced to imagine an isosceles right triangle.

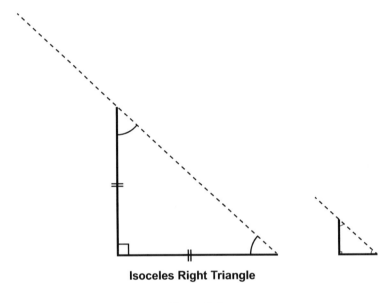

Isoceles Right Triangle

Figure 2.5.

As the illustration directly above shows, it is not simply the large isosceles right triangle that captured Thales's attention—the plane of the pyramid expressed vertically as its height and horizontally as its shadow—but also the smaller isosceles right triangle that was indispensable to the measurement. Without it, he would not have been sure that the time was right to make the measurement. So let's think this matter through. Near the pyramid, or even set up at the extremity of the pyramid's shadow, a stake or gnomon was crucial to provide the confirmation of the exact moment to make the measurement. This means that Thales set up a gnomon—a vertical rod—somewhere nearby. Then, one probable way to proceed was that Thales took a piece of cord whose length was equal to the height of the stick, and placing one end at its base, he swung an arc, that is, he made a circle with the gnomon height as radius. Whenever the shadow cast from the tip of the stick intersected the circle, *that* was exactly the time when every vertical object casts a shadow equal to its length. Here we have the principle of *similarity* displayed in an unforgettable way, a display that, in my estimation, played an enormously important role in Thales's geometrical imagination.

Since we can say confidently that Anaximander, Thales's younger contemporary, had made a seasonal sundial and set up a gnomon, a vertical rod, on the *skiatheria* in Sparta,[40] we can say confidently that our old Milesian friends had a great deal of experience in thinking about shadows and their relations to the vertical objects that cast them. Anaximander, and Thales, would have discovered that in Miletus there was not one time, but rather two times on days between early March and mid-October—but not all days in the year, since the sun never reaches 45° above the horizon in late fall through winter—when every vertical object casts a shadow equal to its height. During the winter months the sun remains too low in the sky to make such a measurement. The only circumstance in which a right-angled triangle can have two legs of equal length is when that triangle is isosceles, and the only circumstance for this case is when the base angles opposite those legs are equal, and this can only be when the opposite angles are 1/2 of a right angle, or 45°. This is one of the theorems attributed to Thales, and it is no lucky coincidence. As we proceed I hope my readers will pay attention to the centrality of mathematical intuitions suggested unmistakably by this theorem with interesting consequences for this "lost narrative" about Thales. We cannot be certain that the Milesians reckoned angles in numbers—though it is not out of the question, since Thales is likely to have been aware of both the Babylonian and Egyptian traditions of a great circle of the year being 360 days and so the division of the circle into four right parts or angles each of 90°—but there are no good reasons to lack confidence that Thales realized the base angles had to be 1/2 a right angle. While it is true that the *proof* that the interior angles of a triangle sum to two right angles is credited to the Pythagoreans,[41] this in no way undermines the reasonable possibility that Thales knew it—without having arrived at a definitive proof, or without being credited for it. I remind readers that in the Introduction we considered Geminus's report that the "ancients" (i.e., Thales and his school) had explored how there were two right angles in each species of triangle. I alert the readers again not to get sidetracked by the important and interesting question of whether and when a formal proof was achieved, and instead focus on the question of what mathematical intuitions would have to be grasped and whether a plausible circumstantial case can be made that Thales had grasped them. Then, our question is whether Thales or Pythagoras of the sixth century could have produced a line of reasoning that would be regarded as persuasive to their compatriots, whether or not it falls short of a proof acceptable to Euclid. The lines of connection are simply these: every triangle reduces to right triangles, and every right triangle is the building block of every rectangle, and every rectangle contains four right angles, and so the division of every rectangle by its diagonal into two right triangles is a fortiori the proof that the angles of every triangle sum to two right angles. In the Introduction I already suggested that this was in RMP problem #51, the identification of "the triangle's rectangle."

It is important to notice that when we place together the diagrams of the theorems of the measurement of the height of the pyramid and the distance of a ship at sea, we begin to see Thales preoccupied with right-angled figures, and especially the isosceles right triangle. Since that triangle is the building block of every square—whereas the other right triangles are the building blocks of every rectangle, with the exception of the square—it is difficult to believe that Thales did not grasp that there were two right angles in a straight "angle" (= line), and four right angles in every rectangle, and indeed in every square that is an equilateral rectangle, and that since every square is composed of two isosceles right triangles, the insight that there were two right angles in every triangle seems inescapable.[42] We will explore this, and its consequences, in more diagrammatic detail later in this chapter.

102   The Metaphysics of the Pythagorean Theorem

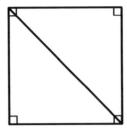

Figure 2.6.

Now, when Thales went to Egypt and stood on the Giza plateau, the problem was vastly more complicated. For while it is true that the sun reaches 1/2 a right angle for *more* days of the year in Giza than in Miletus, the mass of the pyramid is so great that almost always when the sun is at 45°, the shadow never falls off the face of the pyramid itself! Had he sought to measure the height of an obelisk, for example, the problem would have been more simply analogous to the Milesian experiments with a gnomon and sundial. But the Great Pyramid is some 480 feet in height, its equal sides are more than 755 feet in length, the pyramid face is inclined at slightly more than 51°, and since the pyramid is aligned to cardinal directions, for the measurement of pyramid height to be made using *this* technique, the sun had to be at 1/2 a right angle *and* at the same moment due south, or even due east or due west. This condition is illustrated, below:

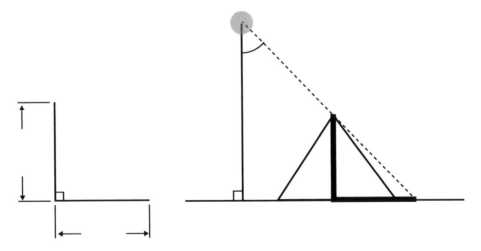

Figure 2.7.

In the span of a whole year, we can confirm also by a solar calculator[43] that there are only *six* times over the course of a mere *four* days when the sun would be at a 1/2 a right angle and either due south, due east, or due west. These are the *only* dates in the course of a year when the sun is at the right height and with the right alignment to make that measurement. The window of opportunity is perhaps as few as twenty minutes long (for east and west orientations, and slightly longer at midday when the sun is due south), and it had better not be cloudy! According to the solar calculator, these rare occasions occurred in 2013 on the dates listed below. While it is true that due to precession, the specific dates some 2,700 years ago would have been different, there is no ancient argument for a date when the measurement was made, and the point of my argument here is simply to make clear the rarity of the times when the conditions are right to do so, had the measurement been made when the shadow was equal to the pyramid's height. The 2013 dates where the solar altitude at the cardinal points is closest to 45° for a latitude of 29° 59' are:[44]

DUE SOUTH				
Date, Time	Altitude	Azimuth	Feet	Royal Egyptian Cubits
10 Feb, 12:09:42	45.81°	180.0°	466.6	271.7
9 Feb, 12:09:41	45.49°	180.0°	471.9	274.8
8 Feb, 12:09:40	45.17°	180.0°	477.2'	277.8
7 Feb, 12:09:38	44.85°	180.0°	482.5'	280.9
4 Nov, 11:38:59	44.56°	180.0°	487.4'	284.3
3 Nov, 11:38:59	44.87°	180.0°	482.2'	280.8
2 Nov, 11:38:59	45.18°	180.0°	477.0'	277.7
1 Nov, 11:39:00	45.50°	180.0°	471.7'	274.6
DUE EAST-WEST				
Date, Time	Altitude	Azimuth	Feet	Royal Egyptian Cubits
Morning				
24 May, 8:37:05	45.32°	90.0°	474.7	276.9
23 May, 8:35:05	44.83°	90.0°	482.9	281.7
19 July, 8:44:35	45.3°	90.0°	475.0	277.1
20 July, 8:44:35	44.8°	90.0°	483.4	282.0
Afternoon				
24 May, 15:07:00	45.45°	270.0°	472.5	275.6
23 May, 15:08:50	44.97°	270.0°	480.5	280.3
19 July, 15:18:10	45.0°	270.0°	480.0	280.0

Table 2.1.

Thus, had Thales made the measurement when the height was equal to the shadow length, as Diogenes Laertius and Pliny both report, he would have depended almost certainly on the formulation of a solar calculator.[45] Whether he could have gotten this information from the priestly records in Egypt tracking the path of the sun, as their great god Re crossed the heavens, or by means of solar records obtained through Babylonian sources, or even by means of his own industry, we cannot say for sure. The doxographical reports that credit Thales with some sort of discovery about the solstices and equinoxes,[46] which Burkert conjectured was the book to which Eudemus had access—as they credit also Anaximander with making a seasonal sundial that must have identified the solstices and equinoxes—make perfectly good sense in this context because they would have been important factors to be considered in reckoning the few opportunities to make the measurement. Had Thales made the measurement by *this* technique, the idea that he just lucked out—a clever fellow who just showed up on the right day—deserves to be dismissed. It deserves to be dismissed because the supposition behind this kind of reasoning is that the height of the pyramid could have been measured more or less on any day, like the height of a stick. This technique of measurement by *equal* shadows could not be carried out in such a way.

How did Thales "measure" the shadow length? One thoughtful writer conjectured that he did so by "pacing" out the distance of each pyramid side, then divided by half to reckon the distance from the pyramid center to the outside perimeter, and then paced off the shadow distance.[47] Thales might well have done the measurement this way, but if so, no one would have any idea if the "answer" proposed was correct (compared to what?), and the pacing would have had to be done with great care even to approximate the number of "feet"—and then to whose "foot" measure was the measurement being referred? However, had he used the surveyor's royal cubit cord, Thales's answer in cubit measures probably astounded his Egyptian hosts. We have good evidence for the routine use of this cord in ancient Egyptian architecture and surveying; consider the examples below:

Figure 2.8. After Egyptian statues of surveyors, Sennehem and Senmut, with measured cord, in the British Museum.

Thus, had Thales made the measurement with the Egyptian royal knotted cord, he would have obtained the result in cubit measures, and the reason the result was so astounding was that, as I have conjectured, it was extremely close to or even matched the result the Egyptians knew by their own method and recorded in the RMP. The successful measurement, then, would have identified a shadow of 102.5 feet (or roughly 60 cubits at 52.30 cm = 1 cubit), which had to have added to it 1/2 the distance from the pyramid side to the middle. And since the building is square, the same result could have been gained by measuring the distance of the shadow protruding from the side of the pyramid and adding to it 1/2 the distance of any side of the pyramid, that is 377.5 feet, to produce the result of 480 feet in height (or 220 cubits), and thus 220 + 60 = 280 cubits for the whole height of the pyramid.

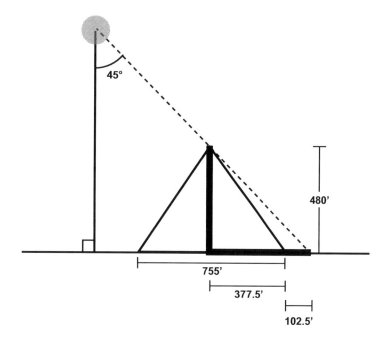

Figure 2.9.

If Thales used this first technique, his success came at a price. By fixating on the isosceles right triangle, he must have reached a surprising conclusion, namely, that whenever the gnomon or stake cast a shadow equal to its height, and that gnomon had a measure expressible as an integer of cubits, that is, a whole number, when Thales stretched the cord from the tip of the shadow to the top of the gnomon (the hypotenuse, or ὑποτείνουσα, literally the "stretching under of the cord," the term used in Plato's *Timaeus* 54D and elsewhere), it would *never* be an integer, that is, a whole number. Proclus observed exactly this point, and each of us today, already knowing the famous hypotenuse theorem algebraically, can confirm by means of the metric formula, that when the length of two sides of a right triangle can be expressed as the *same* integer, the length of the hypotenuse of that right triangle will *never* be an integer.[48] Thus, if the length of the gnomon was 3 cubits, and hence its shadow was 3 cubits, by $a^2 + b^2 = c^2$, then by substitution $3^2 + 3^2 = c^2$, and so $9 + 9 = \sqrt{18}$, and $\sqrt{18}$ is not a whole number.[49] And this is why, I speculate, that Thales, who, by my hypothesis, was looking for the unity that underlies all three sides of a right triangle, did not pursue the numerical or metric interpretation of what became later the famous theorem, algebraically. It is my thesis that he was looking for the underlying unity of the sides of a right triangle because he came to regard that figure as the fundamental geometrical figure, as I shall continue to argue. However, because he focused on the *isosceles* right triangle, he saw with a kind of immediacy that is not apparent with, say, the scalene 3, 4, 5 triangle—the paradigm of the numerical or metric interpretation—that the squares constructed on both sides were equal in area to the square constructed on the hypotenuse.

106　The Metaphysics of the Pythagorean Theorem

To see this point, let us join Thales again in reflecting on the kind of sketch he might well have made on the Giza plateau,[50] or the diagram that had to appear to him in his imagination. Looking for the unity of all three sides of a right triangle, he could not find a numerical connection among the lengths of the sides—there were limitless examples of isosceles right triangles where the legs had whole numbers but not the hypotenuse—but he came to see a unity when squares were constructed out of them, that is, a unity revealed through a composite constructed from them. First, he realized that the isosceles right triangle was the basic building block of every square, below left; and the next step, below right, was to envision a square made on each side of the right triangle. Each square is composed of two isosceles right triangles, and the square made on the hypotenuse is equal to four isosceles right triangles—the sum of the squares on the two sides. I remind the reader that the plausibility of this case rests on grasping that *Thales was looking for an underlying unity* in geometry, as he did in nature, and in this case of all geometrical figures.

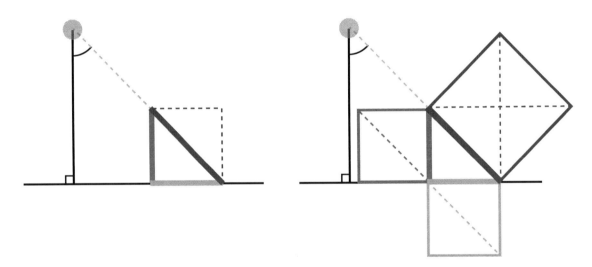

Figure 2.10.

Before moving to consider the second technique with which Thales is credited in measuring the height of the pyramid, let us be very clear that even in this first technique, the method is by means of *similarity*. We have already considered how Thales knew the right time to make the measurement, because he either set up a gnomon whose shadow was observed to be equal to its height, or he measured his own shadow to confirm the equality of heights of vertical object and shadow, though his own shadow would offer a less precise exemplar.[51] But now let us reflect on the proportional thinking required by Thales or anyone else who thinks through the measurement by this method. As his own shadow (or the gnomon's) was to his (its) height, so the pyramid's shadow was to its height.[52] Thus not only was the measurement the result of grasping "similar triangles," but the similarity was established by means of a ratio and proportions—ratios (λόγοι) and proportions (ἀναλόγοι).

>   Thales's height   :   Thales's shadow
>   [Gnomon height   :   Gnomon shadow]
>   as
>   Pyramid's height   :   Pyramid's shadow

Now as we discussed in the last chapter, a ratio is a comparison between the relative size of two magnitudes; it tells us how many copies of the smaller will fit into the larger. So a ratio is implicitly a comparison of all the potential multiples of one magnitude to all the potential multiples of the other. Thus when the ancient Greeks said that two ratios are in the same *proportion* (ἀνάλογον), they meant literally that they both had the same ratio—the term ὃ αὐτος λόγος.[53] Geometric *similarity* is the structure that underlies the macroscopic-microscopic argument. This early vision was of a cosmos connected by continuous proportion, the small world growing into the big world that shared the same λόγος. The small world and the big world share the same structural principles. For Thales, the cosmos originates from ὕδωρ; the whole cosmos is nothing but ὕδωρ. Once one adopts this starting strategy, the structure of the small world is grasped to contain precisely the same structure as the big world.

Whether λόγος was a term used by sixth-century Ionians in this mathematical sense of ratio is subject to controversy and dispute. What is less disputed is that λόγος came to denote "ratio" and ἀνάλογον—"proportion"—to denote ὃ αὐτος λόγος, things in the same ratio. As von Fritz put it well, "*Logos* designates the word or combination of words in as much as they convey a meaning or insight into something. It is this connotation which made it possible in later times to acquire the meaning of an intrinsic law or the law governing the whole world. If *logos*, then, is the term used for a mathematical ratio, this points to the idea that the ratio gives an insight into a thing or expresses its intrinsic nature."[54] For our concerns in the sixth century, regardless of which terms were used, ratios between the gnomon's shadow and height, the pyramid's shadow and height—and the two compared—revealed an intrinsic nature of these things, and revealed something metaphysical about the nature of similar triangles, the same structure that both the small world and large world shared.

B.2 Thales's Measurement of the Height of a Pyramid

Technique 2: When the Shadow Length Was NOT Equal to Its Height

So far, our investigations have shown how rare were the opportunities to have measured the height of a pyramid when the shadow was equal to its height. On merely four days per year, the sun is at 1/2 a right angle and due south, east, or west. Had Thales made such a measurement, almost certainly he had to have made a solar calendar to know when those few windows of opportunity would be. Thus when Thales is credited with a book *On Solstices and Equinoxes* (whether or not he wrote it, he knew it) it seems so plausible, had he measured when shadow equals height. Is the situation more favorable for making the measurement of pyramid height when the shadow is unequal but proportional?

We have a report in Plutarch that Thales measured the height of a pyramid by a technique using ratios and proportions.[55] "Among other feats of yours, he [Pharaoh Amasis] was particularly pleased with your measurement of the pyramid when, without trouble or the assistance of any instrument, you merely set up a stick at the extremity of the shadow cast by the pyramid and, having thus made two triangles by the impact of the sun's rays, you showed that the pyramid has to the stick the same ratio which the shadow has to the shadow" (Plutarch, in his dialogue with Niloxenus speaking).[56] But let us also be clear that the measurement when the "shadow equals the height" is also a proportional technique; it is also a demonstration by *similar triangles*, but in such a case, 1 : 1 proportions. And it is a measurement established by *ratios* (λόγοι) and proportions (ἀναλόγοι). In this second case, he placed a stake (gnomon) at the end of the pyramid's shadow (or nearby) and reckoned that the length of the shadow was to the height of the stake as the length of the pyramid's shadow was to its own height. This was a case of ὁ αὐτὸς λόγος—they both had the same ratio, and thus stood in the same proportion. This is not a measurement at a time of day when the shadow is equal to the height of the object casting it, but rather is unequal to it. This, then, requires that Thales imagined further *proportional* relations between *similar* triangles. Again, in my estimation, he would have used the royal cubit cord to make both measurements—the shadow lengths of the gnomon and pyramid.

Let us be clear, then, about the distinction and relation between equal and similar triangles. Triangles are equal when we could superimpose (ἐφαρμόζειν) one on top of the other and they would fit (more or less) perfectly; thus the three angles would all be the same, as would be the lengths of each side. We might have to rotate one triangle on top of the other for fit, but if they were equal they would align, angle to angle and length to length. Euclid's congruence theorems in Book I of the *Elements* show that equality is obtained when two sides are of equal length and the angle they share between those equal sides is the same (side-angle-side, or SAS) I.4, when all three sides are of equal length (SSS) I.8, and when two angles are the same and the side they share between them is the same length (ASA) I.26. The technical term for what we call "congruence" is derived from the sense of ἐφαρμόζειν, a term that appears in Euclid as Common-Notion 4—"coincide," and it is an idea that Euclid tries to avoid because of the inherent imperfection in all physical measurements and rotations to fit. It is plausible that if Thales had proved that "a circle is bisected by its diameter" (which Euclid includes as definition 17 in Book I, and thus nothing in need of proof at all for him), he would have cut the circle in half (made of some material, possibly clay) and placed one half on top of the other to show they were equal—and this must have been a fundamental feature of "proving" anything in these early stages—and certainly it was the hands-on technique by which monumental architecture advanced in its projects.[57] One of the theorems of equality—ASA—is Euclid I.26: "If two triangles have the two angles equal to two angles respectively, and one side equal to one side, namely, either the side adjoining the equal angles, or that subtending one of the equal angles, they will also have the remaining sides equal to the remaining sides and the remaining angle equal to the remaining angle." Thales, as we have already considered, is sometimes credited with this understanding because it was seen as a principle that enabled him to measure the distance of a ship at sea, a measurement that we shall now explore.

To see this interconnection, let us imagine Thales measuring the distance of a ship at sea. Since we are not informed of just how he did so, we set out to imagine the various ways that he might have. All the techniques require an understanding of similarity. The measurement requires that two triangles share ὁ αὐτὸς λόγος—the same ratio. In one interpretation, Thales is standing on the shoreline at A looking out to an inaccessible ship at B, below. Historians of mathematics such as Heath and others speculate in the scholarly literature that the sighting might rather have been made from an elevated location along shorelines since flat plains inland are in short supply in this geography, and while that might well be, it might just as well have been made from the shoreline, perhaps just outside Miletus, pictured in the diagram, below.[58] Now at a right angle, Thales measures off a distance across the shoreline to point D, and from D stakes out a straight line inland at a right angle to AD of arbitrary length, labeled E. Now, Thales could next bisect line segment AD, by means of compass and straightedge, or even by means of a measured cord as used by the Greek architects when building monumental temples in Didyma and elsewhere, like the ἁρπεδονάπται, and identified the midpoint as M. He might then walk across from point A to D, and then walk inland to some point along line DE, when he would sight the distant ship from some point P. All this could be done on the level seashore that we still see near Didyma. Since he can measure the distance of DP, he now knows the distance to the ship AB. How can he be sure? How might he have replied to his doubting compatriots? He may well have argued along the lines of what later became Euclid I.26, namely, that in the triangles PDM and BAM, side DM = MA, angle D = angle A since they are right by construction, and angle DMP = AMB, since they are vertical angles—a fact Thales was said to have discovered—and, therefore, the remaining corresponding parts of the triangle are equal. In particular, BA = DP, so that by measuring DP, which is on land, BA can be known.

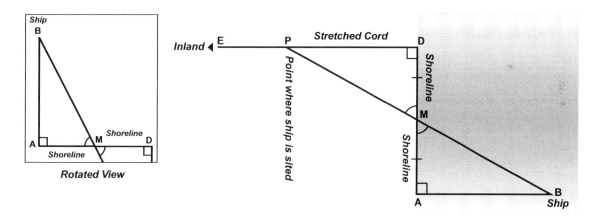

Figure 2.11.

Let me also direct the readers to take another look at the diagram directly above and compare it to the geometry of the pyramid measurement; the very conception of this measurement of distance to the ship can be seen to mirror the Egyptian RMP technique of measuring the height of the pyramid. In the pyramid case, the relation of "pyramid base–pyramid height–*seked*" is analogous to "triangle base, distance to ship, inclined line/angle for sighting," whether or not the sighting is done from the shore or from a tower above.

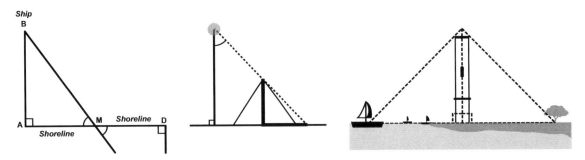

Figure 2.12.

But it is also possible, keeping within the theme of similar triangles that Thales reasoned from the smaller but proportional triangle to the distance of the ship sighted—all the angles are the same and the sides are proportional. Thus, we can suppose that Thales chose M so that DM is some given part of MA, say 1/2. Then, having concluded that triangles AMB and DMP are similar by noting that the angles are equal to those of the other, he would infer that DP should be the same part of AB that DM was of MA, that is, 1/2. So measuring DP he could say that AB is twice the length he measured.

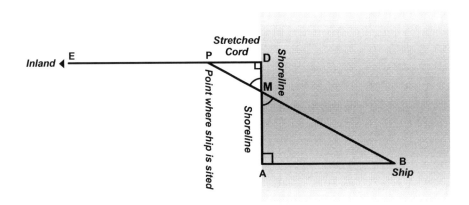

Figure 2.13.

In the diagram, above, Thales knows that both angles at M are equal, which is why he is credited with the theorem that when two straight lines cross, they make the vertical angles equal. Thus, below, the vertical, opposite angles are equal whether or not the angles are right, and one can see instantly that angle AEC and angle AED make two right angles, and so any two angles so connected on a straight line equal two right angles.

Thales and Geometry   111

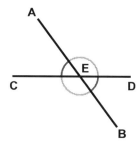

Figure 2.14.

Now let us consider another way that Thales might have imagined the measurement. Below, on the left is more or less the same situation of having a measured cord stretched along the shoreline, making the base of a triangle. The ends of the cord are the markers for the angles of the triangle, and the length of the perpendicular from the ship to the base is the distance of the ship at sea. Next, Thales makes a small model to scale, just as I discussed with the Eupalinion tunnel that is roughly contemporaneous with the latest part of Thales's life. His scale model is ὅ αὐτος λόγος—in the same ratio. And then he reasons that the distance across the shoreline BC is to the base of the model of the triangle EF, as the distance to the ship AD is to the perpendicular in the model GH.

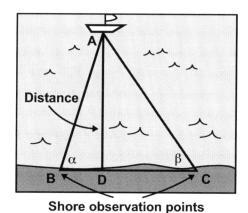 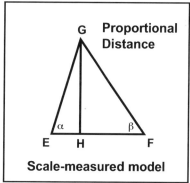

The actual distance is calculated by proportionality. That is:
BC : AD :: EF : GH

Figure 2.15.

112   The Metaphysics of the Pythagorean Theorem

Perhaps Thales made the measurement by another variation. Consider the diagrams below. In this scenario, Thales is standing atop a tall tower. He first looks seaward and determines the angle to the ship, and then looks inland and sights something, say a tree. He selects the tree, or whatever, because it is at the *same angle*—determined by a simple device, as Heath conjectures—a piece of wood with a nail in it that can be rotated. Thus he has imagined an isosceles triangle, with right angles at the base of the tower. Naturally, this same procedure could be done by unequal but proportional triangles; consider the second and third examples, below, as well.

Figure 2.16, 2.17, 2.18.

Heath suggested another possibility, below. In this case, Thales is standing on a platform on segment DE and he sights the ship at C. The little triangle ADE is similar to the large triangle on base BC, ABC.

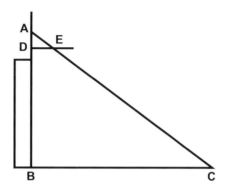

Figure 2.19.

Thus, by trying to connect some of the dots, as it were, a plausible case can be made that Thales grasped the difference and relation between equal and similar triangles. But a point to be emphasized became clear to me and places this discussion, for instance, of the measurement of the distance of a ship at sea, in a new and broader light. Once one begins to play with these diagrams, one begins to discover surprising insights into shapes and lengths. From my experience, it seems more likely, not less, that Thales tried several of these ways, and perhaps more. Once a person sees that geometric formulations offer insights into such matters as distances and heights—like rotating a tile to see if it fits the standard—it would come as no surprise to see Thales trying multiple ways of confirming his insights.

Now, pressing further with the pyramid measurement using the second technique, Thales is imagining a right triangle, but the "base" angles, while not equal to each other (in each triangle), were nevertheless correspondingly equal in both triangles—the one formed by the pyramid, and the one formed by the gnomon. Thus, all three angles were correspondingly the same in both triangles *at any one time*, and only the lengths of the sides were different but stood in the same proportions—ὅ αὐτος λόγος. This is precisely the meaning of *similar* triangles. Once we reflect on the enormity of the task of making the measurement by the first technique when the rare conditions would be met only six times a year on four days when the pyramid casts a measurable shadow equal to its height, and the window of time due east and due west is about twenty minutes, and only slightly longer at midday when due south, this second technique recommends itself. Had we sat with Thales day after day on the Giza plateau waiting for the conditions to be right for the first technique that almost never happened, he would have noticed that there was a time *every* day when the sun crosses the meridian—which we call "local noon"—when the sun is highest above the horizon, *and thus due south*. On a sunny day, there would always be a measurable shadow. Had the cord that Thales used had fractions of a cubit indicated by the multitude of knots, he was in a position to produce an accurate measurement

114   The Metaphysics of the Pythagorean Theorem

every clear day by *proportional* reasoning. Below is such a diagram when the shadow length is twice the object's height.

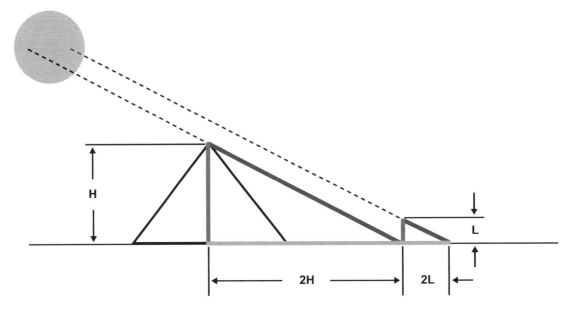

Figure 2.20.

Below is a photograph of the Giza pyramids with proportionally longer shadows. These photos were taken in the afternoon when the sun was low in the western sky, and so the shadows project eastward.

Figure 2.21. Photo by Marilyn Bridges from her book *Egypt: Antiquities from Above*. Boston: Little Brown, 1996, p. 10.

Had Thales grasped this proportional relation in the first technique, he certainly would have grasped its application in the second technique. In both cases, the proportional relations are by similar triangles. He could not have made such a measurement without understanding *similarity*; that is, such a measurement *presupposes* its principles. That includes the central insight—the deep mathematical intuition—that the small world has the same structure as the big world, and the underlying presupposition that endured until incommensurability was discovered, namely, that there was a whole number of times that the little, similar figure (or some smaller similar figure still), would expand to fill up fully the bigger figure. I claim that this geometrical intuition supplied for Thales a vision of *transformational equivalences*; it offered to explain how, while there was only one underlying unity—ὕδωρ—and all appearances were nothing but ὕδωρ *altered without* being changed, it looked so different at different times because the right triangles were combined differently to make them. Thus, from a single underlying principle, all appearances are expressible in terms of ratios from the basic figure of the right triangle. But now, instead of focusing on how all rectilinear figures reduce to triangles and ultimately to right triangles—the way *down*—he is seeing quite literally a model of how the cosmos grows—the way up—and this leads him to see that growth as a continuous proportion, or growth and expansion in terms of mean proportionals (i.e., geometric means), through a series of continuous ratios, and that growth was visualized in terms of areal equivalences.

It is this proportional thinking that also, plausibly, led to his discovery of an aerial interpretation of the famous theorem, according to Tobias Dantzig,[59] an historian of mathematics, who stands almost alone when he suggests that Euclid's VI.31 was probably the line of thought by which Thales "proved" the theorem. As we have seen already, it is a proof that relies on an argument by similar triangles to reach the conclusion of I.47, which is a special case. I am not arguing, however, that Thales "proved" this theorem, but only that he plausibly connected the lines of thought that later were set out rigorously in Euclid's VI.31. When we supply the geometrical techniques that seem to be presupposed to measure the distance of a ship at sea, both from the seashore and from a tower, and to measure the height of a pyramid by both techniques, we can see that Thales had reflected long and hard on the ideas of equal and similar triangles. And by appeal to the geometrical principles revealed in the measurement of a pyramid at the time of day when the shadow length is equal to its height, Thales was forced to imagine an isosceles right triangle. Here was a moment when he thought through the theorem concerning all isosceles triangles, that when "their base angles are 'equal' the sides opposite them must also be equal. Proclus tells us that Thales used the archaic expression γωνίαι ὁμοίας,—"similar angles," not "equal angles" γωνίαι ἴσας[60]—and this suggests further that Thales was aware of similarity in its original and broad sense. In an important way, we get insight into Thales's fascination for searching for underlying unities—perhaps we should say underlying "similarities." In addition, when we consider the pyramid measurement that required Thales to imagine an isosceles right triangle, we get a clue about how he might have come to see the unity of all three sides of a right triangle revealed by the squares constructed on those sides—in terms of areal equivalence. And finally, before moving forward, we should remind ourselves that the discovery that "every triangle inscribed in a (semi-)circle is right angled," attributed to Thales, highlights further his preoccupation with right angles and right triangles, and the interrelation of triangles, squares, rectangles, circles, and semicircles (i.e., diameters) that is suggested *visually*. The reader who will sketch out diagrams of these reflections will begin to come to terms with the visual facts of the matter, that accepting the attribution of *these* theorems to Thales means that Thales had sketched out countless diagrams, inspecting each to discover if any properties of lines and figures could be grasped by means of these inspections. The context I am pointing to here seems never to have been emphasized in the scholarly literature for an understanding of Thales *metaphysical* speculations.

Let me try to emphasize this last point in another way. As soon as one accepts the idea that Thales was reflecting on these matters by making geometrical diagrams, a completely new vista for understanding Thales opens. To grasp it, we must join Thales in making sketch after sketch; when we do, a different kind of mentality appears. Proclus informs us, clearly from Eudemus's book, that "Thales, who had travelled to Egypt, was the first to introduce this science into Greece. He made many discoveries himself and taught the principles for many others to his successors, attacking some problems in a general way and others more empirically."[61] It is not as if Thales gained insight into *four or five* theorems. The point is that he gained insights into dozens and dozens of propositions about space and its articulation in terms of rectilinear figures, and he grasped that the right triangle was the basic or fundamental geometrical figure, *since this is what he was looking for*.

In the Introduction I pointed out that the in RMP problem 51, a triangular plot of land had an area reckoned in comparison to the rectangle in which it was inextricably connected. In his translation and commentary[62] the expression "There is its rectangle" showed Peet that the problem was to be answered "graphically," that is, *visually*. Since the evidence that we do have shows land in Egypt was customarily divided into *arouras*—rectangles—though Herodotus mistakenly reports the land was divided into squares[63]—it is suggestive to see that this method of calculating area was useful because it showed that the area of every rectangle could be expressed as the summation of triangles into which it could be divided. The height of the pyramid, problem 57, in similar fashion could be seen visually as the triangle's square. Traveling in Egypt and measuring the height of the pyramid plausibly allowed Thales the opportunity to learn these things, and I dare say much more, based on the evidence that survives. I will return to some of these diagrams later in this chapter. And among the "successors" to whom Thales taught these principles—whether directly in person, which is my suspicion, or by some members of his school, since this knowledge was current—was the young Pythagoras.

C

Thales's Lines of Thought to the Hypotenuse Theorem

Having already explored Euclid's two proofs of the hypotenuse theorem, I.47 and VI.31, I have suggested that the basic lines of thought for the discovery of VI.31 are to be found in reports about Thales—if we can connect the dots, so to speak—as a consequence of similar triangles. The case I have been exploring urges that we place in front of us geometrical diagrams attributed to Thales; that as we continue to work out diagrams connected to the ones mentioned we recognize that he must have produced and inspected countless more diagrams; and that we then add Aristotle's doxographical report, which sheds light on what Thales was up to—he was looking for, and discovered, the fundamental stuff that underlies everything. When we do this, a new light shines on a series of thoughts that plausibly led to his discovery of an areal interpretation of the famous hypotenuse theorem. To see this connection, we must keep in mind that the metaphysics of the *hypotenuse theorem by ratios, proportions,* and *similarity* is that the right triangle is the fundamental geometrical figure, the microcosmic structure that underlies all appearances, and as the right triangle expands macrocosmically, it blows up or scales up proportionately in geometric means, and the areal relation among the similar and similarly drawn figures on its sides holds—*every figure* scales up in the duplicate ratio of the corresponding sides. All rectilinear figures reduce to triangles, and all triangles reduce to right triangles, and *inside every right triangle are two similar right triangles dissectible ad infinitum*. It was this understanding that led to the search for regular solids, all of which are constructed out of right triangles—a project of Pythagoras inspired by Thales or by the kind of knowledge that Thales had and that was current in Ionia—the combinations of triangles that compose the elemental building blocks of all appearances.

Once we place before us a variety of geometrical diagrams showing "Thales's theorems"—that the base angles of isosceles triangles are equal when the sides opposite those angles are of equal length, that straight lines that cross make the opposite angles equal, that triangles are equal if they share two angles respectively and a side between them, that the diameter bisects the circle, and that *in the bisected circle every triangle is right angled*—and keep in mind Aristotle's testimony that Thales is the first philosopher because he identified the basic stuff that underlies all appearances, we can surmise a direction for Thales's interests. The plausibility of my thesis rests further on looking forward to Plato's *Timaeus*, and then after seeing what is contained there, on looking back to trace this project in its earlier stages of echo—the construction of the cosmos out of right triangles—to Pythagoras who was inspired to think this way by Thales. Why Thales is bothering with geometry is not first of all and most of the time for the purposes of practical problem-solving, though he realized geometrical diagrams offered insight into them, but rather because he is looking, analogously, for the geometrical figure that underlies all others, as he landed upon ὕδωρ as the basic unity that underlies everything. Thus, the overall framework that I am suggesting Thales provided is that ὕδωρ is the ἀρχή, geometry is the structure, and I will only mention here that condensation/compression is the process by means of which this underlying unity appears so divergently.[64] These are the essential ingredients to understand the project of transformational equivalences—how, given an underlying unity that alters without changing, things can appear so differently. This ultimate unity is suitable to combine and reconfigure itself to appear as our marvelously diverse world. How does it do so? Geometry offers a window into the structure, though not the process, of how the right triangle, the basic geometrical figure, combines and recombines to produce what Pythagoras went on to explore and show, were the basic elements—fire (tetrahedron), air (octahedron), water (icosahedron), and earth (hexagon)—out of which all other things are made. The discovery of the regular solids is a tradition that traces itself back to Pythagoras, but in my estimation, the originator of the project of building the world out of triangles was Thales.

The path to Thales's discovery can be traced through the report from Diogenes Laertius's report; Pamphile says πρῶτον καταγράψαι κύκλου τὸ τρίγωνον ὀρθογώνιον, καὶ θῦσαι βοῦν, that "[Thales was] the first to describe on a circle a triangle (which shall be) right-angled, and that he sacrificed an ox (on the strength of this discovery)."[65] Heath remarked that "This must apparently mean that Thales discovered the angle in a semicircle is a right angle (Euclid III.31)."[66] As Heath sorts through the difficulties of these reports, he is almost able to see that,

118   The Metaphysics of the Pythagorean Theorem

if we accept this attribution, Thales might well have discovered an areal interpretation of the hypotenuse theorem, and though neither Allman nor Gow attribute explicitly the discovery of the hypotenuse theorem to Thales (though Gow supposes that Thales may have known the contents of the first six books of Euclid, and a fortiori must have then had knowledge of the hypotenuse theorem), they go much further to work out the connecting themes (as we shall consider). It seems to me that the reason Heath cannot see this is that, apparently, he does not provide any indication that he grasps why Thales is dabbling in geometry in the first place, seeing instead Thales's preoccupation with practical problems or with formulating proofs and demonstrations as if he were in a math seminar.

Proclus claims that Thales demonstrated (ἀποδεῖξαι, 157.11) that the diameter bisects the circle (= Euclid I. Def. 17) but only notices (ἐπιστῆσαι, 250.22) and states (εἰπεῖν, 250.22) that the angles at the base of an isosceles triangle are equal. Proclus says that Eudemus claims that Thales discovered (εὑρημένον, 299.3) that if two straight lines cut each other, the vertically opposite angles are equal, but he did not prove it scientifically (Euclid I.15). And as we have considered already, Eudemus claims that Thales must have known that triangles are equal if they share two angles, respectively, and the side between them (= Euclid I.26), because it is a necessary presupposition of the method by which he measured the distance of a ship at sea (Εὔδημος δὲ ἐν ταῖς γεωμετρικαῖς ἱστορίαις εἰς Θαλῆν τοῦτο ἀνάγει τὸ θεώρημα. Τὴν γὰρ τῶν ἐν θαλάττῃ πλοίων ἀπόστασιν δι'οὗ τρόπου φασὶν αὐτὸν δεικνύναι τούτῳ προσχρῆσθαι φησιν ἀναγκαῖον. 352,14–18).

What could it mean to say that Thales *demonstrated* that the diameter of a circle bisects it, when Euclid does not prove it at all but rather accepts it as a definition?[67] Heath supposes that it might have been suggested to Thales by diagrams on Egyptian monuments, or even wheeled vehicles in Egypt—multispoke wheels—where circles are divided into sectors by four or six diameters, a subject I have discussed at length elsewhere in a chapter on ancient wheel-making.[68] Heath provides the following diagram to illustrate what he means:[69]

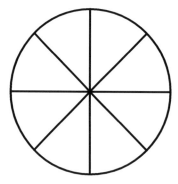

Figure 2.22.

Whatever visual clues might have suggested it, a "proof" in the sixth century almost certainly began by superposition (ἐφαρμόζειν). After the visible showing has been presented, Thales might well have noticed the recurring conditions—the similarities—that accompanied the equalities he pointed out.

Let us now consider Thales's grasp that the base angles of an isosceles triangle are equal. Some have thought that Thales's lines of thought follow the proof by Aristotle in the *Prior Analytics* at 41b13ff, a sequence of thoughts that depends on the equalities of mixed angles.[70] What recommends the "mixed angles" approach is that it combines "the diameter of a circle" theorem with "the angle in a semi-circle theorem" in an effort to prove the isosceles triangle theorem. The point is that they are all interconnected.

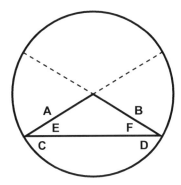

Figure 2.23.

Suppose the lines A and B have been drawn through the center. If then one should assume that the angle AC is equal to the angle BD [that is, the angle between the radius A and the circumference at C and the angle between the radius B and the circumference at D], without claiming generally that angles of semicircles are equal; and again if one should assume that the angle at C is equal to the angle at D, without the additional assumption that every angle of a segment is equal to every other angle of a segment, and further if one should assume that when equal angles are taken from the whole angles, which are themselves equal, the remainders E and F are equal, he will beg the thing to be proved, unless he also states that when equals are taken from equals the remainders are equal.[71]

The proof that the base angles of an isosceles triangle are equal appears in Euclid's Book I.5:

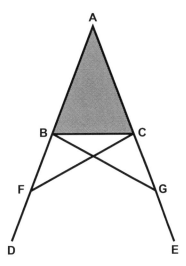

Figure 2.24.

120   The Metaphysics of the Pythagorean Theorem

We begin with triangle ABC, and the proof is to show that base angles ABC and ACB are equal. First, AB is made equal to AC, and then AB is extended to some arbitrary point D, and another arbitrary point F is placed between points B and D. Then, AC is extended to some arbitrary point E, and from the longer AE the segment AG is cut off equal to AF. Then, F is joined to C, and G is connected to B, thus AF = AG (by construction) and AB = AC (by assumption), angle A = angle A, therefore, triangles ACF and ABG are equal, so that FC = BG, angle AFC = AGB, and angle ABG = angle ACF. Moreover, since AB = AC and AF = AG, BF = AF − AB = AG − AC = CG, and base FC is equal to base GB. And since the two sides BF and FC equal the two sides CG and GB, respectively, angle BFC equals CGB, while the base BC is common to them, and CBG is equal to BCF, and thus triangle FBC is equal to triangle GCB. Thus the whole angle ABG is equal to the whole angle ACF, and since FBC is equal to GCB, when we subtract FBC from the whole angle AFC, and subtract GCB from the whole angle AGB, the angles ABC and ACB remain. And when equals are subtracted from equals, the remainders are equal. And so the base angles of the isosceles triangle are equal.

Could Thales have made a proof like this?[72] Proclus says that Thales notices (ἐπιστῆσαι, 250.22) and states (εἰπεῖν, 250.22) that the angles at the base of an isosceles triangle are equal; he does not use either the terms ἀποδεῖξαι (demonstrate/proved) or εὑρημένον (discovered). Very well, then, what might it have meant for him to "notice"? What is the construe of ἐπιστῆσαι here? This is a term probably used by Eudemus, and it conveys an understanding. So what might Thales have understood, had Eudemus used this word in an appropriate sense, distinct from "proved" or "discovered?" The deep intuition seems simply to be this: *there is no way for two sides of a triangle to be of equal length without the angles opposite them being equal.* Angles and sides are so inextricably connected that we cannot have one without the other.

Sorting out the problems of Pamphile's claim, reported by Diogenes Laertius, offers us the chance to gain a penetrating insight into Thales's connection with the hypotenuse theorem. Here again, I begin by following Heath to show where we are led well and where he leaves us hanging. Then I will reflect on earlier approaches by Allman and Gow, whose works influenced Heath, to review this matter further. It is clear that Euclid proves III.31, that every angle in a semicircle is a right angle, by means of I.32, and that the sum of the angles of a triangle equals two right angles; this is sometimes why I.32 is also referred to as "Thales's theorem."[73] The proof, which rests on the parallel theorems of Euclid Book I, is this (diagram, below): construct a triangle ABC and extend BC to D; from point C make a line CE parallel to AB; now when a straight line BD falls on two parallel lines, it makes the exterior ECD and the interior and opposite angle ABC equal; then again, when a straight line AC falls on parallel lines AB and CE, it makes the alternate angles BAC and ACE equal. Now there are two right angles in every straight line (= straight angle), and angle ECD is equal to angle ABC, and angle BAC is equal to angle ACE, and thus when we add angle ACB to angles ACE and ECD (= the whole angle ACD), we get a straight line of two right angles. And since angle ECD is equal to ABC, and angle ACE is equal to angle CAB, then when we add ACB to both, we get the same result, namely, the sum of the angles of the triangle is equal to two right angles.

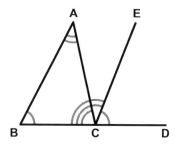

Figure 2.25.

Now the problem is, as we already mentioned, that Proclus states explicitly that the first to *discover* (εὑρεῖν, 379.3–4) as well as to prove (δείκνυμι, 379.5) that the sum of the angles of a triangle is equal to two right angles were the Pythagoreans. And Heath thoughtfully grasps that, if this is so, this cannot be the method by which Thales reached this understanding. Since the proposition that "Every triangle in a (semi-)circle is right angled" presupposes the proposition that every triangle contains two right angles, had Thales grasped the latter, he had to have grasped the former. So, how might Thales have come to this understanding? Heath contended that had Thales proved the semicircle theorem in some other way, "he could hardly have failed to see the obvious deduction that the sum of the angles of a *right-angled* triangle is equal to two right angles." Here, Heath seems to me to be exactly on target, but the understanding (ἐπιστῆσαι) was already Thales's.

As Aristotle shows, though Heath does not mention him in this context, the proof of what later became III.31 in Euclid is accomplished in a different way, and this might well hearken back to Thales. The diagram with the isosceles right triangle might well point back originally to Thales because areal equivalence is immediately obvious in that case.

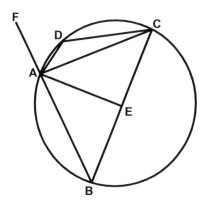

Figure 2.26.

122   The Metaphysics of the Pythagorean Theorem

A review of the medieval manuscript diagrams, even if schematic, shows that the isosceles right triangle is more often displayed (Codex P, B, b, V).[74]

Proposition III.31

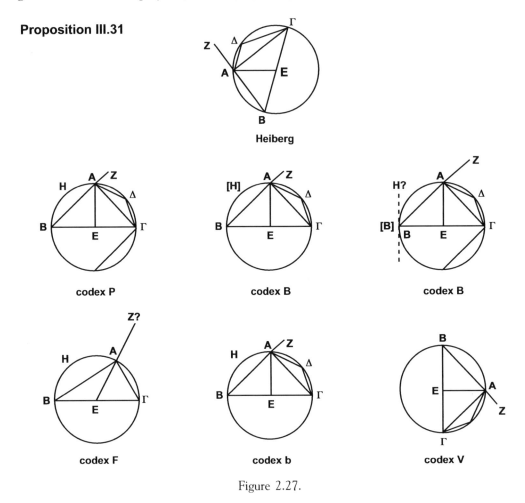

Figure 2.27.

The proof relies on dividing any angle inscribed in a semicircle into two isosceles triangles—this approach resonates with what we already know, that Thales focused on isosceles and right-angled triangles, and that when two sides of a triangle are of equal length, the base angles across from them must be *similar*. In the diagram below, a triangle ABC is inscribed in a semicircle, then a line is drawn connecting A to the center of the circle D. The first version (below left) is with the isosceles right triangle, and mirrors the diagrams in Heiberg's text, Codices P, B, p, b, and V; the second (below, right) is the variation when the triangle, though still right, is divided differently, as in Codex F.

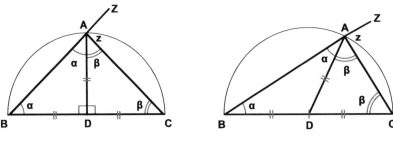

Figure 2.28.

The result in either case is to produce two isosceles triangles since DB is equal in length to DA, since both are radii, and thus triangle ADB is isosceles; DC is equal to DA since both are also radii of the circle, and thus triangle ADC is isosceles. This makes each triangle have two base angles α and two base angles β. And the task of this proof is to show that angle α + β = 1 right angle. The outline of the proof in Euclid is as follows: Let's call z angle CAZ. Then since z is the exterior angle of triangle BAC, it is the sum of the interior opposite angles α + β. Therefore, the angle BAC = CAZ. Therefore, by the very definition of a right angle (Def. 10), BAC is a right angle. This proof sequence has the advantage of relying on the definition of a right angle, and the ones that were plausibly known to Thales.

But had Thales also known that there were two right angles in every triangle, whether or not he had produced a formal proof (perhaps he simply noticed it?), Thales's train of reasoning might have gone like this (Fig.ure 2.28, left): He started with the most obvious case, the isosceles right triangle. Since angles ADB and ADC are supplementary angles, they sum to two right angles. Since there are two right angles in every *right* triangle, then 2α + 2β + 2 right angles = 4 right angles. Therefore, 2α + 2β = 2 right angles, and finally, α + β = 1 right angle, which is precisely what is there to prove. This angle equivalence is most immediately apparent in the diagram on the left—the isosceles right triangle—when the angle in the semicircle is half a square, with the angle at A (= α + β already known to be a right angle). Then, the angle at B and the angle at C must each be 1/2 a right angle because each is the angle of a square (= right angle) bisected by a diagonal, which is the circle's diameter. Then, we could say that 2α + 2β + 2 right angles = 4 right angles; subtracting the sum of two right angles (angles ADB and ADC), what remains is 2α + 2β = 2 right angles, and thus α + β = 1 right angle. OED. The isosceles right triangle, which so absorbed Thales in the first pyramid measurement, obviously has its angle at A equaling a right angle, now divided into two 1/2 right angles.

Thus, our line of argument is that it is plausible that Thales became convinced reasonably that the angle in a (semi-)circle is right angled by appeal to the other theorem he is credited with knowing—that the base angles of an isosceles triangle are equal (whichever version or versions of that connection he may have embraced). For had he known that angle BAC is right, he would know that the two base angles in the isosceles right triangle BAC are equal (the angle at B and the angle at C), and for the same reason the two angles in isosceles triangle ADB are equal (angles DAB and DBA), and also the two angles in ADC are equal (DAC and DCA). Consequently, angles DAB and DAC are equal to angles DBA and DCA (when equals are added to equals the wholes are equal), and since we know that angles DAB and DAC are equal to a right angle, then the equal angles ABD and ACD are equal to a right angle, and thus the sum of all four angles (so partitioned) that make up the triangle BAC are equal to two right angles.

Furthermore, if the angles of a right triangle are equal to two right angles—we have already considered the diagram of a square that contains four right angles, and draw the diagonal dividing the square into two isosceles right triangles, hence each right triangle contains a right angle and two 1/2 right angles that together make the second right angle—then, as Allman pointed out and both Gow and Heath follow, "it is an easy inference that the angles of any triangle are together equal to two right angles" because every triangle can be divided into two right triangles by dropping a perpendicular from the vertex to the base.[75] Note that this is the diagrammatic frame of VI.31, though Heath mentions nothing about it here—it is principally the same diagram in the codices, with the (semi-)circle removed.

124 The Metaphysics of the Pythagorean Theorem

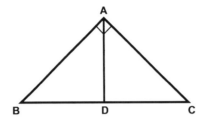

Figure 2.29.

Drawing the perpendicular, AD, divides triangle ABC into right triangles ADB and ADC, and together they must contain four right angles. "Now the angles of triangle BAC are together equal to the sum of all the angles in the two triangles ADB, ADC less the two (right) angles ADC and ADC. The latter two angles make up two right angles; therefore the sum of the angles of triangle BAC is equal to four right angles less two right angles, i.e., to two right angles."[76] Thus, Heath's argument about knowing that all triangles contain angles equal to two right angles rests on the understanding that every triangle can be divided into two right-angled triangles by dropping a perpendicular from the vertex to the base, and from there on, our understanding of all triangles reduces to our understanding of the right triangles to which one and all of them reduce. This is, of course, fundamental to the proof of the "enlargement" of the hypotenuse theorem at VI.31. Here we find yet another suggestion of how Thales could be connected to the hypotenuse theorem by way of similar triangles, and the understanding that every triangle in a (semi-)circle is right, as Pamphile attests.[77]

The proposition that the sum of angles in every triangle is equal to the sum of two right angles might well have been proved in yet another way.[78] Eudemus tells us, according to Proclus,[79] that the Pythagoreans discovered that the sum of the angles of a triangle is equal to two right angles. But we have the testimony of Geminus, who reported that while the Pythagoreans *proved* that there were two right angles in every triangle,[80] the "ancients" (who could be none other than Thales and his school) *theorized* about how there were two right angles in each species of triangle—equilateral, isosceles, and scalene. There are several ways this may have been investigated by Thales (though no more details are supplied by our ancient sources), but all of them could have been revealed to have two right angles by dropping a perpendicular from any vertex to the side opposite and then completing the two rectangles, as below; I argued in the Introduction that this may very well have been the way that the Egyptian RMP problems 51 and 52 (= MMP problems 4, 7, and 17) were solved *visually*—the problem of calculating a triangular plot of land, and a truncated triangular plot of land (i.e., a trapezium)—and that Thales learned or confirmed the *triangle's rectangle* from the Egyptians. By dropping a perpendicular from the vertex to the base, the rectangle can be completed: This is the triangle's rectangle.

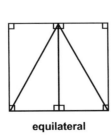
equilateral

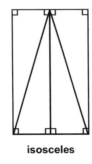
isosceles

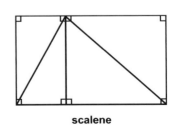
scalene

Figure 2.30.

It is fascinating to note that the same diagrams that reveal that each species of triangle contains two right angles, at the same time reveals that every triangle reduces to two right triangles.

In addition, Allman conjectures that had Thales known that every triangle inscribed in a semicircle is right, he deserves credit for the discovery of *geometric loci*—that if one plots all the points that are vertices of a right-angled triangle whose base is the diameter of a circle, all those points define the circumference of the circle. In this case, Allman is reflecting on some of the conditions that stand out when one further inspects this discovery—if Thales discovered that every triangle in a (semi-)circle is right, he would surely have realized also that the vertex of every right triangle in that (semi-)circle defines the circumference of the circle. I find Allman's speculations enticing but I would describe this insight differently, placing it in the context of metaphysical speculation rather than a mathematics seminar. When one discovers that every triangle inscribed in a (semi-)circle—whose vertex is at A and whose base is diameter BC—is right, one has discovered *all* the right-angled triangles that fill up and are contained within the circle. The circle is filled completely with and by right-angled triangles, the right angles at A mark out completely the circumference of any circle. Thales's investigations that led to this realization about right angles in the (semi-)circle came about as he was exploring what was the basic geometrical figure—circles, rectangles, triangles—and his conclusion, as we shall now examine further through VI.31, was that it was the right triangle. *Even the circle is constructed out of right triangles—that within every circle are all possible right-angle triangles!*

Thus ALL the possible right angles are potentially displayed in the half circle, and by placing in sequence all the right triangles, a circle is created by connecting their vertices at the right angle.

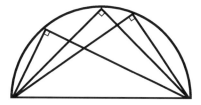

Figure 2.31.

And consequently, if ALL the right angles are continued in their mirroring opposite half of the (semi-)circle, then ALL possible rectangles are displayed in the circle, including of course, the equilateral rectangle or square.

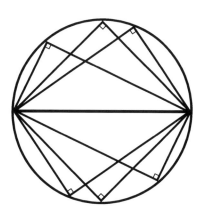

Figure 2.32.

126 The Metaphysics of the Pythagorean Theorem

And if one removes the semicircle, one is left to gaze on almost exactly the same diagram as VI.31 (without the figures drawn on the sides, and not necessarily having a radius but rather any line connecting the diameter to the semicircle). In the ancient manuscripts, unlike Heiberg's later version, the diagram at VI.31 is an isosceles right triangle, containing no figures drawn on the sides (sc. the right angle box supplied for clarity).

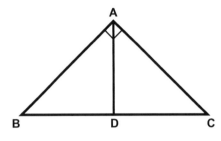

Figure 2.33.

First, VI.31 requires a right triangle. From the diameter of the semicircle we have a method that guarantees that the triangle inscribed within it is right. Next, a perpendicular is dropped from the right angle to the center of the circle. Did Thales know that the perpendicular from the right angle was the *mean proportional* or *geometrical mean* between the two sides, and moreover, its corollary, that the perpendicular divided the original triangle into two similar triangles, each of which shares a mean proportional with the largest triangle? This is precisely what an understanding of *right-triangle similarity* implies. Had he understood it, as I have come to be persuaded he did, he was aware of the relation among three proportionals, and hence he knew that the length of the first is to the length of the third in duplicate ratio as the area of the figure on the first is to the area of the similar and similarly drawn figure on the second. This is all a way of expressing what he was looking for: how the right triangle dissects indefinitely into right triangles, indeed collapses by mean proportions, and how the right triangle expands so as to produce areal equivalences on its sides again in mean proportions. This was the key to Thales's grasp of *transformational equivalences*, the project of the cosmos growing out of right triangles. Pythagoras is credited with the knowledge of *three means*: arithmetic, geometric, and harmonic, and if we accept this, it is certainly not too much of a stretch to realize that Thales understood the idea of a geometric mean contemporaneously and earlier. The geometric mean or mean proportional was the grasp of the continuous proportions that structures the inside of a right triangle and at once describes how the right triangle grows and collapses.

How might Thales have discovered this or convinced himself of it? Recall the discussion, earlier, of VI.31 and the geometric mean. Let's try to follow the diagrams. The mean proportional is an expression of length and areal ratios and equivalences. When the hypotenuse of the right triangle is partitioned equally and unequally by the perpendicular depending on where on the semicircle the vertex is located, nevertheless all form right triangles. When the perpendicular divides the hypotenuse unequally, the longer part stands to the shorter part in a duplicate ratio; of course it does so as well in the isosceles right triangle, where the partition is 1 : 1. Let us follow through the equal and unequal divisions. We can imagine a rectangle made from both lengths as its sides—a rectangle whose sides are the lengths of both extremes. The area of this rectangle is equal to the area of the square drawn on the perpendicular that divides them, and thus the area of the square on the first side stands in duplicate ratio to *both* the square on the perpendicular and the rectangle composed from the first and third side lengths, below:

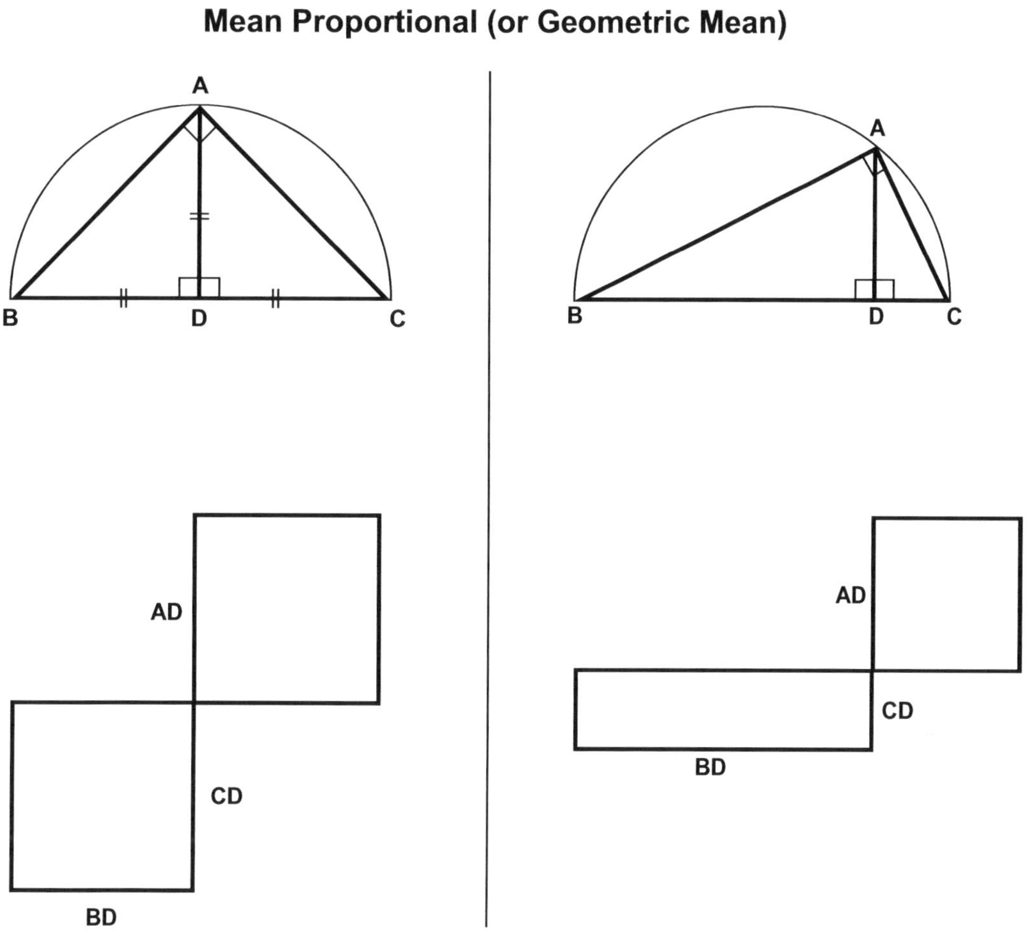

Mean Proportional (or Geometric Mean)

Figure 2.34.

When the series of connected ideas is presented in Euclid Book VI, the series begins with VI.8, that a perpendicular from the right angle to the hypotenuse always divides the right triangle into two right triangles, each similar to the other and both to the whole largest triangle into which they have been divided, and (porism) that the perpendicular is the mean proportional. Next, VI.11, to construct a third proportional, three lengths that stand in continuous proportion, then VI.13, the mean proportional of the lengths into which the hypotenuse has been partitioned is constructed by any perpendicular from the diameter of the semicircle to its circumference; the perpendicular is extended upward from the diameter, and the moment it touches the circumference, it *is* the mean proportional between the two lengths into which the diameter is now partitioned. When we connect that circumference point to the two ends

128   The Metaphysics of the Pythagorean Theorem

of the diameter, the triangle formed must be right. Next, VI.16, if four lines are proportional, the rectangle formed by the longest and shortest lengths (= extremes) is equal in area to the rectangle made of the two middle lengths (= means). Next, VI.17, *if three lines are proportional, the rectangle made by the longest and shortest lengths (= extremes) is equal in area to the square made on the middle length (= means)*—here is the conclusion to the mean proportional, the three making a continuous proportion. Next, VI.19, similar triangles are to one another in the duplicate ratio of their corresponding sides, and finally VI.20, all polygons reduce to triangles, and similar ones are to one another in the duplicate ratio of their corresponding sides.

Let me set these out again in the context of how we might imagine Thales's discovery. My case is that in the distant past, Thales stumbled on these ideas in their nascent form as he was looking for the fundamental geometrical figure. Once he realized that it was the right triangle, he explored further to see how it grows into all other appearances. The interior of the right triangle dissolves into the same structure into which it expands. Here was his insight into *how*, structurally, a single underlying unity could come to appear so divergently: this is what I believe Aristotle is referring to when he states that the Milesians claimed that all appearances alter without changing. I am not arguing that Euclid's proofs mirror Thales, or that his project got further into constructing the regular solids that are built out of right triangles, but rather that it is plausible that Thales had connected a series of mathematical intuitions that supported a new metaphysical project—the construction of the cosmos out of right triangles, and an idea about transformational equivalences to offer an insight into how ὕδωρ could alter without changing.

Let us continue by recalling the diagram that Thales was forced to imagine in measuring a pyramid in Egypt at the time of day when vertical height equals the horizontal shadow (left), and then add to it a matching isosceles right triangle making the larger one into which both divide. Or, since they are similar, simply rotate the same triangle into a different position (right). Again, this is the diagram that appears most in the ancient manuscripts for VI.31.

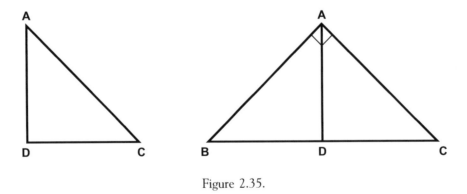

Figure 2.35.

The *mean proportional* or geometric mean is immediately obvious—if you are looking for areal equivalences in a right triangle—in the case of the *isosceles right triangle*. The square on the perpendicular is equal to the rectangle formed by the two segments into which the perpendicular partitions the hypotenuse. It is immediately obvious because the segments—AD,

BD, and CD—are all radii of the circle and hence equal, and so the rectangle made by BD, CD, is an equilateral rectangle, or square.

But the next step for Thales, having discovered that every triangle in the (semi-)circle was right, was to wonder whether this areal equivalence held when the perpendicular was begun at some other point along the circle, as in the right column, 2.34 above. If this areal relation held, then the smaller square on AD should still be equal in area to the rectangle BC, CD. He may have suspected this, but what diagrams might have persuaded him and his retinue? By working through the diagrams I am proposing a line of reasoning by which Thales may have thought through the idea of the mean proportional.

Once Thales saw the areal equivalence between square on the perpendicular and rectangle made by the two parts into which the hypotenuse was divided, he investigated further to see if there were other areal equivalences. And he found them. Finding them—the two other cases of mean proportionals—was tantamount to discovering an areal interpretation of the Pythagorean theorem. First he thought through the isosceles right triangle that had been so much his focus since the pyramid measurement:

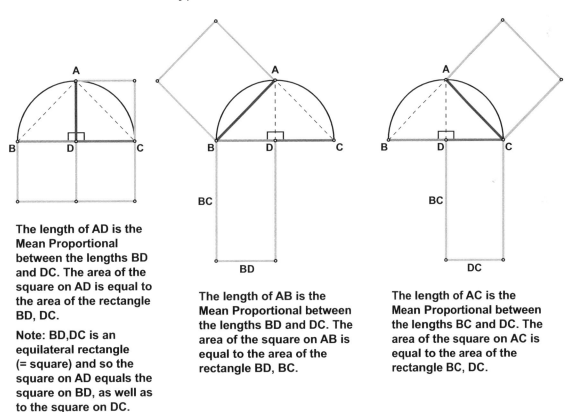

Figure 2.36.

Then, he thought it further with regard to *all* right triangles. While it is certainly true that this intellectual leap is an extraordinary one, I argue that these features of a right triangle were explored *because* Thales had already concluded—or suspected that—the right triangle was the

130   The Metaphysics of the Pythagorean Theorem

basic building block of all other appearances, that all other appearances could be imagined to be built out of right triangles:

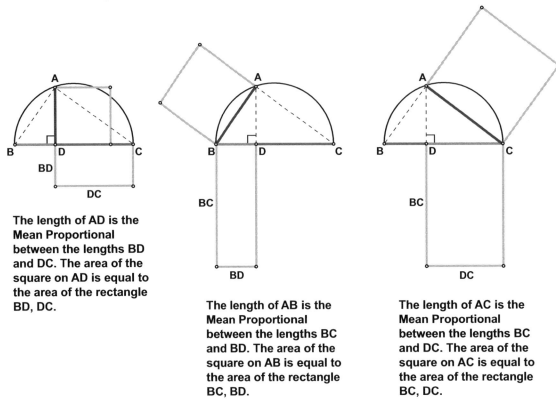

The length of AD is the Mean Proportional between the lengths BD and DC. The area of the square on AD is equal to the area of the rectangle BD, DC.

The length of AB is the Mean Proportional between the lengths BC and BD. The area of the square on AB is equal to the area of the rectangle BC, BD.

The length of AC is the Mean Proportional between the lengths BC and DC. The area of the square on AC is equal to the area of the rectangle BC, DC.

Figure 2.37.

And thus, what he had before him, when the second and third diagrams in each row are combined, is an areal interpretation of the Pythagorean theorem:

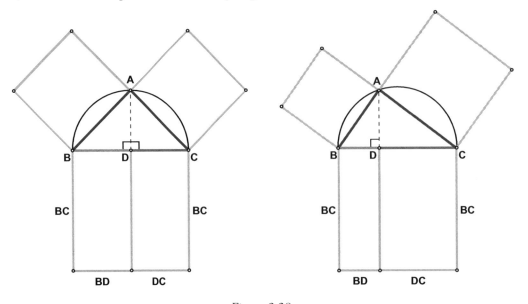

Figure 2.38.

Next, as I imagine it for Thales and his retinue, to be sure about areal equivalences, he divided the squares by diagonals in the case of the isosceles right triangle (left), and probably at first by measurement, he confirmed the areal equivalences between square and rectangle for scalene triangles (right).

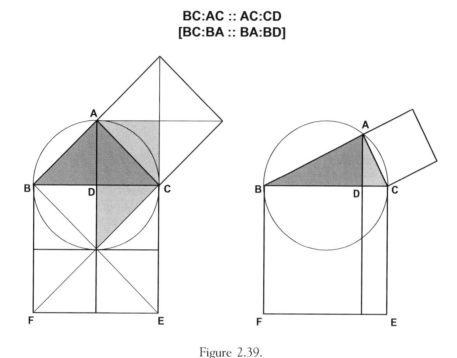

Figure 2.39.

Stated differently, Thales realized that in addition to perpendicular AD being the mean proportional between lengths BD and CD, he realized that length AB was the mean proportional between diameter BC and segment BD, and thus AC was also the mean proportional between diameter length BC and segment CD. The recognition of these geometric means—"means" of lengths between extremes—amounted to the realization that the area of the square on AB was equal to the rectangle BC, BD, and the area of the square on AC was equal to the rectangle BC, CD. The case of the isosceles right triangle was most immediately obvious (above, left), but dividing the figures into triangles by diagonals allowed him to confirm areal equivalences between squares and rectangles.

Once you know that ABD is *similar* to CAD, you know that BD : AD :: AD : DC, for the ratios of corresponding sides must be the same. Stated differently, you know that BD : AD : DC is a continuous proportion. So *by the very definition* of a "mean proportional," AD is a mean proportional between BD and DC. The conclusion of Euclid VI.16, 17 is that, as a consequence, the rectangle formed by BD, DC is equal to the square on AD. The porism at VI.8 is in effect the analysis for the problem whose synthesis is given in VI.13—every perpendicular from the diameter of a semicircle to the circle's circumference is a mean proportional, or geometrical mean, of the lengths into which it divides the diameter. In my estimation, the

132   The Metaphysics of the Pythagorean Theorem

diagram below with an isosceles right triangle was the gateway to his realization of the metaphysics of the fundamental geometrical figure that he was looking for.

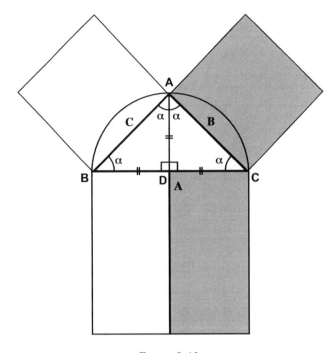

Figure 2.40.

Thus, the areal relation of the similar and similarly drawn figures constructed on the sides of a right triangle stand in such a relation that the figure on the hypotenuse can be expressed as a square, divisible into two rectangles, and equal in area to the sum of the squares constructed on the two sides, expressed as rectangular parts of the larger square—regardless of the shape of the figures. It applies to *all* shapes, to *all* figures, similar and similarly drawn.

In a review of the scholarly literature I could find no one but Tobias Dantzig who expressed the position that I, too, adopted that Thales knew an areal interpretation of the Pythagorean theorem. Of course, Dantzig acknowledged that his reasons were speculative, and did not find popular support. He put it this way: "With regard to the Pythagorean theorem my conjecture is that at least in one of its several forms the proposition was known before Pythagoras and that—and this is the point on which I depart from majority opinion—it was known to Thales. I base this conjecture on the fact that the hypotenuse theorem is a direct consequence of the principle of similitude, and that, according to the almost unanimous testimony of Greek historians, Thales was fully conversant with the theory of similar triangles."[81] And consequently, Dantzig came to believe that "the similitude proof of [Euclid] Book Six bears the mark of the Founder (of geometry), Thales."[82] Long before I happened on Dantzig's book, I reached the same conclusion, thinking through similitude, but at that point I was still not focused on the fact that there were in Euclid *two* proofs of the hypotenuse theorem, not one. As I continued to gaze on the diagrams of measuring the height of the pyramid at the time of day when every object casts a shadow equal to its height—and hence focused on isosceles right triangles—having concluded that Thales was looking for the unity that underlies all three sides of the right triangle, it seemed to me that Thales had already come to some clarity about isosceles triangles, right and otherwise, and that the squares on each of the three sides, each divided by its diagonal, showed immediately that the sum of the areas on the two sides was equal to the area of the square on the hypotenuse. Thales's cosmos is alive, it is *hylozoistic*,[83] it is growing, and collapsing, and it is ruled and connected by geometrical similitude as the structure of its growth. The little world and the big world shared the same underlying structure. And when I read Dantzig's thoughts on Thales, they made an enormous impression because it was from him that I realized the metaphysical implications of this areal discovery, which are almost never even considered in the vast secondary literature on the Milesians and early Greek philosophy, and even ancient mathematics.

In discussing the Pythagorean theorem and its metaphysical implications, Dantzig claimed: "To begin with, it [the theorem] is the point of departure of most metric relations in geometry, i.e., of those properties and configurations which are reducible to magnitude and measure. For such figures as are at all amenable to study by classical methods are either polygons or the limits of polygons; and whether the method be *congruence, areal equivalence* or *similitude*, it rests, in the last analysis, on the possibility of resolving a figure into triangles."[84] Thus Dantzig articulated what I, at that stage, had not formulated adequately. The famous theorem is equivalent to the parallel postulate, as we have already considered, which Dantzig notes as well.[85] To grasp one is to grasp the other, though it might not be obvious at first that each implies the other. I am not claiming that Thales grasped the parallel postulate and all of its implications, but rather that he grasped an array of geometrical relations that are exposed when straight lines fall on other straight lines, and how those lines reveal properties of figures, especially in triangles and rectangles. Thus, Thales came to imagine the world as set out in flat space, structured by straight lines, and articulated by rectilinear figures. All rectilinear figures can be reduced to triangles, and every triangle can be reduced ultimately to right triangles. It is from his inquiries into geometry that Thales came to discover and to grasp an underlying unity, an unchanging field of relations that, remarkably, gave insights into the ever-changing world of our experience. This is the legacy of the *practical* diagram. Thales claim that there was an ἀρχή announced that there was an unchanging reality that underlies our experience; his insight into spatial relations through diagrams revealed the structure of it.

Chapter 3

Pythagoras and the Famous Theorems

A

The Problems of Connecting Pythagoras with the Famous Theorem

In the last fifty years, the scholarly consensus on Pythagoras has largely discredited his association with the famous theorem that bears his name, and indeed with any contribution to what we might today call "mathematics." While this consensus was built throughout the twentieth century, the avalanche of scholarly sentiment came finally with the publication of Walter Burkert's *Lore and Wisdom in Ancient Pythagoreanism* (1962, 1972), which subsequently became the foundation of modern Pythagorean studies. The core of Burkert's thesis was that there was no reliable, secure testimony datable to the fifth and fourth centuries BCE connecting Pythagoras with the famous theorem, or indeed with mathematics in general. And with that verdict, the connection between Pythagoras and the theorem was disengaged. All the reports that connected Pythagoras to the theorem and mathematics were judged too late to be trustworthy. But this is not to deny that the Greeks were aware of the hypotenuse theorem, or even members of Pythagoras's school, though the usual dating is placed in the fifth century.

Recently, Leonid Zhmud published *Pythagoras and the Pythagoreans* (2012), a major revision of his 1997 work *Wissenschaft, Philosophie und Religion im frühen Pythagoreismus*. These works developed the thesis that Zhmud had announced already in 1989, arguing that perhaps Burkert had been too hasty in discrediting "Pythagoras the mathematician." Zhmud's most recent version of the argument has two threads, one on securing testimony dating to the fifth and fourth centuries BCE, and the other making a plausible case based on the very internal development of geometry. Diogenes Laertius reports the words of a certain "Apollodorus the arithmetician" that Pythagoras proved that[1] in a right-angled triangle the square on the hypotenuse was equal to the sum of the squares on the sides adjacent to the right angle,[2] and Zhmud argues that this is the same writer known as Apollodorus Cyzicus, dating to the second half of the fourth century BCE, a follower of Democritus.[3] Zhmud argues that writers from Cicero to Proclus refer to this testimony of Apollodorus Logistikos—the very same Apollodorus—and thus these late reports echo testimony from the fourth century BCE.

Ἡνίκα Πυθαγόρης τὸ περικλεὲς εὕρετο γράμμα
κεῖν᾽, ἐφ᾽ ὅτῳ κλεινὴν ἤγαγε βουθυσίην

When Pythagoras discovered the famous figure (γράμμα),
He made a splendid sacrifice of oxen.

While it is true that the earliest explicit mention of Pythagoras and the hypotenuse theorem dates to Vitruvius of the first century BCE,[4] Zhmud traces the source to Callimachus of the first half of the third century BCE, who mentions Thales's drawing a geometrical figure in the sand by the temple of Apollo at Didyma but claiming he learned it from Euphorbus, an earlier (re)incarnation of Pythagoras.[5] In any case, once again we have an early mention connecting Pythagoras with geometrical figures, and, Zhmud thinks, with the famous theorem itself. Moreover Zhmud's claim is this: If, in the fourth century, some ten authors mention Pythagoras and mathematics, in the third only Callimachus does so, making use of the fourth-century tradition; it was on this that Cicero and Vitruvius relied in the first century; and Diodorus Siculus cited Hecateus of Abdera. The late mentions connecting Pythagoras and the theorem trace back to testimony from the fourth and fifth centuries BCE; they are echoes of this earlier tradition.

In addition, there is Zhmud's case that the logic of development of geometrical thought places the discovery of the hypotenuse theorem not later than the first quarter of the fifth century BCE, before the achievements of Hippasus, and who else to better credit than Pythagoras?

> There is no doubt that the pupils and followers of Pythagoras from Hippasus to Archytas were engaged in *mathēmata*. It is therefore quite natural to suppose that these studies were launched by the founder of the school. It is true that this natural supposition could have been made in antiquity, indeed even had Pythagoras actually not engaged in science. The logic of the development of Greek mathematics, which makes up in part for the acute lack of reliable evidence, permits an escape from this circle of suppositions. Between Thales, to whom Eudemus attributes the first geometrical theorems, and the author of the first *Elements*, Hippocrates of Chios (c. 440), from whom came the first mathematical text, there passed a century and a half, during which geometry was transformed into an axiomatic and deductive science. Although we shall never be able to establish the authors of *all* the discoveries made in that period, in a number of cases a combination of historical evidence and mathematical logic makes it possible to do so sufficiently reliably. If Hippocrates makes use of Pythagoras' generalized theorem for acute and obtuse-angled triangles (II, 12–13), it is clear that an analogous theorem for right-angled triangles had been proved before him. Further, tradition connects the discovery of irrationality with Pythagoras' pupil Hippasus, and the Pythagorean proof that the diagonal of a square is incommensurable with its side, i.e. irrationality $\sqrt{2}$, preserved at the end of book X of the *Elements*, is based on Pythagoras' theorem. Clearly, it was proved before Hippasus. Finally, Apollodorus the arithmetician, probably identical with Apollodorus of Cyzius (second half of the fourth century), attributes the discovery of the theorem to Pythagoras, to which virtually all the

authors of antiquity who wrote about it assent. Is it worthwhile to reject the attribution, the history of mathematics having not yet proposed a single worthy alternative to Pythagoras?[6]

Thus, Zhmud's second argument, the second main thread connecting Pythagoras himself with the famous theorem, is that the logic of the development of Greek mathematics places the discovery of the hypotenuse theorem before the time of Hippasus, and the *only* name connected with it is Pythagoras.[7] I am very sympathetic to Zhmud's arguments and conclusions but I became convinced by a rather wholly different approach, and that was to think through the geometrical diagrams connected to a series of claims that are at least plausible—for Thales—and directing attention to the metaphysical project, the metaphysical context, that underlies them. Accordingly, my point of departure will be to invite my readers to move beyond Zhmud's thoughtful words about Hippocrates and Hippasus by turning to the diagrams that illustrate their foci and achievements.

B

Hippocrates and the Squaring of the Lunes

Zhmud's claim is that Hippocrates of Chios, the author of the first *Elements* (ΣΤΟΙΧΕΙΟΝ), made use of the hypotenuse theorem, and so it was certainly known before his time. Hippocrates is credited with solving the problem of the quadrature of the lunes. This problem is usually associated with ancient attempts to square the circle. One way to understand this problem is to see that what is at stake is to determine the relation between the area of a square and the area of a circle, the area of a square being understood long before the area of a circle. We have already traced a panoply of geometrical diagrams that I have argued are plausibly connected with Thales; these diagrams formed a background of interests more than a century before Hippocrates's achievements. The lines of thoughts that precede this investigation are the ones that seek to determine the relation between a square and a rectangle. This relation is central to VI.31, the line that, I have been arguing, allowed Thales to grasp the hypotenuse theorem by similar figures. And it will be the same line of thought through which Pythagoras followed and expanded upon Thales. Here, I follow Heath in his diagrammatic presentation of Hippocrates and the quadrature of the lunes:

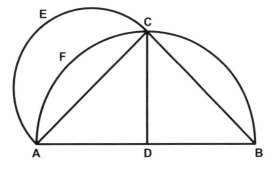

Figure 3.1.

138   The Metaphysics of the Pythagorean Theorem

Thus, once again we meet a semicircle and the isosceles right triangle inscribed in it, the very figure that I have suggested was the route by which Thales, and Pythagoras after him, arrived at the hypotenuse theorem through similar triangles, and the areal relations of similar figures that stand in proportions to the side lengths. Euclid's diagram may very well point back to Hippocrates himself, and the tradition that preceded him. This time, however, instead of constructing rectilinear figures—squares or rectangles—we begin by constructing semicircles; for example, on the hypotenuse CA of right triangle CDA, we draw semicircle CEA. And as I pointed out in the last chapter, an understanding of the "enlargement" of the hypotenuse theorem at VI.31 shows that the areal relation holds for any figures, similar and similarly drawn, and hence semicircles as well.

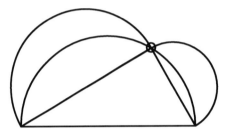

Figure 3.2.

The argument is that Hippocrates of Chios knew the theorems later recorded by Euclid as II.12 and 13; these were required for, and presuppositions of, his quadrature of the lunes. Euclid II.12 and 13 are a different kind of generalization of Pythagoras's theorem from the one in Book VI. Here, one considers the squares on the sides of an angle in a triangle that is not right: in II.12, the angle is obtuse, and the squares on the sides of the angle fall short of the square on the side subtending the angle, below.

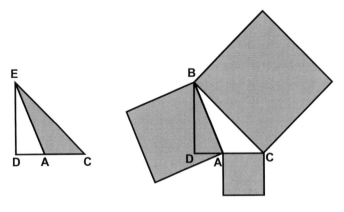

Figure 3.3.

By contrast in II.13 the angle is acute, and the squares on the sides exceed the square on the subtending side (below). The amount by which the squares fall short in the one case and exceed in the other is the rectangle formed by one of the sides and the piece cut off by the perpendicular.

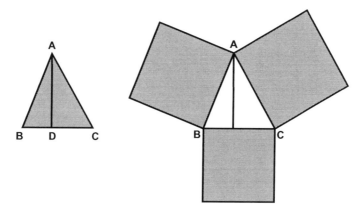

Figure 3.4.

Thus in the case where the angle is right, the perpendicular meets the vertex so that there is neither excess nor deficiency. The central point for our considerations is that *both* proofs suppose the hypotenuse theorem; it is used to prove these theorems at II.12 and 13. In Eudemus's account of Hippocrates (and since it is a later account it might include details that were not part of the earlier formulation),[8] there is one point where Hippocrates argues that because the sum of squares on the sides around a certain angle exceeds the square on the side subtended by the angle, the angle must be acute (from which he concludes that the lune is contained by more than half the circle). So, this can be contrived as a reference to these theorems in Book II. Zhmud's claim, then, is that if Hippocrates had used these theorems—had he reached his conclusions by appeal to these propositions—he must have had a proof for Pythagoras's theorem (indeed, in Euclid, the proofs for II.12–13 rely on I.47).

Heath discusses Hippocrates, expressing his work in terms of geometrical algebra as "[$BD^2 < BC^2 + CD^2$]. Therefore the angle standing on the greater side of the trapezium [<BCD] is acute."[9] I think this is the spot where one can say Hippocrates is using a version of II.13, in fact its converse, namely, that if the sum of squares on two sides of a triangle is less than the square on the third side (II.3 of course gives a precise value for the difference), then the angle must be acute. Similarly, Heath adds, "Therefore $EF^2 > EK^2 + KF^2$, so that the angle EKF is obtuse . . . ,"[10] which is a version of II.12, again, the converse.

The broad problem Hippocrates was investigating was that of quadrature of a circle, also called squaring a circle, which is to find a square (or any other polygon) with the same area as a given circle. Note that a problem that earned the focus of early Pythagoreans was to find a square with the same area as a rectangle (i.e., a rectilinear figure), which was a central underlying theme of what later became Euclid Book II proved in the last theorem 14, and which we shall explore later in this chapter. But the quadrature of the rectangle is also achieved in VI.17 and in principle by VI.31. Hippocrates did not solve that problem, but he did solve a related one involving lunes. A *lune* (also called a *crescent*) is a region of nonoverlap of two intersecting circles. Hippocrates did not succeed in squaring an arbitrary lune, but he did succeed in a couple special cases. Here is a summary of the simplest case.

140   The Metaphysics of the Pythagorean Theorem

> Draw a square *ABCD* with diameters *AC* and *BD* meeting at *E*. Circumscribe a semicircle *AGBHC* about the right isosceles triangle *ABC*. Draw the arc *AFC* from *A* to *C* of a circle with center *D* and radius *DA*. Hippocrates finds the area of the lune formed between the semicircle *AGBHC* and the arc *AF* as follows.

Note that there are three segments of circles in the diagram; two of them are small, namely, *AGB* with base *AB*, and *BHC* with base *BC*, and one is large, namely, *AFC* with base *AC*. The first two are congruent, and the third is similar to them since all three are segments in quarters of circles.

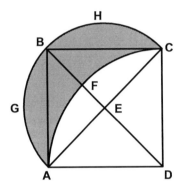

Figure 3.5.

Hippocrates then uses a version of this proposition VI.31—generalized so the figures don't have to be rectilinear but may have curved sides[11]—to conclude that the sum of the two small segments, *AGB* + *BHC*, equals the large segment *AFC*, since the bases of the small segments are sides of a right triangle while the base of the large segment is the triangle's hypotenuse.

Therefore, the lune, which is the semicircle minus the large segment, equals the semicircle minus the sum of the small segments. But the semicircle minus the sum of the small segments is just the right triangle *ABC*. Thus, a rectilinear figure (the triangle) has been found equal to the lune, as required.

Note that at Hippocrates's time, Eudoxus's theory of proportion had not been fully developed, so the understanding of the theory of similar figures in Euclid Book VI was not as complete as it was after Eudoxus. Also, Eudoxus's principle of exhaustion, preserved in Euclid XII, and anticipated especially in proposition X.1 for finding areas of curved figures, was still to come, so there may have been little consensus about reckoning the area of curved figures.[12] Such a situation has proved to be common in mathematics—mathematics advances into new territory long before the foundations of mathematics are developed to justify logically those advances.

The problem of calculating the area of a lune stands a long way down the line from the earliest geometrical diagrams in this tradition. And these earliest stages may trace a lineage from Thales; Thales discovered, proved, or observed that the base angles of an isosceles triangle were "similar." Which geometrical diagram did he propose to explain his insights? There is a proof in Aristotle concerning this theorem that Zhmud conjectures might well be the one that traces back to Thales; it is a proof by *mixed angles*. I turn to explore this briefly, because this offers an early anticipation of the kinds of things Hippocrates addresses in his problem much later. It is certainly not the proof at Euclid I.5, and Heath took this to suggest both that this proof was probably proposed much earlier, and that the proof at I.5 was more likely the work of Euclid himself.[13] And it is not the proof at VI.2 where the same principle is reached by ratios and proportions, and that is sometimes also referred to as "Thales's theorem."[14] In this case, triangle ABC is isosceles—AB is equal to BC—and its vertex lies in the center of the circle at B, below. If we see that VI.31 follows the lines of thought through III.31—the triangle in the (semi-)circle—we have yet another example of how Thales might have given much thought to curved and straight angles.

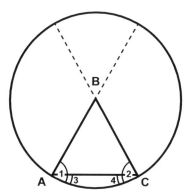

Figure 3.6.

Angle 1 is equal to angle 2 since both are angles of a semicircle, and angles 3 and 4 must be equal because they are angles of a segment of semicircle. If equal angles are subtracted from equal angles—3 is subtracted from 1, and 4 is subtracted from 3—then angles BAC and BCA are equal.

When we entertain this diagram—using mixed angles—and place this side by side with the ones concerning angles in the semicircle, a rich picture forms that suggests how theorems concerning the diameter in a circle, angles in an isosceles triangle, and triangles in a semicircle may have been grasped in an interconnected way very early in this geometrical tradition. Moreover, it suggests a way that these interconnected diagrams had long been the focus of the geometer's gaze and reflections by the time of Hippocrates.

C

Hippasus and the Proof of Incommensurability

The second event in the internal development of geometry upon which Zhmud's argument relies, that the knowledge of the hypotenuse theorem must have predated the time of Hippasus and hence the early first quarter of the fifth century or even earlier, is the proof of incommensurability of the diagonal of a square and its side. Aristotle refers to this proof twice in the

Prior Analytics.[15] The proof used to appear as Euclid X.117, but Heiberg regarded it as a later interpolation and removed it to Appendix 27 for Book X.

The problem, and solution, of doubling the volume of a square was later presented in Plato's *Meno*. It seems to me that some basic form of this argument, as it appeared in Euclid, certainly goes back to the Pythagoreans;[16] in any case, the "Pythagorean theorem" predates Hippasus's proof. Below is the proof of Euclid X.117 (now in the Appendix to Book X, 27):[17]

Let it be proposed to us to prove that in square figures the diameter is incommensurable in length with the side. Let ABCD be a square, of which AC is the diameter. I say that AC is incommensurable in length with AB.

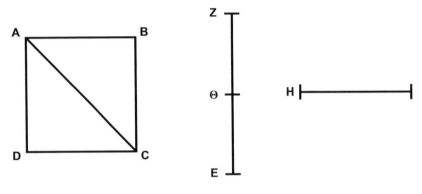

Figure 3.7.

For if possible, let it be commensurable. I say that it will follow that the same number is odd and even. Now it is manifest that the square on AC is the double of that on AB. Since AC is commensurable with AB, then AC will have the ratio [*logon*] to AB of one number to another. Let these numbers be EZ and H, and let them be the least numbers in this ratio. Then EZ is not a unit. For if EZ is a unit and has the ratio to H which AC has to AB, and AC is greater than AB, then EZ is greater than H, which is impossible. Thus EZ is not a unit; hence it is a number [NB: in Greek mathematics a "number" is a collection of units]. And since AC is to AB as EZ is to H, so also the square on AC is to that on AB as the square on EZ is to that on H. The square on AC is double that on AB, so the square of EZ is double that on H. The square on EZ is thus an even number; thus EZ itself is even. For if it were odd, the square on it would also be odd; since, if an odd number of odd terms is summed, the whole is odd. So EZ is even. Let it be divided in half at θ. Since EZ and H are the least numbers of those having this ratio, they are relatively prime. And EZ is prime; so H is odd. For if it were even, the dyad would measure EZ and H. For an even number has a half part. Yet they are relatively prime; so this is impossible. Thus H is not even; it is odd. Since EZ is double Eθ, the square on EZ is four times the square on ET. But the square on EZ is the double of that on H; so the square on H is double that on Eθ. So the square on H is even, and H is even for the reasons already given. But it is also odd, which is impossible.

Hence, AC is incommensurable in length with AB. This was to be proved. In commentaries, this proof is often explained by resorting to algebraic notation. But, the Greeks did not have algebra, and to place it in these terms as a way to make clearer what they meant is perfectly misleading because it misrepresents how they reached their conclusions, the geometrical sense in which they thought through this matter. So, let us review the proof in the form in which it is presented, and in the context of what incommensurability meant.

Let us go back to Thales and his measurement of the pyramid when the shadow is equal to the height of a vertical object casting it:

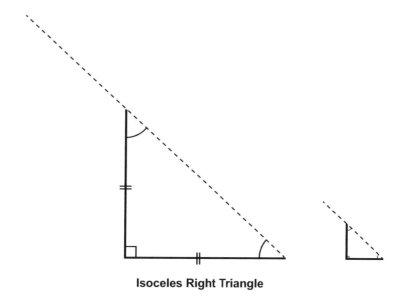

Figure 3.8.

The gnomon height and its shadow length are equal, and thus the time is right to infer that the pyramid's height is equal also to its shadow. The little world and the big world have the same geometrical structure—similarity—and the *ratio* of shadow to height proposes the relation of how many of the small world fit into the big world. If the little and big world are *proportional*, the ratio offers the number of times the little world scales up or blows up into the big world. This matter can be stated differently: the small world might not simply scale up to fill up evenly the big world, but there might be a module or least common denominator that is a factor of both lengths, so that, reduced to this smallest common module, the big world incorporates the small world wholly, fully. If the little world scales up a whole number of times—an integer— to fill up the big world, then the Greeks understood that the little world and big world were *commensurable* (σύμμετρος). The deep intuition had been remarkably beautiful, that everything measurable was commensurable. And this is precisely what is at stake in the proof. If the side of a square and its diagonal are commensurable—each being measurable—then there should be a common module, a common denominator of the ratio, so that the smaller and bigger stand in a whole-number relation to one another, number to number—a common denominator of the ratio. The proof shows that there is not, on the supposition that, if it were the case, then one and the same length had to be a number both odd and even. In my estimation, Thales never understood this result, though he must have been confounded by the fact that in every isosceles right triangle, when the two sides had lengths describable as whole numbers, the length of the hypotenuse was *never* a whole number but rather a fraction more than some whole number. Stated differently, while there could be found squares of *different* sizes whose sum was another square such as 3 and 4 (9 + 16 = 25), you could never find two *equal* squares whose

144   The Metaphysics of the Pythagorean Theorem

sum was a square (3 and 3: squares of 9 + 9 = square root of 18—the hypotenuse is never a whole number!).

To understand the proof we have to think also of "odd" and "even" numbers, how they were grasped, and that the relation between them was contradictory: every number was either even or odd, never both. Zhmud follows Becker, Reidemeister, von Fritz, van der Waerden, and Waschkies, crediting Pythagoras himself with the theorems on even and odd numbers preserved in Euclid IX, 21–34.[18] Zhmud's list is shorter than the one proposed earlier by van der Waerden but includes:

IX 21. The sum of even numbers is even.
IX 22. The sum of an even number of odd numbers is even.
IX 23. The sum of an odd number of odd numbers is odd.
IX 24. An even number minus and even number is even.
IX 25. An even number minus and odd number is odd.

Becker and van der Waerden also include, among others, the theorems:

IX 30. If an odd number measures an even number, it will also measure the half of it.
IX 32. A number that results from (repeated) duplication of 2 is exclusively even times even.

Before turning to consider the proof, let's try to get clearer about the distinction between even and odd numbers. Numbers were visualized as shapes. The numbers that are "even" (ἄρτιοι) are divisible into two equal parts, and line up uniformly in two parts, while the numbers that are odd (περισσοί) quite literally "stick out." They seem to have been imagined, in the context of oppositions, as two rows of pebbles (ψῆφοι)—it is striking that they appear as if they were representing a plan view of a temple stylobate, the pebbles perhaps indicating where a column was to be installed. Perhaps the idea developed first from inspecting and reflecting on architectural plans. First, below left, here is an example of the visualizing of the "even" number 10, and the "odd" number 9, right:

Figure 3.9.

The shape of oddness, then, was something like this, with some part "sticking out":

Pythagoras and the Famous Theorems   145

Figure 3.10.

In the context of bifurcation, numbers were imagined to fall into one of two groups. Those numbers that preserved a basic symmetry came to be called "even," and those that did not came to be called "odd." And the sense of συμμετρία was extended so that it came to refer to numbers attached to lengths—"commensurable" (σύμμετρος), and consequently to lengths that had no numbers—"incommensurable" (ἀσύμμετρος).

If the diagonal AC of square ABCD was commensurable with its side AB, then one and the same number must be both even and odd. The proof starts with the assumption that the relation is commensurable; thus the proof is indirect. If this supposition leads to a contradiction, then we may regard the assumption as false, and hence to be replaced by its contradictory, namely, that the relation is incommensurable. The square on diagonal AC is double that on AB; dividing each into isosceles right triangles, the square on AB is made of two while the square on AC is made of four, each of the same area.

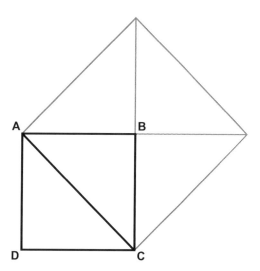

Figure 3.11.

Now the ratio of AC to AB will be in the lowest possible terms, and those will be identified—as numbers—EZ and H. Since diagonal AC is greater in length than AB, and AB is considered 1 unit, then AC is greater than 1. If the ratio is the least possible, then AB must be odd, otherwise it will be divisible further and thus not the least possible. And since the square on

146   The Metaphysics of the Pythagorean Theorem

AC is double the area of the square on AB, its area is reckoned by multiplying the smaller area by 2. Now every number multiplied by 2 is even because whether a number is odd or even, when that number is multiplied by 2 the result is always even. And since the square on AC is even, then AC must be even, because even times even results in an even number. Had AC been odd, the square would be odd as well, because odd times odd is always odd. And so, if the ratio of AC : AB is the least possible, and the square on AC is 2 times the area of the square on AB, then AC must be even, and AB must be odd.

And since the square on AC is to AB as the square on EZ is to H, then just as the square on AC is double that on AB, so the square on EZ is double that on H.

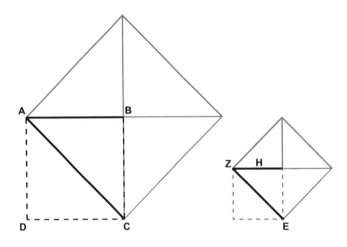

Figure 3.12.

Consequently, the square on EZ must be an even number, and thus EZ must be even—for if it were odd then the square on it would be odd because an odd number times an odd number is always odd. Thus, if EZ were odd, then the square on EZ would be an odd number; this means that EZ would be added to itself an odd number of times (namely, EZ). A visual image is helpful here:[19]

Figure 3.13.

Each row is EZ (i.e., EZ = X) an odd number. There are EZ rows, again an odd number. An odd sum of odd numbers is odd; therefore, the square on EZ would have to be odd if EZ were odd.

Now, since EZ has been established as even, the proof then divides EZ at θ, so that twice Eθ is EZ. Now, H must be odd because if it were even both EZ and H would be measured by 2; however, this is impossible because EZ : H was said to be "the smallest of all ratios (of a number to a number) which are in the same ratio as AC : AB."

Now, this means that the square on EZ is 4 times the square on EΘ, and 2 times the square on H (Therefore, twice square EΘ is equal to square H [i.e., 2 times square EΘ = square H].)

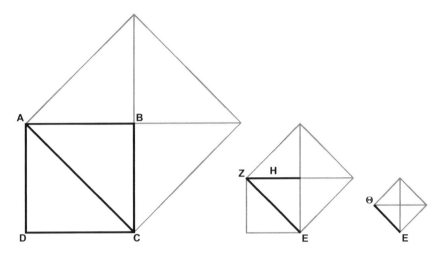

Figure 3.14.

Hence, by the same argument that showed EZ to be even, H must be even. But H is odd, which is impossible (ὅπερ ἐστὶν ἀδύνατον)—but there is no other way for the ratio of EZ to H to be the smallest possible. Therefore, AC and AB cannot be commensurable, which is what was to be shown.[20]

Zhmud had argued that *this* proof of the irrational is rightly attributed to Hippasus in the first half of the fifth century BCE, and that it presupposes the hypotenuse theorem, and thus the hypotenuse theorem must have been proven before it. But let us be clear *how* this proof presupposes the hypotenuse theorem. The proof is a continued reflection on just the sort of diagram that must have been made by Thales when he measured the height of a pyramid at the time of day when the shadow length was equal to its height—and the height of all other vertical objects as well. Thales's measurement required a deep reflection on the isosceles right triangle—half a square, of course. The square on the diagonal was twice the square constructed on any of its sides. This is the case where the hypotenuse theorem is most immediately apparent, and hence the areal unity among the sides of a right triangle, where the square on the hypotenuse of an isosceles right triangle is equal to the sum of the squares on the two other sides—each area dissected to isosceles right triangles. And it is this reflection that continued to turn up a sore point, namely, that in every isosceles right triangle, when the length of the two equal sides could be expressed as whole numbers (i.e., integers), when the measured cord was "stretched under" (ὑποτείνω, ὑποτείνουσα) the tip of the shadow to the top of the gnomon, the length of the hypotenuse was *never* a whole number. And this must have surely troubled Thales and his retinue. At all events, for Thales this meant that the length of the hypotenuse was always a fraction greater than the length of a side, though even with a measured cord one could identify the length exactly without clearly being able to supply a number. Identifying the exact length by the measured cord was sufficient for architectural purposes, but the relative rarity in which the lengths of all three sides were whole numbers—so-called Pythagorean triples[21]—must surely have been a source of intellectual consternation. The investigation of numbers and their divisions into odd and even, were part and parcel of trying to discern a pattern where these rare instances of "Pythagorean triples" could be identified.

Let me try to summarize and integrate the two arguments that the diagonal of a square is incommensurable with its side—the one concerning areal equivalences, the other concerning numbers.[22] The temple builders and tunnel engineers, probably like Pythagoras and the early Pythagoreans, regarded all things measurable as commensurable. For in practical skills that use counting and measuring, the only numbers that are meaningful are integers and rational fractions. There may not be a whole number to correspond to the hypotenuse of a right triangle whose sides are of equal length, but the length can be measured exactly with a knotted cord. Thus the discovery of incommensurability, and hence irrational numbers, was no obstacle for the architects, builders, and engineers because the rational numbers were the only ones of practical use. The overturning of the principle, beautiful in its simplicity, that everything measurable is commensurable, occurred not because it contradicted experience, but because it was incompatible with the axioms of geometry. To describe the ancient argument in more modern terms:

> Indeed, if the axioms of geometry are valid, then the *Pythagorean theorem holds* without exception. And if the theorem holds, then the square erected on the diagonal of a square of side 1 is equal to 2. If, on the other hand, the *Pythagorean dictim held*, then 2 would be the square of some rational number, and this contradicts the tenets of rational arithmetic. Why? Because these tenets imply, among other things, that *any fraction can be reduced to lowest terms*; that at least one of those terms of such a reduced fraction is *odd*; and that the square of an even integer is divisible by 4. Suppose then that there existed two integers, x and R such that $R^2/x^2 = 2$, i.e., that $R^2 = 2x^2$. It would follow that R was even, and since the fraction is in its lowest terms, that x was odd. One would thus be led to the untenable conclusion that the left side of an equality was divisible by 4, while the right side was not.[23]

D

Lines, Shapes, and Numbers: Figurate Numbers

When Thales measured the height of a pyramid, he had to supply a number. I have already suggested that the number was probably in royal cubits, thanks to the use of the Egyptian royal surveying cord. Had he measured the height of the Great Pyramid, the number should have been very close to 280 royal cubits. Had Thales's measurement not been in cubits, neither the Egyptians nor the Greeks would have been in a position to appreciate the accuracy of this new technique of measuring shadows; there would not have been good reason to marvel, and the story would not have been memorable as an astonishing feat.[24] For more than a thousand years before Thales reached the Giza plateau, the Egyptians knew, as the Rhind Mathematical Papyrus shows, how to reckon the height of a pyramid.

Problem 57 in the Rhind Papyrus shows the calculation of the height of a pyramid given the length of the square base and the *seked* (or, angle of inclination of the pyramid face). The answer of pyramid height, its *peremus*, in this case 93 1/3 royal cubits:

> 57. A pyramid 140 in its ukha thebt (i.e. the length of its square side), and 5 palms, 1 finger in its seked. What is the peremus (i.e., height) thereof? You are to divide 1 cubit by the seked doubled, which amounts to 10 1/2. You are to reckon with 10 1/2 to find 7 for this one cubit. Reckon with 10 1/2. Two-thirds of 10 1/2 is 7. You are to reckon with 140, for this is the ukha thebt. Make two-thirds of 140, namely 93 1/3. This is the peremus thereof.

We have already considered in the Introduction how the pyramid was grasped geometrically as a triangle and inseparable from both its square and rectangle, and thus how it might have been solved graphically.

When Eupalinos, a contemporary of Pythagoras, reached an hypothesis for the length of each tunnel half in Samos, he too had to supply a number, though he probably relied on the sum of the horizontal distances between stakes placed up and along the hillside, not shadows (as we shall consider shortly). How did the Greeks of the time of Thales and Pythagoras grasp the law-like relation between line lengths, areas, and the numbers attached to them?

The story following the early Pythagoreans requires a focus on figurate numbers. Numbers were themselves conceived as shapes, and so, for example, triangular numbers were distinguished from square ones, pentagonal ones, and so on. Beginning with the fundamental number—3 for triangular numbers, and 4 for square numbers, probably represented by pebbles (ψῆφοι)—additional rows were added to produce a sequence of triangular or square numbers, and so on. Below are figurate numbers.

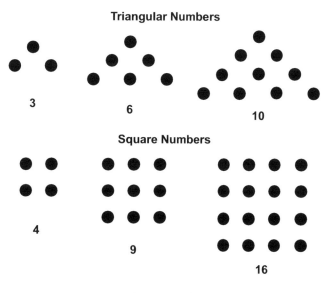

Figure 3.15.

The production of the series of figurate numbers offered a vision of how numbers *grow*. Just as areas grow in some rule-defined process as line length grows, so do numbers. The growth of areas found expression in the duplicate ratio of the corresponding line lengths (in the square of the linear scaling). How should we understand the growth of numbers?

Let us consider this process with focus on square numbers. Quite obviously, they are patterned into squares.

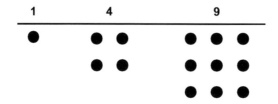

Figure 3.16.

In the case of square numbers, the growth from one square number to another requires the addition of pebbles aligned along the right side and bottom of those already assembled. In short, the addition to the next square number is achieved by what the Greeks referred to as a *gnomon*—an L-shaped area surrounding an area already defined (though the shape of the L, appears more like a backwards ⌐, shown in gray.

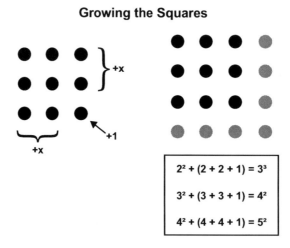

Figure 3.17.

While at 4 (2 × 2), we can jump to 9 (3 × 3) with an extension: we add 2 (right) + 2 (bottom) + 1 (corner) = 5. And consequently, 2 × 2 + 5 = 3 × 3. And when we're at 3, we get to the next square by pulling out the sides and filling in the corner: Indeed, 3 × 3 + 3 + 3 + 1 = 16. Each time, the change is 2 more than before, since we have another side in each direction (right and bottom).

The sequence of the first square numbers is 4, 9, 16, 25, 36, 49, and so on. To find the difference between consecutive squares, we observe that:

1 to 4 = 3
4 to 9 = 5
9 to 16 = 7
16 to 25 = 9
25 to 36 = 11
36 to 49 = 13

The sequence of odd numbers is sandwiched between the squares.

The γνώμων may have been imported originally from Babylon to describe a vertical rod used to cast shadows and hence connected with time-telling; Anaximander had been credited with introducing the gnomon into Greece.[25] The term was also used to refer to a carpenter's square, an L-shaped instrument used by architects and builders both as a set-square to check that a course of stones was level, and to guide the making of a right angle.[26] The L shape may explain how it came to refer geometrically to a shape formed by cutting a smaller square from a larger one. Euclid extended the term to the plane figure formed by removing a similar parallelogram from a corner of a larger parallelogram. Thus, the gnomon came to be identified as the part of a *parallelogram* that remains after a similar parallelogram has been taken away from one of its corners. Consider the diagram, below:

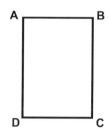
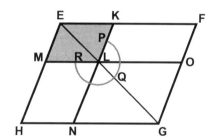

Figure 3.18.

On the left is a rectangle, literally a right-angled parallelogram, ABCD. The rectangle is contained by AB,BC (or BC,CD, and so on). On the right is parallelogram EFGH: there is a diameter EG with a parallelogram LNGO about it and the two complements KLOF and MHNL, and these three parallelograms together make up the gnomon. This means that a gnomon is an L-shaped tool with its two sides at right angles. And gnomons are illustrated in modern editions of Euclid by arcs of circles around the inner vertex—in this case the gnomon is ⌒ PQR—whether or not he did so himself.[27]

For the Pythagoreans, the gnomons came to be identified also with the odd integers. Square numbers could be generated, starting with the monad, by adding this L-shaped border that came to be called a gnomon, and thus the sum of the monad and any consecutive odd number is a square number. Thus, stated differently, the sum of the monad and any consecutive number of gnomons (i.e., odd numbers) is a square number.

152   The Metaphysics of the Pythagorean Theorem

$$1 + 3 = 4$$
$$1 + 3 + 5 = 9$$
$$1 + 3 + 5 + 7 = 16$$
$$1 + 3 + 5 + 7 + 9 = 25$$

We can observe clearly in visual presentation that the sequence of odd numbers is sandwiched between the squares.

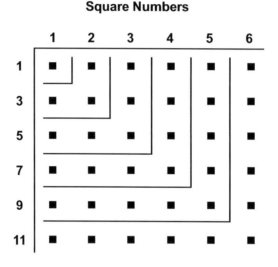

Figure 3.19.

Notice also that the jump to the next square is always odd since we change by "2n + 1" (2n must be even, so 2n + 1 is odd). Because the change is odd, it means the squares must cycle even, odd, even, odd, and this should make sense because the integers themselves cycle even, odd, even, odd; after all, a square keeps the "evenness" of the root number (even times even = even, odd times odd = odd). The point to emphasize here is that these insights about odd and even numbers, their squares, and their "roots" all are discoveries of properties hiding inside a simple pattern. By following the diagrams, it offers us a way to grasp plausibly how Pythagoras and his retinue first came to *visualize numbers* as shapes.

If we consider an odd square, for example 25 = 5 x 5, then since 25 = 2 x 12 + 1 this can be thought of as a gnomon formed by two pieces of length 12 and a corner stone. Thus, placing this gnomon over a square of side 12, one obtains a square of side 13. In other words, $12^2 + 5^2 = 13^2$. Thus one can obtain in this way a limitless series of Pythagorean triples (it does not exhaust them, however; there are others that cannot be formed in this way—for example $8^2 + 15^2 = 17^2$ is not obtainable by this procedure).

Let me add also in this discussion of the visualization of numbers that Aristotle uses an example of a square gnomon in *Categories* 15a30–35 to show how a square can increase in size without changing its form. "But, there are some things that are increased without being altered; for example a square is increased (i.e., enlarged) when a gnomon is placed round it, but it is not any more "altered" (i.e., changed in shape) thereby, and so in other similar cases." While it is true that the term γνώμων has a history we can trace from the sixth century BCE that includes the pointer of a vertical rod, as suggested by Anaximander's sundial; that Theogonis refers to as an instrument for drawing right angles like a carpenter's square; that Oinopides refers to a perpendicular as a line drawn gnomon-wise (κατὰ γνώμονα), and "the Pythagoreans use[d . . .] the term to describe the figure, which, when placed round a square produces a larger square,"[28] this Pythagorean sense is the one presented by Aristotle in the *Categories*; it deserves to be seen as a legacy from Thales and his compatriots.

E

Line Lengths, Numbers, Musical Intervals,
Microcosmic-Macrocosmic Arguments, and the Harmony of the Circles[29]

One of the important contributions that Zhmud's new study makes to our understanding of the early Pythagoreans is to remind us how diverse were their ideas, and thus how difficult it is to be confident that we know Pythagoras's specific view on many questions. There is a view familiarly identified with Pythagoras and the Pythagoreans that "All is number." Riedweg summed up the view that scholars who conclude that before Philolaus, the Pythagoreans did not go beyond a kind of philosophical arithmological speculation *and* those who saw a serious metaphysical exploration of numbers all concurred that the general doctrine of the Pythagoreans was that "All is number."[30] What Zhmud argues is that the doctrine "All is number" is a double reconstruction, from later commentators who sought to follow Aristotle's often contradictory assertions, and Aristotle's own reconstruction based on often inconsistent evidence.[31] If we take Zhmud's position on this matter sympathetically, where does this leave us in terms of the place of "numbers" in Pythagoras's thought? What I would like to suggest by the end of this section is that, contrary to Zhmud's position, Aristotle had the Pythagoreans more or less right, that in his own terms, the Pythagoreans, and I suggest Pythagoras himself, held "number" to be the ἀρχή ("principle"). And this helps us to see how Pythagoras took what he learned from Thales and his retinue, and came to see that number structures all the other appearances—fire, air, and the rest.

The pictures of the Pythagoreans of the late sixth and fifth centuries BCE share some common investigations without establishing a doctrine of teachings. What survives about Alcmaeon, for instance, suggests that he subscribed to no doctrine of principles, though Hippasus held that the ἀρχή is fire; this suggests that Hippasus was a monist, while the Italian philosophers after Pythagoras seem to have been dualists. Similarly, on the doctrine of the nature of the soul, there seems to be no single "Pythagorean" doctrine, though Pythagoras himself seems to have been convinced of transmigration. But the lack of continuity within the ranks of the early Pythagoreans does not undermine the chance to identify some doctrines with Pythagoras himself. None of this, of course, is incontestable, but Zhmud sums up the big picture plausibly: "The discovery of the numerical expression of concords, the first three proportions, the theory of odd and even numbers, and Pythagoras's theorem all correspond fully with the stage *mathêmata* (i.e., μαθήματα, what can be learned or taught, and hence more generally, education) had reached before Hippasus. Hence nothing was attributed to Pythagoras which could not *in principle* have been his, or, in more cautious terms, could not have belonged to a predecessor of Hippasus."[32]

A central part of my argument is that when we see the metaphysics of the hypotenuse theorem—following the ratios, proportions, and similarity in VI.31—line lengths, numbers, the first three proportions (arithmetic, geometric, harmonic), the theory of odd and even numbers, musical intervals, microcosmic-macrocosmic argument (whose structure is the principle of geometric similarity), and the harmony of the spheres, all appear together in an interconnected light—*the project was the building of the cosmos out of right triangles*. We have now emphasized how by the middle of the sixth century BCE there was a deep appreciation of the relation between the length of a line and the area of a figure constructed on it. As the lengths of the lines of a figure increased, the area increased in the *duplicate ratio* of the corresponding sides. In modern terms we would say that as the line lengths of a figure increase, the area increases in the square of the linear scaling. When Thales measured the height of a pyramid or the distance of a ship at sea, he had to supply numbers—the pyramid is so many cubits in height, the distance to the ship is so many cubits, feet, or ells. We have already considered a Pythagorean connection through figurate numbers—line lengths and areas were represented as *numbers visualized as shapes*.

About Pythagoras and the mathematical treatment of music, Zhmud writes: "The Pythagorean origin . . . is not open to serious doubt. The establishment by Pythagoras of a link between music and number led to the inclusion of harmonics in *mathêmata*."[33] After reviewing the ancient evidence and scholarly debates, Zhmud concluded that "Given an unbiased approach to the fifth and fourth century sources, we can state with justifiable confidence that Pythagorean harmonics *must* go back to Pythagoras himself."[34] According to Nicomachus, on the authority of Xenocrates, "Pythagoras discovered also that the intervals in music do not come into being apart from number"; and it is from Nicomachus that we learn explicitly that Pythagoras studied the intervals of the octave, fifth, and fourth, and made his discovery of them by experiment.[35] While it is true that the story linking Pythagoras's discovery of the musical intervals to hearing varying pitches as he walked by the blacksmiths' shop has been discredited,[36] there is no good reason to discount the discovery from other kinds of experimentation. The anecdote from the blacksmiths' shop claimed that Pythagoras was taken by the different pitches emitted by hammers of different weight as they struck the anvil, but it happens that the anvil, not the hammers, emits the sound, and furthermore, pitch is not a function of the weight of the hammers in any case. However, the variety of stringed instruments and pipes offered a plethora of suitable resources to experiment with ratios and proportions to discover harmonic intervals.

Andrew Barker set out with exemplary clarity the simple musical facts behind Pythagorean intervals. First, "to a given musical interval between two notes or pitches, there corresponds a specific ratio between two lengths of a true string," a true string being one that is consistent throughout its length in tension, thickness, and material constitution—quantities depending upon qualities. Secondly, "that to the three intervals standardly called *symphônai*, concords, there corresponds an orderly set of three strikingly simple numerical ratios."[37] And these are whole-number—integer—intervals. Moreover, about these three cordant intervals, "The lengths of two sections of a true string which give notes at the interval of an octave are in the ratio of 2 : 1, the ratio corresponding to the interval of a fifth is 3 : 2, and that of an interval of a fourth is 4 : 3."[38] The special significance attributed to these ratios is indicated by the fact that only these three were considered "cordant." Consider the diagram below, where the mark ∧ indicates the point where one would stop the string to produce the desired pitch above the fundamental (= fund.) or lower pitch.³⁹

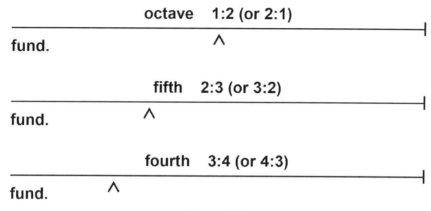

Figure 3.20.

The relation between consonant sounds was discovered as expressible in whole-number ratios: 2 : 1, 3 : 2, 4 : 3. It seems likely that it was here that Pythagoras was inspired also to see a relation between line lengths and numbers.

Árpád Szabó, in Part II of *The Beginnings of Greek Mathematics*, set out to show how the pre-Eudoxian theory of proportions initially took place in the Pythagorean theory of music.⁴⁰ Szabó supported this thesis with an analysis of the Greek technical terms of the theory (διάστημα, ὅροι, ἀνάλογον, λόγος) and their recognition in the supposedly Pythagorean experimental practice of a string stretched across a ruler, the so-called "canon," divided in 12 parts.⁴¹ Here the Pythagoreans found three main consonances: the octave (διπλάσιον, 12 : 6), the fourth (ἐπίτριτον, 12 : 9 or 8 : 6, 4 : 3), and the fifth (ἐμίολιον, 12 : 8 or 9 : 6, 3 : 2). This musical theory was explicitly connected by Archytas with the means: "Now there are three means in music . . .": arithmetic $a - b = b - c$, geometric $a : b = b : c$, harmonic or subcontrary $(a - b) : a = (b - c) : c$.⁴² The point is that the *geometric mean* or *mean proportional* is the natural relationship between the different octaves and hence seems even a sort of natural consonance,⁴³ and that Aristotle recognized instead that only the arithmetical and the harmonic means were related to musical harmony.⁴⁴

The earliest mention of what is called familiarly in the secondary literature "harmony of the spheres" survives in Plato's *Republic*—but there it is the "harmony of the circles," not spheres.⁴⁵ The correlation between astronomy and music is first introduced in Book VII (530c)—they are sisters—and then the theory is presented at the very end of the *Republic* in Book X. There, Socrates relates the Myth of Er; there have been debates about to what degree this is a Pythagorean tale, for some of the details there is no Pythagorean evidence. But the tale involves the transmigration of souls and a correlation between astronomy and music, and for these we have evidence connecting us back to Pythagoras himself. When offering to explain the process by which reincarnation happens—the souls approach *Anangke's* spindle of necessity—Ἀνάγκης ἄτρακτον—the Sirens stand on circles of the planets, each uttering a single note, and all eight tones together produce a single harmony: ἕνα τόνον—ἐκ πασῶν δὲ ὀκτὼ μίαν ἁρμονίαν ξυμφονεῖν.[46] The whirling of the cosmic bodies produces a cordant sound.

Aristotle, in discussing and criticizing Pythagorean teachings, confirms, without mythological guise, the idea of the heavenly harmony. He explains that the Pythagoreans held the theory that music is produced by the movements of planets and stars, because the sounds they make are harmonious. They reasoned that these large heavenly bodies, moving so rapidly, must produce enormous sounds just as our own experiences show that large bodies moving rapidly produce sounds. And because the Pythagoreans reckoned the relative speeds of these heavenly bodies to be a function of their distances, and those distances were claimed to be in the ratios of musical intervals, the sound that these cosmic bodies emit is cordant.[47] Here, then, we have explicated the kernal of the doctrine of the harmony of the circles, and thus the interconnectedness of number and spaces made possible by shared, interpenetrated structure. By virtue of understanding geometric similarity, Thales and Pythagoras shared the view that the little world has the same structure as the big world, and thus we could discover things about the big world—infer them—by attending to the details of the little world to which we had access. This is the structure underlying the microcosmic-macrocosmic argument. The relation between the musical intervals and the harmony of the circles is another display of just this projection of structural *similarity*. The musical intervals exhibit cordant sounds in whole numbers; the varying distances—lengths—between the heavenly bodies are claimed to be in these same ratios, and hence Pythagoras came to *the vision of the harmony of the "spheres"* (i.e., the harmony of the "circles"). This picture accords well with Proclus's report on the authority of Eudemus that "Pythagoras transformed mathematical philosophy into a scheme of liberal education."[48] The overlap and interpenetration of arithmetic, geometry, astronomy, and music, then, resulted in the vision that *there is geometry in the humming of the strings; there is music in the spacing of the spheres.*[49]

Aristotle sums up the Pythagorean teaching on the correlation between these μαθήματα in the *Metaphysics*:

> They said too that the whole universe is constructed according to a musical scale. This is what he means to indicate by the words "and that the whole universe is a number" because it is both composed of numbers and organized numerically and musically. For the distances between the bodies revolving round the center are mathematically proportionate; some move faster and some more slowly; the sound made by the slower bodies in their movement is lower in pitch, and that of the faster is higher; hence these separate notes, corresponding to the ratios of the distances, make the resultant sound concordant. Now number, they said, is the source of this harmony, and so they naturally posited number as the principle on which the heaven and the whole universe depended.[50]

The Pythagoreans studied number, according to Aristotle, and came to see that its principles are the principles of everything, the structures of the little world and big are joined microcosmically-macrocosmically, and the relation is further delineated by means of number.

F

Pythagoras and the Theorem: Geometry and the Tunnel of Eupalinos on Samos

Could the experiences of watching the architects and builders, in addition to the musicians and instrument makers, have provided Pythagoras with a breakthrough about metrical or numerical relations relevant to the hypotenuse theorem? Could the project of digging the famous tunnel in Samos—the tunnel of Eupalinos—have offered Pythagoras a way to see, or to confirm further, the numerical formula behind the lengths of the sides of a right triangle?

We know Herodotus's opinion that the three greatest engineering feats in the Greek world were in Samos.[51] They were (i) the temple of Hera, the greatest temple ever to be attempted in Greece, (ii) the enormous sea mole that protected the fleet from storms and surging sea and so allowed the Samian economy to thrive by controlling the sea lanes along the west coast of modern Turkey; and (iii) the tunnel of Eupalinos that brought fresh water to ancient Samos. Herodotus had this comment about the tunnel:

> [A]nd about the Samians I have spoken at greater length, because they have three works which are greater than any others that have been made by Hellenes: first a passage beginning from below and open at both ends [i.e., double-mouthed], dug through a mountain not less than a hundred and fifty fathoms in height; the length of the passage is seven furlongs and the height and breadth each eight feet, and throughout the whole of it another passage has been dug twenty cubits in depth and three feet in breadth, through which the water is conducted and comes by the pipes to the city, brought from an abundant spring: and the designer of this work was a Megarian, Eupalinos the son of Naustrophos. This is one of the three; and the second is a mole in the sea about the harbor, going down to a depth of as much as twenty fathoms; and the length of the mole is more than two furlongs. The third work which they have executed is a temple larger than all the other temples of which we know. Of this the first designer was Rhoikos the son of Philes, a native of Samos. For this reason I have spoken at greater length of the Samians. (Herodotus III.60)

If we place Pythagoras's birth at around 570 BCE, the first monumental stone temple to Hera, Dipteros I, was already underway around the time of his birth.[52] The architect was Theodorus, who had connections with Egypt, where perhaps the great temple of Luxor inspired the double-peristyle design.[53] By around 550 BCE, a second temple to Hera—Dipteros II—was begun about forty-three meters to the west. This temple must be the one that Herodotus saw, credited to the architect Rhoikos, perhaps assisted by the aging Theodorus. It now seems that the reason Dipteros I was dismantled and the new, even larger plan for Dipteros II was developed, was not fire, as had been conjectured earlier, but rather the first temple beginning to list under its own immense weight.[54] Here, then, in Samos in the mid-sixth century BCE, was a veritable experimental laboratory of ancient building technologies—planning, quarrying stone, transporting the monoliths, finishing the architectural elements, and installing them. Geometrical principles were displayed—ratios and proportions of lengths and areas—for anyone who cared to look for them. The overall proportions of the archaic temples in Samos were 1 : 2 : 4—indeed not only in Samos but also the other great archaic Ionic temples in Ephesus and Didyma, probably begun around 560 BCE. They were all twice as wide as they were tall, and twice as long as they were wide. The building followed the geometrical progression, central also to Egyptian mathematics of doubling: 1, 2, 4, 8, 16, 32, 64, etc. Coulton conjectured that in the architects' early prose books, contemporaneous with the first philosophical book in prose by Anaximander, "a description of the buildings, both in terms of absolute size and *the rules of proportion* used, was probably included."[55]

Let us try to be clear about the kinds of things involved in the rules of proportion. As I have discussed elsewhere,[56] Vitruvius informs us that the success in building monumentally begins with the selection of a module. In Ionic architecture, Vitruvius reported that "column diameter" was the selected module;[57] the module is a single element of the building in terms of which the other architectural elements are reckoned as multiples or submultiples. In an important sense, it serves as the One over Many; the module is the underlying unity and the other elements are microcosmic or macrocosmic expressions of it. Once column diameter is assigned a measure (i.e., so many feet or ells), then intercolumniation is calculated to separate one column from another east to west, and north to south. The sizes of the temple parts—*pronaos*, *adyton*, and *opisthodomos*—are all reckoned as multiples or submultiples of the module, as are the upper orders of architrave, pediment, *geison*, and *kyma*. This one underlying unity is concealed in all the details but revealed upon inspection; stated within our central metaphysical theme, there is an underlying unity that alters without changing. When Anaximander expressed that the size of the cosmos could be reckoned in column-drum proportions—literally column diameters—he not only made use of an architectural technique of modular design, he appealed to exactly the module that the Ionic architects consciously selected. He appealed to column diameter, the architects' module, because he came to imagine the cosmos as cosmic architecture, an architecture built up in stages like the temple. The temple, through its symbolic elements and design and through its proportions, sought to make the invisible, divine powers visible; the proportions revealed the power and harmonious balance by which the supersensible and invisible power appears.

Now Eupalinos was an architect and engineer. Herodotus calls him an ἀρχιτέκτων[58] and tells us that he comes from Megara—that is, mainland Megara, west of Athens, and yet he was selected by the leaders of the eastern Greek island of Samos for this extraordinary task of tunneling through a limestone hill simultaneously from two sides. How shall we make sense of this? We have no additional information from Herodotus or other ancient sources. We have no archaic evidence for a tunnel driven from two sides anywhere in the Greek world before this time. We know the central motivating factor for the tunnel was the creation of a water channel to carry fresh water to the blossoming town, and that new a source of water was discovered in the modern town of Ajades, about 900 meters from the north entrance to the tunnel, outside the city walls and on the far side of the hill. Now, the Samians already had extensive experience quarrying limestone for both the Heraion and city wall. The quarry for the Heraion is located in the hill just north of the north side of Mount Castro overlooking the north entrance of the tunnel. Perhaps Eupalinos had moved to Samos some time before the tunnel project and distinguished himself among the building teams of the temple of Hera, or even the city wall? He might very well have been a quarry foreman, because a visit to the quarry shows very deep tunneling into the hill, one of the skills indispensable to the successful tunnel project. The experience of working through the deep tunnels in the hill just north of the north entrance might have provided much-needed experience, and the topography and geology were comparable to those of the nearby quarry.

Figure 3.21.

But recent evidence has now appeared from mainland Megara showing that enormous underground water channels were dug through the limestone there.[59] Perhaps the fact that Eupalinos came from Megara was not simply a coincidence but rather indicative of long or recent experience that the Megarians in general, and Eupalinos in particular, had with such water-channel projects? Whatever Eupalinos's background and previous experience was, credit for the successful projects rests with him. The planning and execution of the tunnel takes places within the timeframe when Pythagoras was still in Samos, not yet having emigrated to southern Italy. The dating of the tunnel continues to be a subject of debate, and some recent reflections on the clay pipes that make up the water channel have suggested a date significantly earlier than the 530s that is sometimes conjectured, perhaps closer to the midcentury and before the time of Polykrates. Within this timeframe, we have young Pythagoras, who plausibly had already met with Thales and Anaximander and/or members of their school—because both Milesians were dead shortly after the turn of midcentury—and consequently was plausibly familiar with geometrical thinking, diagrams, and model-making. Anaximander had already been credited with making a cosmic model, perhaps an early kind of planetarium, and a book on geometry. Could the problems to be solved dealing with applied geometry have played some role in Pythagoras's reflections on what he learned from Thales, Anaximander, or members of the Milesian school? The problems posed by digging the tunnel are multifold. The successful results show that Eupalinos was able to start both tunnel entrances at almost exactly the same level, even though they were not visible at the same time. He was able to control the floor level with such accuracy that, at the meeting point, some 616 meters from the north entrance and more than 420 from the south entrance, he was off by not more than 60 cms. The successful meeting of the tunnels proves also that he controlled the bearing—the direction of digging—of both tunnels with remarkable accuracy. And the completed tunnel reveals other challenges that show the remarkable ingenuity of applied geometrical skills by which Eupalinos nevertheless succeeded. When we consider some competing hypotheses for the applied geometrical techniques Eupalinos knew and employed, we will be in a position to see how Pythagoras might have brought to bear what he learned from Thales, and so clarified or confirmed the hypotenuse theorem, or even further resolved the metric interpretation of it. For our project, there are three principle questions to answer:

1. How did Eupalinos manage to establish that both tunnel entrances were on the same level?

2. How did Eupalinos come up with an hypothesis for the length of each tunnel half?

3. How did Eupalinos find his way back to the originally planned meeting point under the ridge when, because of soft stone and too much natural groundwater, he was forced to leave the straight line in the north tunnel?

There has been some consensus that Eupalinos established that both tunnel entrances were on the same level by the use of a *chōrobatēs* (χωροβάτης), a simple water-leveling device; the supposition that he made use of a dioptra has been largely ruled out as anachronistic.

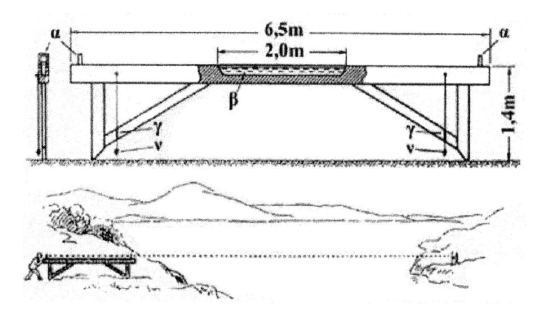

Figure 3.22.

The χωροβάτης works by enabling a clear sighting of level at a distance. The instrument is a small bathtub-like vessel filled with water up to a uniformly marked level. The vessel rests on legs, and perhaps had two plumb lines alongside the vessel, above the legs, to allow the surveyor to be assured that the instrument itself was level. At the front and back of this bathtub-like container there are sighting devices, perhaps making use of animal hairs, both of which are adjusted to be at the same level themselves. By means of this instrument, a stake with a horizontal board placed some distance away is lined up with the two hairlines until the board is seen at the same level. Thus the level is transferred over distance, setting up and re-setting-up the χωροβάτης all around the hillside, as pictured below, after Kienast.

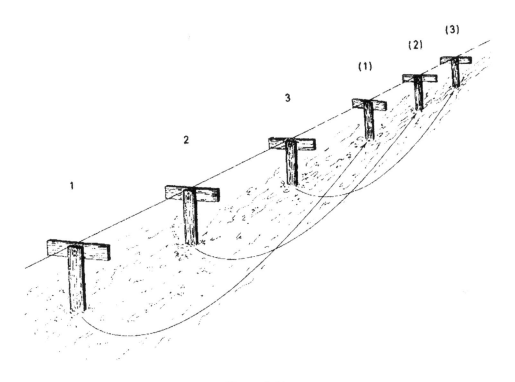

Figure 3.23.

Thus, the process of establishing that both entrances were at the same height above sea level began with a decision about the height of one entrance and then by means of a χωροβάτης that level was secured all around the hill until it reached the site of the other proposed entrance.

The process of reaching an hypothesis about the tunnel length for each side has also achieved broad consensus; the technique was to "stake out the hill," and the argument for it is that the two tunnels met—and were planned to meet—directly under the ridge.[60] For the question can surely be asked, "Where should these two tunnels meet?" The decision to dig from two sides was certainly motivated to cut in half the duration of the project, and perhaps the urgency of gaining access to a sufficient supply of natural fresh water. But the tunnel halves do not meet in the arithmetical middle, at the arithmetic mean; since the total tunnel length is 1,036 meters, the tunnels might have met halfway at 518 meters. Instead, the north tunnel is almost two hundred meters longer than the south tunnel. The unequal lengths of the tunnel

162   The Metaphysics of the Pythagorean Theorem

halves invites us to think about why the tunnels meet where they do, that is, directly under the ridge.

   According to Kienast's hypothesis, as the stakes ran up each side of the hill, Eupalinos could add up the horizontal distances between the stakes and infer that their sum was the length of each tunnel half: this is why the tunnels were planned to meet under the ridge, and why they did. This process of staking required that poles of a standard size be placed up and along the hillside, the horizontal distance between poles being the crucial ingredient for determining the measure of the tunnel. The white dots up along the south side of the hill indicate the possible placing of the stakes.

Figure 3.24.

The white dots, above, running up and along the south side of Mount Castro, allow one to calculate the sum of the horizontal distances between the stakes (i.e., between the dots), the proposed length of the tunnel. To make clear why this is so, consider the illustration below. Had anyone made a geometrical diagram, as we must imagine Thales and his retinue to have done many times, it would show how geometry is applied to the problem of tunnel length so that each of the horizontal distances, from stake to stake, expresses the projected tunnel length for each tunnel half.

Pythagoras and the Famous Theorems 163

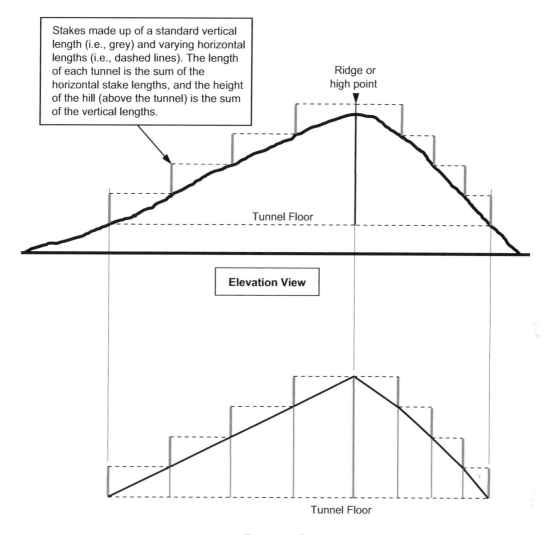

Figure 3.25.

Now, before we bring these engineering issues together to see how they might have been relevant to a youthful Pythagoras, let us turn to consider the third question, concerning a problem inside the tunnel itself. The technique of staking out the hill can provide crucial information to Eupalinos about the length of each tunnel half, but only if the digging can proceed uninterruptedly in a straight line. The summation of all the horizontally staked lengths is the proposed length of the tunnel, but if Eupalinos is forced to leave the straight line, as he was in the north end, the technique loses its security. Among the greatest challenges for tunnel diggers is the prospect of collapse. The two most important factors in this regard are the softening or crumbling of the limestone, and the profuse presence of natural groundwater and mud. Both of these factors loomed large as Eupalinos dug in the north tunnel. So long as the workers could look backward and still see daylight at the entrance, there was reasonable assurance that they were at least working in a straight line. But once he was forced to leave the straight line, curving outward toward the west and replicating

164   The Metaphysics of the Pythagorean Theorem

the same curve inward to regain his place in the originally planned straight line, both Eupalinos and the workers were proceeding quite literally in the dark.⁶¹ This is an important point, because it highlights the importance not only of planning, but most especially of the use of a *diagram and/or scale model* by which Eupalinos could keep track of where he was in the tunnel progress. It was only by recourse to such diagrams and models that Eupalinos could control the project. Once again, this means that the little world—the diagram and/or model—had the same structure as the big world, the tunnel itself. By focusing on the small world, one could discover the structure of the big world. Here, once again, we have archaic evidence of microcosmic-macrocosmic reasoning, a grasp of the structure of things through an argument by *analogia*—by means of ratios and proportions.

In the process of excavating the north tunnel, fearing collapse, Eupalinos was forced to make a large detour, and thus leave the straight line. Now he was faced with the problem of how to do so, to leave the straight line and yet find his way back to the original meeting point under the ridge; he realized that the stratigraphy folded to the northeast, and so he dug to the west where there was less risk of collapse. The difficulty highlighted that whatever would be the exact solution, the length of the tunnel would have to be prolonged by making such a detour. Eupalinos's solution, according to Kienast, was to dig at an angle of ≈22.5° to the west, and when he reached a point where the stone was suitable again, he dug back at the *same* angle of ≈22.5° to regain his position in the straight line—that is, his solution was to construct an isosceles triangle. As Kienast reconstructs the evidence, one of the five marking systems, painted in red along the western wall, displays the Milesian system of numeration. Eupalinos used this marking system to keep track of where he was in the tunnel, that is, to determine how close he was to the planned meeting point. Consider the diagrams, below, whereby Kienast conjectures how Eupalinos solved the problem using a scaled-measured diagram:⁶²

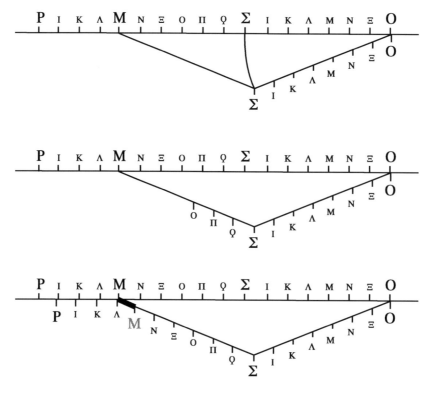

Figure 3.26.

Kienast's conjecture is that there was a geometrical diagram and perhaps also a three-dimensional model, constructed to scale. In the diagrams above, Kienast conjectures how Eupalinos's scaled-measured plan looked—while no such diagram or model survives, its existence in inferred from the letters painted on the wall itself. First, he set out the straight line, then planned the triangular detour because he had to leave it. The straight line has the original Milesian numeration—counting in tens—with Greek letters P, M, Σ, and O highlighted, and O marks the return to the original straight line. Starting at M, Eupalinos began the detour; his plan was to continue in a straight line westward until the conditions for collapse no longer presented a significant risk. When he reached that point, knowing the stratigraphy sufficiently well, he planned to turn back *at the same angle* to the original line. With a compass, and the point on either (or both) M or O—since the proposed plan was based on an isosceles triangle, it did not matter which letter since the length of both sides would be equal—Eupalinos transferred the point Σ from the planned straight line to the vertex of the triangle in the detour. Kienast's claim is that the placement of Σ on the wall itself at the vertex of the triangular detour reveals that Eupalinos worked from a scaled-measured plan—a lettered diagram. Why? Because the Σ should not have appeared *there* unless the plan was to arbitrarily move Σ to the vertex and then start counting backward. From the Σ, he proceeded to transfer the remaining distances on the straight line—I, K, Λ, M, N, Ξ—to reach the original straight line at O. This is the reason that the letters on the wall itself of the detour are all *shifted inward*. Before the detour begins, the letters on the straight line, like the ones in the south tunnel, are all placed at 20.59 meters apart, starting from outside the tunnel and counting inward. But in the north tunnel, while M should appear on the straight line just *before* the detour, it is now in the detour itself. Eupalinos knows the distance left to dig from Σ to O, the second leg of the detour created by transferring Σ to the vertex; now he counts backward on the first leg of the triangular detour from Σ—Ϙ, Π, O, Ξ, N—to reach M and so discover the prolongation of the tunnel. This is why M now appears shifted inward. The distance from where M appears to where it should have appeared (before the detour) is exactly the additional digging required.

Käppel studied the elongation of the tunnel, paying careful attention to the measurements, and proposed an understanding of a remarkable feature in the north tunnel, an interpretation that has been adopted subsequently by Kienast.[63] On a retaining wall, built *after* the completion of the tunnel (and so it could not play any role in the construction process itself), and not far from the north entrance, the word "PARADEGMA" appears, in red paint with letters 30 cm in height. The inscription is bordered by two measuring lines and the distance from one measuring line to the other is (almost) exactly the extra distance of the tunnel elongation.

Figure 3.27.

Thus Käppel's conjecture, embraced by Kienast, is that this was how Eupalinos celebrated his calculation of the elongation of the tunnel. "The inscription is . . . the engineer's legacy, a message to those who might one day attempt to decipher the plan of the tunnel."[64]

Now, let us try to place this daring engineering project in the context of Samos shortly after the middle of the sixth century BCE. The techniques and applied geometry of determining the height above sea level of the tunnel entrances, the hypothesis that tunnel lengths were determined by staking out the hill, the control of the bearing or direction of the digging and the level of the tunnel floor by other marking systems, the use of diagrams and models—all are evidence of what was available to anyone in Samos who might have been interested in watching Eupalinos and reflecting on these techniques.

Now suppose that the Samian council sought among its citizens to confirm Eupalinos's hypothesis of tunnel length. Let us keep in mind the realities of a tunneling project. The digging would have taken years to complete. No one could be sure that the project was on course until the sound of the hammering could be heard from both tunnel halves. This means that if the work tunnel (as opposed to the water channel, which could be completed only *after* the level tunnel was done) took about ten years—based on conjectures from the rates of excavation at the silver mines in mainland Laurion, with comparable iron tools and through similar limestone hills—no one would be in a position to know if the project was on target until labor had continued for a decade or more! Had the level been lost or the bearings significantly diverged, the tunnels might never have met and the reasons why might never have been discovered.

Surely, circumspect supervisors would have felt the practical need to confirm independently Eupalinos's hypothesis. Or if this sensed need was found to be uncompelling, it still might have challenged Pythagoras to find another way to check the hypothesis of the architect-engineer. From the picture we have been developing of Pythagoras, it is hard to resist the idea that this daring tunnel project would be of interest to him.

Let us note, then, that the plan was to have the two tunnel halves meet directly under the ridge. This was a consequence of staking out the hill, which gave Eupalinos the horizontal distances from each entrance to the meeting point. From either the south or the north entrance, he could not see the stakes on the *other* side of the hill, which introduced an element of error into his calculation of the bearing. Thus, each tunnel half was conceived separately, the distance to an ideal meeting point being calculated by adding up the horizontal distances between the stakes and so arriving at a distance *directly under the ridge* of the hill, and the target had to have a width great enough to accommodate some element of error. Once we absorb these claims, and the geometrical picture of staking out the hill, it should be clear that the tunnel project *can* be imagined as two right-angled triangles sharing in common the line from the meeting point to the top of the hill, as seen below:

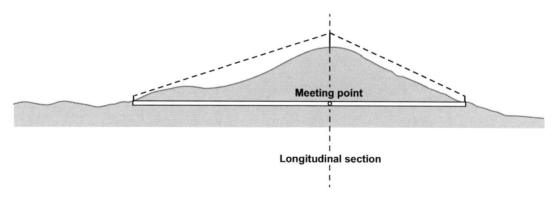

Figure 3.28.

Anyone who watched the staking of the hill would have known two things. They would have known, of course, the sum of all the horizontal lengths, and hence Eupalinos's hypothesis of tunnel lengths. But, by adding all of the *vertical lengths* they would have been able to calculate the height of the hill—the side that both tunnel halves shared in common—for the precise vertical height could be discovered by the same technique that secured horizontal lengths. And while the stakes stood up and along the slopes of Mount Castro, the hypotenuse of each right triangle could have been measured directly. The entrances are both established on the same level, at 55 meters above sea level. The height of the hill at the ridge is 225 meters, and so the vertical line from the meeting point to the ridge is some 170 meters. By means of a knotted cord, the hypotenuse of the south triangle would have measured some 453 meters, and for the north triangle some 639 meters. Here we have, then, the numbers to describe the lengths of the legs of the right triangles. From Thales or his retinue, young Pythagoras would have known already the areal relations revealed by ratios and proportions, and the squares or rectangles that could be drawn on each side could have found expression in figurate numbers.

By attending to the marking systems in the tunnel, Kienast concluded that the basic increment of measurement was not a standard Samian foot or ell, nor any measure we know from the mainland; instead, the unit-measure was 2.059 meters, Eupalinos's tunnel measure. This choice of selecting what seems to be his own *module* fits well in the context of other contemporaneous engineering concerns, such as those in temple building, and the search for the One over Many characteristic of the early Greek philosophers. As we have already considered, to control the monumental temple as it is being built, and to secure the intended appearance as it had been planned, the architect-engineer selected a module in terms of which all the other dimensions were reckoned as multiples or submultiples. According to Vitruvius, the Ionian architects selected column diameter as their module.[65] The fact that Anaximander identified the size and shape of the Earth with a column drum, and then reckoned the size of the cosmos in earthly proportions (i.e., column drum proportions, and thus column diameters) shows not only that he embraced an architectural technique but adopted precisely the same module by which the architects worked in Ionia. He was inspired to appeal to architectural techniques, as I have discussed elsewhere, because he came to grasp the cosmos as built architecture.[66]

We have no evidence, no reports, that connect Pythagoras with the tunnel, but the planning and excavation of it took place some time around the middle of the sixth century, say 550–540 BCE, when Pythagoras was still a young man. The picture of Pythagoras that we have been developing suggests that he would have been aware, from Thales and Anaximander or members of their school, of a range of geometrical diagrams and claims, and a metaphysical project that connects him to the evidence we plausibly do have. The success of Eupalinos's tunnel offers an exemplar that numbers reveal the hidden nature of things, because it was by control of them that the project succeeded. Here we have another contemporaneous experiment in applied geometry that reveals a special interconnection to line lengths and areal relations revealed and confirmed with right triangles.

G

Pythagoras, the Hypotenuse Theorem, and the μέση ἀνάλογος (Mean Proportional)

The narrative I shall continue to tell is the one that follows from the foundation set by Thales. Pythagoras and his retinue took up the project of building the cosmos out of right triangles, having learned from Thales or a member of his school the hypotenuse theorem—along the lines of ratios and proportions in VI.31, not I.47. Their discovery of the theorem was that the right triangle was the fundamental geometrical figure, that within every rectilinear figure are triangles, that all triangles reduce to right triangles, and those right triangles reduce indefinitely to right triangles: to adapt a saying, it is right triangles all the way down. The theorem also meant to them that when the right triangle expands or unfolds, contracts or collapses, it does so in continuous proportions; the principle of the right triangle unfolding into other figures is structured by the geometric mean or mean proportional (μέση ἀνάλογος).

From this platform, Pythagoras and his associates achieved a series of developments. When he tried to understand metaphysically how some fundamental underlying unity was nonetheless able to appear so diversely, he produced a geometrical program to express these transformational equivalences—how fire, air, water, earth, and all the intermediate combinations of them, were nevertheless only this one unity, altered but unchanged (= substance monism). Pythagoras's project of transformational equivalences is his "application of areas" theorem, reported by Plutarch. Through numbers, ratios and proportions in musical intervals, and the projection of this structure by analogy with geometrical similarity, he came to imagine the harmony of the spheres. This was Pythagoras's development of the microcosmic-macrocosmic argument. The little world was analogous to the big world, it shared the same proportional relations, and it exhibited the same geometric structure of similarity. The process of explicating how the little world becomes the big world—exhibiting the same underlying structure—led Pythagoras to another intermediary explanation of how a single underlying unity could nevertheless appear so divergently—the construction of the five regular solids. The narrative I shall reach by the end of this chapter is how Pythagoras plausibly came to the insight that right triangles—the fundamental building block of all appearances—can be repackaged and combined to form the regular solids—the basic blocks out of which all these other appearances, in turn, are structured.

The case connecting Pythagoras to the famous theorem is framed, then, by the convergence of a series of ideas that point plausibly to Pythagoras himself. Although surrounded by legend, there is broad agreement, or ought to be, about a few claims; the relevant ones for my case I list here: (i) Pythagoras was familiar with ratios and proportions; (ii) he knew three proportional relations—arithmetical, geometrical, and harmonic; (iii) Pythagoras knew the theorems about odd and even numbers; (iv) he discovered or knew the musical ratios of octave (2 : 1), fifth (3 : 2), and fourth (4 : 3)—that euphonic sounds could be expressed by whole-number relations; (v) he projected these euphonic musical ratios onto the distances between the heavenly bodies and so described the "harmony of the spheres"—this is, structurally, the microcosmic-macrocosmic argument, and that argument is analogous to geometrical *similarity*; (vi) he discovered or knew the application of areas theorem(s); and (vii) he knew that the application-of-areas theorem had something to do with the "putting together" of the regular solids—whether he was credited with three or all five cases, he was preoccupied with them [NB: I am not arguing that he "proved" them, but that the "discovery" of them plausibly belongs to him and his school].[67]

The case I have already set out, then, follows from Thales's grasp, discovery, and/or proof that every (tri)angle in a semicircle is right-angled, a proof for him that followed from his proof that the base angles of every isosceles triangle are equal, and thus that the sides opposite the angles are equal, and also that the diameter of a circle bisects it—these all fit together. But it also needs to be set in the context of the other theorems that are reflected in the measurements of pyramid height and the distance of a ship at sea. This theorem that every (tri)angle in a semicircle is right-angled is credited to both Thales and Pythagoras, and I am suggesting that rather than this sharing of credit being a confusion among reports in the doxographical tradition, instead it suggests that both Thales and Pythagoras shared the same lines of mathematical intuitions—a line of thought that Pythagoras learned from Thales.

The Pythagoreans, and in my estimation Pythagoras himself, deserve to share credit for grasping and proving that there are two right angles in every triangle. Proclus attributes this theorem to them, and provides the following diagram:[68]

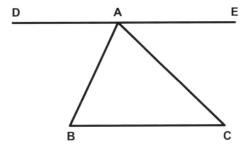

Figure 3.29.

Lines DE and BC are parallel, and thus the alternating angles DAB and ABC are equal. Accordingly, the alternating angles EAC and ACB are equal. Now add the common angle BAC so that the angles DAB, BAC, and EAC together make a straight line or two right angles, and see that since DAB is equal to the base angle of the triangle ABC, and angle EAC is equal to the base angle ACB, and angle BAC is common, thus the sum of all three angles of the triangle ABC, BAC, and ACB equals two right angles.

170   The Metaphysics of the Pythagorean Theorem

But let us follow the diagrams to plausibly connect the steps by which this proof was achieved:

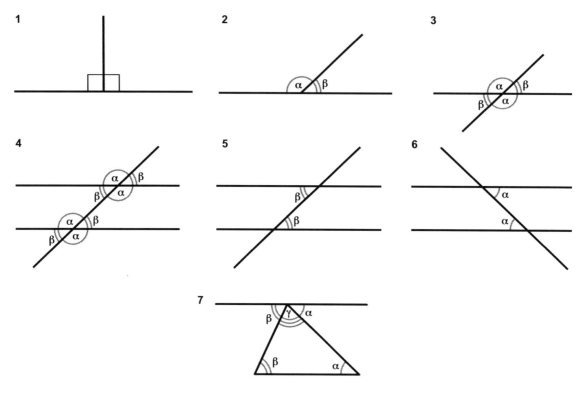

Figure 3.30.

In 1, we see that every straight line (angle) contains two right angles. And in 2, no matter where the straight line falls on the straight line, it creates two angles α and β that also sum to two right angles. Now 3 shows that continuing the vertical line falling on the horizontal straight line makes the opposite angles equal, and that any two angles sharing the same vertical side sum to two right angles. Now 4 shows that when the vertical line is extended through another straight line horizontal to the first, the interior angles along the vertical straight line (α + β) sum to two right angles, and the equivalence of the alternating angles becomes immediately apparent, so that in 5 and 6 the alternating and opposite angles β and the alternating and opposite angles α are then equal. And therefore, in 7, if angle β is equal to its alternate and opposite angle β, and angle α is equal to its alternate and opposite angle α, then since α + β + γ equal two right angles, every triangle must contain exactly two right angles.[69]

It seems clear to me that in some form, Thales understood this, whether or not he "proved" it, and this is plausibly what Pythagoras also learned from Thales or his retinue. As I have pointed out already, if one sets out this theorem (below, left) in a diagram, and removes the semicircle, one is left, almost, with the diagram in the family of VI.31 (minus only the rectangles drawn on each side, and the perpendicular rather than radius from the base, following Heiberg); and as we already considered in the ancient manuscript tradition, the predominant diagram is one pictured on the left, below, but without the details and semicircle—just an isosceles right triangle.

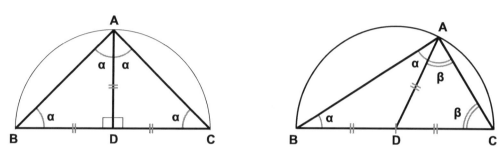

Triangles in Semi-Circle are always Right

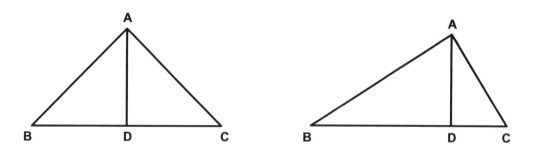

Euclid VI.31 "Enlargement" of Pythagorean Theorem

Figure 3.31.

When we review the lines of thought in VI.31, *similarity*, *ratios*, and *proportions* play key roles in the demonstration. In the diagram on the right, above, VI.31, a right-angled triangle is divided into two right-angled triangles by the perpendicular dropped from the right angle to the hypotenuse. This divides the triangle into two right triangles, both of which are similar to each other and the largest triangle that has now been divided. All three triangles share the same angles, while only the lengths of the lines are different but proportional—thus, the three triangles are all similar. The discovery of the diagram for VI.31 is the discovery also of the isosceles triangle in the (semi-)circle. *It is the discovery that inside every right-angled triangle there is a pair of right-angled triangles that dissect indefinitely into pairs of similar right triangles. The pattern of growth—expansion and collapse—is by geometric mean proportions.*

172   The Metaphysics of the Pythagorean Theorem

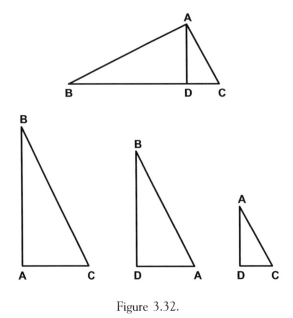

Figure 3.32.

What VI.31 captures is a series of ideas that reveal the correlation of line lengths and the areas of figures constructed on them.

Geometric Mean

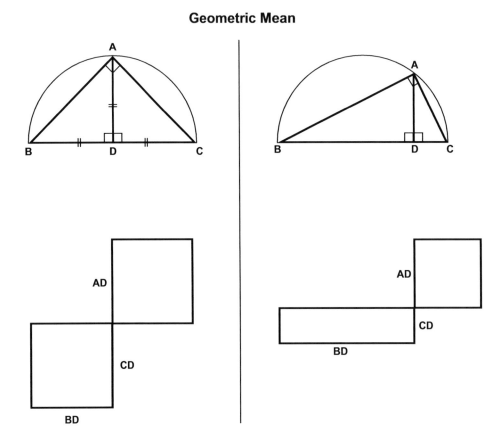

Figure 3.33.

Additionally, the same ratios applied to the relation of the three similar right triangles, both to the lengths of the respective sides and to the areas. And this means that not only do the lengths of the sides in this three-term proportional stand in duplicate ratios to one another, but so do the areas of the triangles.

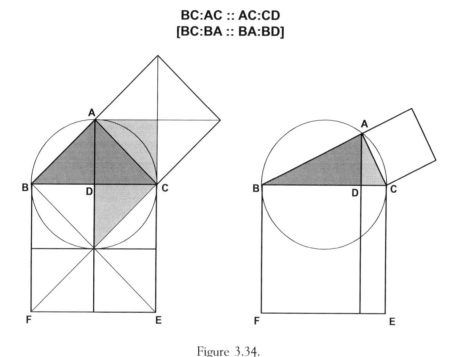

Figure 3.34.

Now the following analysis applies to both triangles, directly above, but is clearer with the one on the right because the one on the left has legs of equal length. With AC as the mean proportional between BC and DC, the ratio of BC to DC is the same ratio as the area of triangle ABC to the area of triangle DAC; in other words, with AC as the mean proportional, the ratio of the areas is the same as the ratio of the longest side of the large triangle, BC, to the shortest side of the small triangle, DC. Thus, as the sides stood in these relations, so did the areas: as the ratio of the length of the first was to the third, so the ratio of the area of the figure on the first was to the second. The realization of the famous hypotenuse theorem by *similarity*, then, emerged for Thales by grasping the ratios and proportions of lengths and areas, and grasping the principle of similarity. This is part of what young Pythagoras plausibly learned from Thales or his retinue.

174   The Metaphysics of the Pythagorean Theorem

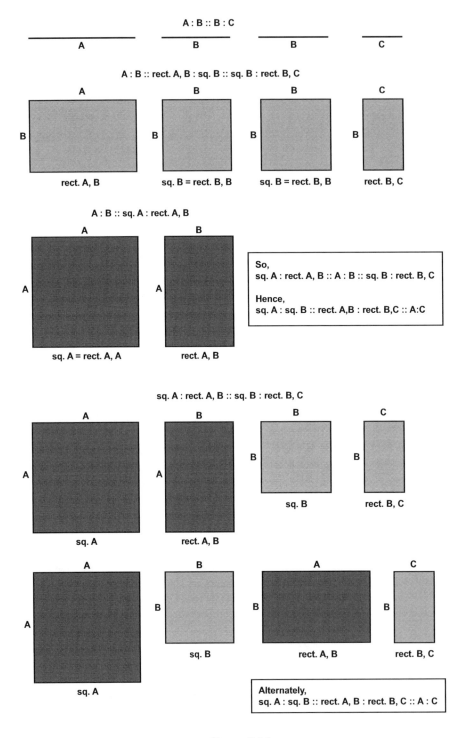

Figure 3.35.

Now, following through this idea of duplicate ratio, line length A is to C in the same ratio as the square on A is to the square on B. Thus, the ratio of the square on A is to rectangle A,B as the square on B is to rectangle B,C. Alternately, the square on A is to the square on B, as the rectangle A,B is to the rectangle B,C.

Line length C grows to length B by the same factor as B expands to A. Thus C expands to A in the duplicate ratio of C's expansion to B. Commensurately, the rectangle BC grows to square B by the same factor—the same increment of growth—by which the square of B grows to rectangle AB, in the duplicate ratio of the corresponding sides. And thus the ratio of line length A is to line length C as the square on A is to the square of B. The ideas of duplicate ratio, mean proportionals, and areal equivalences become so much clearer when you explore these ideas *visually*. My claim is that this was how Thales and Pythagoras imagined these geometrical matters—through geometrical *diagrams*—and this approach is virtually absent in the secondary literature exploring their possible connection to the famous hypotenuse theorem.

Heath had conjectured that Pythagoras probably proved the hypotenuse theorem in the following way[71]:

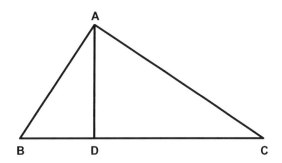

(i) AB/BC = BD/AB hence AB² = BC x BD;

(ii) AC/BC = DC/AC hence AC² = BC x DC;

(iii) Adding these we obtain: AB² + AC² = BC (BD + DC)

Figure 3.36.

When we look at Heath's proposal, the problem it presents is this. This formulation, while making use of proportions, presents it in a form that we have come to call "geometric algebra." And while Book II of Euclid, for example, *can* be expressed as "geometric algebra" (and often is presented this way), it is not the form that Euclid presents. Indeed, there are no numbers anywhere presented in Books I–VI, though some of the theorems are amenable to such a presentation; indeed, *no square is reached by multiplying a side by a side*, (though later in the arithmetical books VII–IX of the *Elements*, square *numbers* are obtained by multiplying a number by a number). So I propose now to review Heath's conjecture of the proof, but placing his form of the proof in geometric, not algebraic, terms. And the connections to Thales are obvious from the lines of thoughts I have proposed. After the geometric argument is set out, the numerical presentation is in order because, at all events, Pythagoras's own achievements probably include arithmetical expressions as well.

176   The Metaphysics of the Pythagorean Theorem

The first claim, (i) is that AB : BC so BD : AB—the ratio of the short side of the biggest triangle to its hypotenuse is the same as the ratio between the shortest side of the smallest triangle is to its hypotenuse. The ratio of lengths implies a claim of areal equivalence: the area of a rectangle whose sides are BC and BD is equal to the square of AB. And it is important to note that this is what it means to establish that AB is the mean proportional between BC and BD.

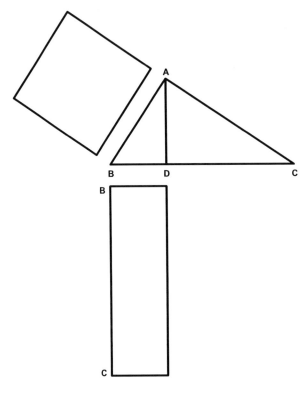

Figure 3.37.

Then, next claim, (ii), is that AC : BC as DC : AC, and thus the ratio of the longest side of the biggest triangle is to its hypotenuse as the ratio of the longest side of the medium-size triangle is to its hypotenuse. Again, this proportion of ratios has another geometric interpretation, namely, as an equality of areas, so that the rectangle whose sides are BC and DC is equal in area to the square of AC. And it is important to note that this is what means to establish that AB is the geometric mean between BC and DC.

Pythagoras and the Famous Theorems 177

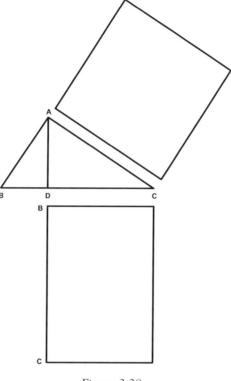

Figure 3.38.

Placing these two diagrams together (below), we can see that the square of the hypotenuse, composed of two rectangles, is equal in area to the sum of the squares constructed on the two sides:

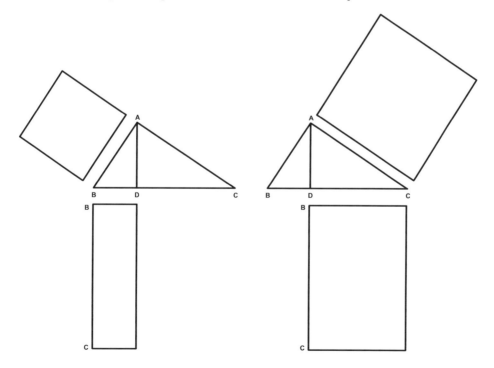

Figure 3.39.

And when we integrate these two diagrams into a single whole, the mathematical intuitions of Euclid I.47 are startlingly presented, *but* the route here is by proportions and similarity. Thus, following Heath, but presenting his insights geometrically rather than algebraically, the proof follows from grasping the mean proportional, one of the three principle means with which Pythagoras himself is credited by Nicomachus and supplanted by Iamblichus's commentary on Nicomachus.[72] In Iamblichus, the names Hippasus, Archytas, and Eudoxus appear, and the fragment in Archytas confirms these details.[73] Pythagoras knew all three proportions; I contend that Thales knew at least the arithmetical and geometrical; the crucial one for the proof is the geometrical mean.

Proclus credits Euclid with the proof of I.47; that proof shows that the "square" or "rectangle" that shares the same base within the same parallel lines with equal (i.e., congruent) triangles is twice the area of each of those congruent triangles. In Heath's reconstruction, the right triangle is divided into two similar right triangles, and both are microcosms of the largest right triangle. This means that the *areas* of the similar figures similarly drawn on the first and second are in the same ratio as are the *lengths* between the first and third. All this is another way of describing that the ratio of lengths of the first to the third projects a rectangle that has the same area as the square drawn on the second, because the square on the second is the mean proportional between these two proportional lengths, first and third. The geometrical mean is an average, like the arithmetical mean, but it is a multiplicative average; it is the result of scaling up, or of using a stepping stone (if you please), to the next proportional scaling up. The microcosm scales up to ever-larger macrocosms by a factor, and the factor is recognized to be a *proportional* scaling or blowing up. The cosmos is interconnected by and through the *similarity* of structures, and the scaling factor is revealed by the ratios expressible by comparative lengths and areas. The deep problem that lies enmeshed in these reflections is to grasp how figures can be similar without being equal; this understanding is presented in the hypotenuse theorem. *Similarity* is the principle of the structure of the microcosmic-macrocosmic argument—expansion and contraction, or perhaps we might even say "compression and decompression."[74] The harmony of the spheres is the macrocosmic projection of the musical ratios; it is the analogy and proportion, the *analogia*, of similarity. Once one sees the harmony of the spheres in this context, attributing the knowledge of the hypotenuse theorem to Pythagoras fits within this broad context. The cosmos is vastly interconnected; the structures that fill it are blow-ups or scale-ups of the right triangle and the areal relations it unfolds. The cosmos that results is a "putting together" (σύστασις) of triangles, to make the regular solids—the elements—out of which the entire cosmos is constituted.[75] This is why Pythagoras was focusing on the regular solids, whether or not he discovered or proved only three or all five.

Pythagoras and the Famous Theorems   179

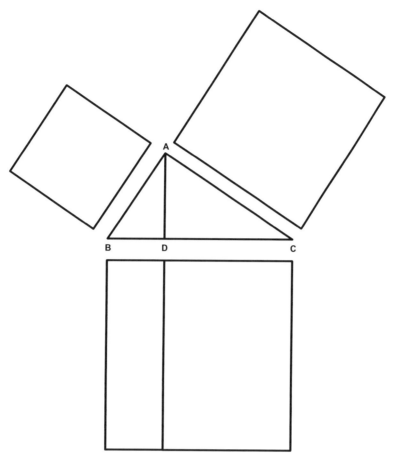

Figure 3.40.

Here, I should like to add one more point that is pivotal, and in this regard follows Heath exactly: "It must not be overlooked that the Pythagorean theory of proportions was only applicable to commensurable quantities. This would be no obstacle to the use of proportions in such a proof so long as the existence of the incommensurable remained undiscovered."[76] Nothing I am arguing here requires the supposition that Pythagoras grasped the existence of incommensurables at the time he learned first, and possibly developed metrically, the hypotenuse theorem. For in my estimation, this foray into geometry gripped his mind at least from the time he met Thales and Anaximander, or members of their schools, an encounter that not implausibly occurred in around 550 BCE. Pythagoras was a young man, and he might well have come to realize the existence of incommensurables later in his life, but the case I am constructing here does not suppose it. Once incommensurables were discovered, however, there would be a need, as Heath put it, "to invent new proofs independent of proportions in place of those where proportions were used."[77] This proof that Heath speculated might well have been the form in which Pythagoras proved the famous theorem, and that I have set out geometrically, is by proportions. And so what Euclid achieved in Book I.47 was to construct a proof of comparable mathematical intuitions—showing that the square on the hypotenuse is equal in area to two rectangles—*without* appeal to proportions.[78]

Thales's interest in geometry, spawned from and invigorated by practical problem-solving that led him to see that the world of our experience could be illuminated intelligibly by means of geometry—that there was an intelligible and stable pattern to things to be discerned behind or underlying the appearances—was driven to find the geometrical figure that underlies all others. If everything is ὕδωρ, then how can this unity nevertheless appear so differently and yet remain always and only ὕδωρ? *This problem is somewhat analogous to the deep meaning of the Pythagorean theorem: how can figures be similar without being equal?* Somehow, ὕδωρ must be expressible as some *structure* that is capable of morphing into different forms (that is, by some process, perhaps felting or compression),[79] into different shapes—similar but unequal. It is my considered opinion that Thales's intuition was that ὕδωρ found expression as a fundamental geometrical figure and that by combination and transposition it comes to appear divergently. As Thales named the basic stuff ὕδωρ, the only reality that underlies and truly *is* all appearances of it, he looked for the geometrical figure that was to the basic stuff, as the basic stuff was constitutive of all other things. As all things are fundamentally different expressions or different appearances of ὕδωρ, all appearances are imagined—similar but unequal—as articulations of flat surfaces; volumes are blow-ups or scaled-up three-dimensional projections of these flat surfaces, and all these surfaces are expressed as rectilinear figures, and every rectilinear figure is reducible to or dissectible into triangles, and all triangles are reducible to right triangles.

Thales's measurement of the height of a pyramid at the time of day when every vertical object casts a shadow equal to its height required that he fixate on isosceles right triangles. Only when the sun is inclined at half a right angle are the conditions right for two of the sides of the right triangle to be equal, because he noticed and stated that only when the base angles are equal are the sides opposite them equal in length. As he stood in Egypt, ready with the Egyptian surveyors holding the knotted cord to measure the length of the pyramid shadow, he stood at the end of the pyramid's shadow, or even nearby, with a gnomon or stake of a cubit measure in a whole integer (e.g., 3 cubits) and a marking to show precisely the same time when it casts a shadow equal to its length. When his gnomon, or his own shadow, was equal to its (or his) height, the pyramid's shadow was equal to its height. And however many times he connected the knotted cord from the tip of the gnomon's shadow to its top—so long as his gnomon was a whole number of cubits in length, 2, 3, 4, 5, 6 whatever, the measure of the hypotenuse, the length of the cord, was never a whole number. Because Thales was looking for the unity underlying the sides of a right triangle, and because he could not discover a metric or numerical connection or unifying rule, he looked for a different connection, a different underlying unity. The isosceles right triangle whose hypotenuse was never a whole number when the two sides of equal length were, had the peculiar advantage that the areal connection was apparent with an immediacy not shared while inspecting right triangles other than the isosceles, such as the 3, 4, 5 "Pythagorean triple" that we considered earlier. If one constructs squares on the three sides of the isosceles right triangle one sees rather immediately—if one is looking for connectedness—that while the squares on the two sides of equal length each reduce to two equal isosceles right triangles, the square on the hypotenuse reduces to four: the sum of the areas on the two legs is equal to the area of the square on the hypotenuse.

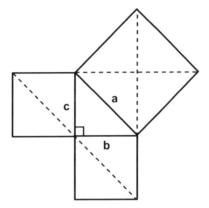

Figure 3.41.

Thales plausibly learned or confirmed everything about the areal connections among triangles, rectangles, and squares from his Egyptian hosts. But he did not yet see the metric connection, the numerical formula or rule, though he was almost certainly aware from the architects and builders that sometimes whole number relations fit a pattern as in the case of so-called "Pythagorean triples." I am not arguing that Thales imagined geometric diagrams without consideration of numbers, as we find later in Euclid. Euclid's *Elements*, through the first six books, contains no numbers at all—a point I have repeated several times for emphasis—and this seems doubtless because Euclid sought to geometrize arithmetic in light of the discovery of the irrational. The discovery of the irrational was the awareness that there were "lengths"—magnitudes—to which no number could be assigned, neither whole numbers nor fractions.[80] One key to its discovery was unlocking further the meaning of the mean proportional, for it is the length of the side of a square whose area is equivalent to the rectangle made up of the lengths of its "extremes."

I am not arguing that Thales was aware of irrationality in this important sense (though both he and Pythagoras were aware of the mean proportional between two lengths), nor is it part of this argument concerning Pythagoras and the hypotenuse theorem. Moreover, the measurements of the height of the pyramid or the distance of a ship at sea were practical problems for which *a number must have been proposed* as an answer. Thus, using the royal knotted cubit cord, Thales answered the question of the pyramid height in cubit measures, the very same cubit measures that the Rhind Mathematical Papyrus shows allowed the priests to reckon its height at least a millennium earlier. And the measurement of the distance of the ship at sea had to yield a number, in whatever stipulated units—Greek feet, ells, cubits, or whatever. But while it seems clear that in addition to geometrical reflections on magnitudes that Thales was forced to assign numbers to those lengths—and so the correlation of lengths and numbers were connected inextricably in practical problem-solving—his fixation on isosceles right triangles prohibited him from seeing the metric or numerical interpretation of the hypotenuse proposition.

This account, then, is the background of what Pythagoras learned from Thales or those in his "school," and that he formulated in diagrams with which he is credited; this knowledge was current contemporaneously with Thales in Ionia, and probably had circulated for decades. Thales may have received the title "*sophos*" as early as c. 582 BCE,[81] and the prediction of the solar eclipse, rightly or wrongly attributed to him, can be fixed plausibly in 585 BCE. Thus Thales was already renowned *before* the birth of Pythagoras. Did Pythagoras meet with Thales, and Anaximander? The reports of the meeting are late, as is so much of the evidence, but they are entirely plausible. When Pythagoras was more or less the age of our undergraduates, Thales was long considered one of the "seven sages." Heraclitus regards Pythagoras as among the "much-learned" ones, though for Heraclitus this counted as an insult rather than a compliment.[82] Had the young Pythagoras shown intellectual acumen as a young man, it seems likely that he would have visited Thales and Anaximander, who were nearby and easily accessible. And the doubts about Pythagoras's discovery *because* of the claim that he sacrificed an ox (or oxen)—from Cicero to Van der Waerden and Burkert—is entirely at odds with his being a vegetarian,[83] and needs to be reassessed in light of *when* he may have celebrated his discovery. As an elder statesman, vegetarian, and believer in reincarnation to include animal species, the objection has weight. But, as I am proposing, had Pythagoras's proof been celebrated as a young man, a *kouros* in Samos, animal sacrifice at the great altar of the temple of Hera would have been how a member of that community would have shown respect and gratitude for divine inspiration. For the celebration of Pythagoras's proof of the famous theorem was not simply the discovery of yet another proposition in a mathematics seminar, but the metaphysical discovery of the basic building block of the whole cosmos.

H

The "Other" Proof of the Mean Proportional: The Pythagoreans and Euclid Book II[84]

The geometrical problem of grasping the mean proportional is to understand how a rectangle can be equal in area to a square—squaring a rectangle. The line of proof—the lines of thought—that convinced Thales, and initially Pythagoras, were plausibly by ratios and proportions; we have explored this idea in detail in both the chapter on Thales and earlier here on Pythagoras. *Visualizing* areal equivalence between square and rectangle is different from *proving* it. In Thales's case, "proving" the mean proportional—that a square was equal to a rectangle—might well have been achieved empirically by use of a caliper or compass. A *geometrical proof* of this equivalence is shown in Euclid Book II.14.

It is customary to attribute the theorems of Euclid Book II to the Pythagoreans, and this certainly seems to be sound, but this program traces back to Pythagoras himself and Thales before him if we follow the diagrams. If readers will review the fourteen theorems in this shortest book of Euclid, perhaps they too will detect a single underlying theme, which I propose motivated initially the whole book, namely the *proof* for the construction of a square equal to a given rectangle—the search for the principle of the mean proportion, the geometric principle by which the right triangle expands and collapses in the cosmos. What I wish to emphasize

Pythagoras and the Famous Theorems   183

is that in Book II, the process of constructing a square equal to a rectangle is exactly how to construct a mean proportional *without the theory of proportions*. For, given any rectangle, a square figure that is equal to a rectangle has for its side a length that is the geometric mean between the two unequal side lengths. Here we have the underlying key to continuous proportions and the underlying principle by which transformational equivalence finds expression in figures of different shapes but the same area, all reducible to triangles (and hence right triangles). It is my surmise that this problem motivated what later became Book II in Euclid. In the proof of II.14, it relies on an understanding of the Pythagorean theorem of I.47. But this should be understood to mean that they mutually imply each other; the knowledge of the one presupposes the knowledge of the other. Analogously, in Book I, proposition 47 proves that "*In right-angled triangles the square on the side opposite the right angle equals the sum of the squares on the sides containing the right angle,*" while in proposition 48, the converse is proved, that "*If in a triangle the square on one of the sides equals the sum of the squares on the remaining two sides of the triangle, then the angle contained by the remaining two sides of the triangle is right.*" The knowledge of the one implies the knowledge of the other.

Before setting out to explore this proof sequence in Euclid II.14, I wish to mention two important theorems relevant to it that are achieved just before at II.5 and II.11. At II.5 it is proved that *If a straight line is cut into equal and unequal segments, then the rectangle contained by the unequal segments of the whole together with the square on the straight line between the points of section equals the square on the half.* Consider its diagram, below:

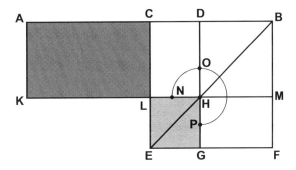

Figure 3.42.

The proof shows that when line AB is cut into two equal segments at C and unequal segments at D, the rectangle AL plus CH plus the square on LG is equal to square CF. For rectangle AL is equal to rectangle CM, since both are constructed on one half the line length.

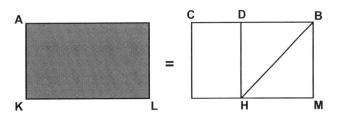

Figure 3.43.

184 The Metaphysics of the Pythagorean Theorem

And since the diagonal EB cuts square CF into two equal halves of which the complements CH and HF are equal (I.43), therefore rectangle AL plus rectangle CH is equal to rectangle CM plus rectangle HF,

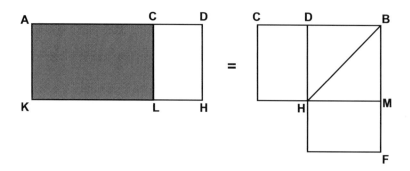

Figure 3.44.

and thus therefore the rectangle contained by AD,DB together with the square on CD is equal to the square on CB.

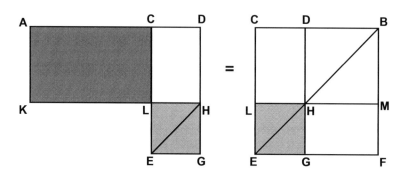

Figure 3.45.

Then at II.11 is a proof of this statement: *To cut a given straight line so that the rectangle contained by the whole and one of the segments equals the square on the remaining segment.* This proof is, in effect, a construction of the "golden section" or the "extreme-mean" ratio, which contains hints of the microcosmic-macrocosmic argument, for we have a line cut into two parts so that the ratio of the smaller to the larger is the same as that of the larger to the whole. This construction is pivotal, later, to the construction of the regular solids in Book XIII. *Elements* II.11, below, has the equivalent statement to VI.30 (*To cut a given finite straight line in extreme and mean ratio; definition VI.3: A straight line is said to have been cut in extreme and mean ratio when, as the whole line is to the greater segment, so is the greater to the less.*) that a line is cut so that the rectangle formed by the whole line and the small segment is equal to the square on the larger segment. At VI.30 the proof is by ratios and proportions, but not here.

\

Pythagoras and the Famous Theorems   185

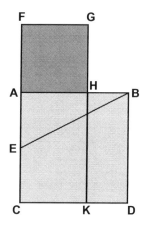

Figure 3.46.

We will not go through the whole proof here, but I wish to note that the proof requires that a line segment AB be cut at some point H so that AB is to AH as AH is to HB. The proof that these line lengths stand in extreme and mean proportion is to show areal equivalences, that the rectangle contained by the whole AB (= BD), which is the rectangle BD,HB is equal to the square on AH. And this offers us yet another way to see, in a book whose authorship traces back to the Pythagoreans, that the focus is on showing how and that a square is equal to a rectangle when a line is cut to produce a series of three proportions in extreme and mean ratios.

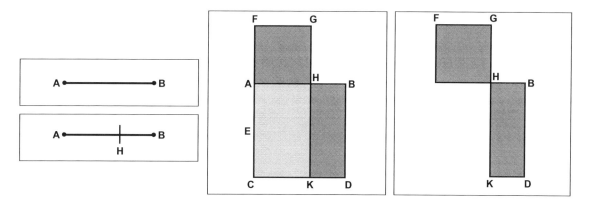

Figure 3.47.

And this proof supposes the Pythagorean theorem at I.47. Here I wish the reader would note that by appeal to visible and invisible lines, as Saito has discussed,[85] the relation between line lengths is illuminated by figures that can be imagined, projected on these line lengths.

186   The Metaphysics of the Pythagorean Theorem

Now we turn to II.14 to help the readers see more than the visualization, but the actual *proof* of how a square can be constructed equal to a rectangle, a mean proportional without an argument by ratios and proportions. Let us consider the very last theorem in Book II: *To construct a square equal to a given rectilinear figure.*

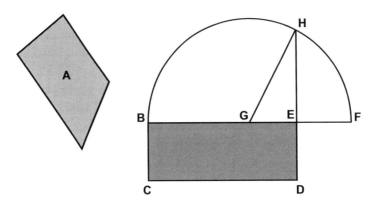

Figure 3.48.

I am now going to adapt this proof to my narrative, which requires shortening the last proof of Book II whereby a given *rectilinear* figure is expressed as a rectangle of areal equivalence and then that rectangle is expressed as a square of equal area. I truncate that proof and instead begin with a rectangle BD, with sides BE,ED, below. But the reader can see that the strategy is to create a mean proportional by constructing a semicircle and then extending a perpendicular from the diameter of the (semi-)circle to the circumference. It is a continued reflection on grasping that every triangle in a (semi-)circle is right, from Thales through Pythagoras and his school, but this proof makes no reference to similarity.

The proof is to construct a square so that the figure will contain sides BE and ED and yet have the same area as the original rectangle, below:

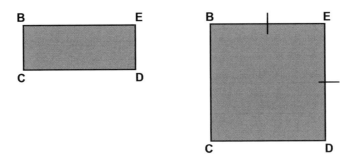

Figure 3.49.

We extend BE to F, below, so that EF is equal to ED; thus the extension EF is the width of ED, and then the whole length BF we bisect at point G, below:

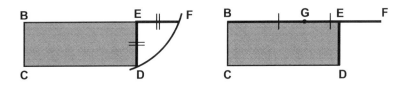

Figure 3.50.

Then, with center G and distance either GB or GF (they are equal), let the semicircle BHF be described (below left), and then let DE be produced to H. We will connect the width of the rectangle to the point on the semicircle H (below middle), and then we join center point G to H, below right:.

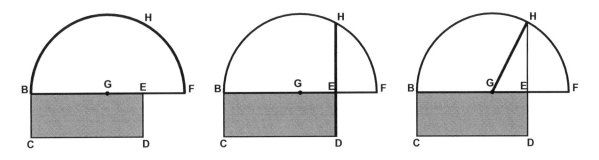

Figure 3.51.

Now, since the straight line BF has been cut into equal segments at G, and into unequal segments at E, the rectangle contained by BE, ED, together with the square on GE, below left, is equal to the square on GF, below right (this was proved at II.5):

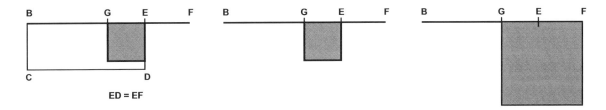

Figure 3.52.

188   The Metaphysics of the Pythagorean Theorem

But GF is equal to GH, since both are radii of the (semi-)circle, and therefore rectangle BE, EF (NB : EF = ED) together with the square on GE is equal to the square on GH, below:

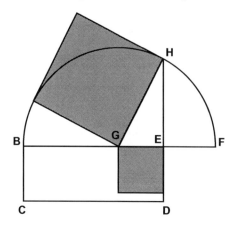

Figure 3.53.

But we know from the hypotenuse theorem that the squares of HE and EG are equal to the square on GH, below:

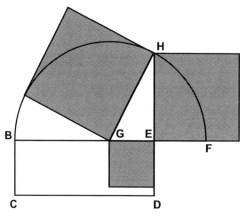

Figure 3.54.

Therefore, rectangle BE, EF, together with the square on GE, below left, is equal to the squares on HE, EG, below right:

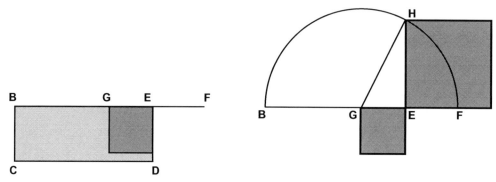

Figure 3.55.

Now let the square GE be subtracted from each; therefore the rectangle contained by BE, EF that remains is equal to the square on EH, below:

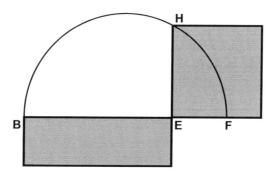

Figure 3.56.

The squaring of a rectangle played a pivotal role for the Pythagoreans in the transformational equivalence of appearances; it displayed an areal understanding of the mean proportional, the way the cosmos—both little and big—grows, that is, expands and collapses.

I

Pythagoras's Other Theorem: The Application of Areas

I.1 The Application of Areas Theorems at Euclid I.42, 44, and 45

From Plutarch we get a report that Pythagoras is connected with a theorem concerning the application of areas.[86] Knorr doubts that Plutarch's report could be correct since he places the construction of such a proof toward the end of the fifth century BCE.[87] It is not my argument that such a formal *proof* was constructed in the sixth century BCE, but rather that the general outlines of such a project were plausibly begun then. Once one realizes that the right triangle is the basic building block of all rectilinear figures, because all those figures dissect to triangles and all triangles contain within themselves right triangles, the project of imagining *how triangles can be transformed into figures of other shapes but with the same area* takes form. It is my contention that this is how the application of areas theorem was first grasped.

Plutarch regarded this theorem as a more important achievement than the hypotenuse theorem. Suppose then that Pythagoras was connected with it. Let us consider what it could have meant to him in the sixth century, long before there were any "Elements" of which we know.[88] To do so we shall explore the later, perfected results in Euclid, keeping in mind that this general project was, in nascent stages, plausibly embarked on long before.

The answer, then, seems rather straightforward once placed within the broad metaphysical project. The theorems, as we find them in Euclid—I.42, 44, and 45, and VI. 25, 28, and 29—all share the same foundation. They show how triangles can be expressed as parallelograms with equal area, how all rectilinear figures can be expressed as parallelograms of the same area, and how a triangle can be expressed as a similar triangle. If one focuses on the underlying metaphysical project, beginning with triangles as the basic building block of all appearances, they are part and parcel of a program to account for what I have been referring to as "transformational equivalences," how from a single underlying unity the world of diversity spawns structures of different shapes but *sharing the same area*, whose structures are all ultimately reducible to the right triangle. My case agrees with Heath on the conjecture that the theorem to which Plutarch was referring traces back plausibly to Pythagoras and his associates in what Euclid preserves as VI.25.

The proof sequence follows proposition I.42: *To construct a parallelogram equal in area to a given triangle.*[89] Parallelogram FB is the result having the same area of the triangle in any given angle. The technique for this construction was already shown earlier in the sequence from I.34 (below left) onward in chapter 2, and is separated out for detail, below right:

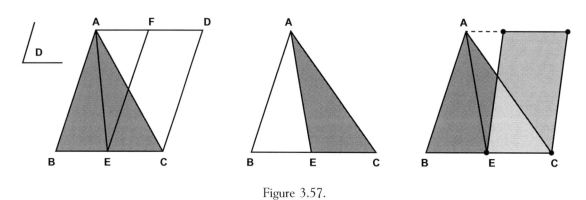

Figure 3.57.

The construction at I.42 begins with a specified triangle, in this case ABC, above left. The base BC is bisected at E and AE is joined dividing the original triangle into two equal triangles. Then, through A line AFD is drawn parallel to BC. Now, on the same base, EC, the parallelogram FC is constructed, and we already know that a parallelogram constructed between the same two parallel lines—for the sixth century it would have been enough to claim between *straight lines*—and on the same base as a triangle, the area of the parallelogram double that of the triangle is made (I.41). And since triangle AEC is half of the original triangle ABC, and on the same base, the parallelogram FC, so constructed, must be equal in area to the whole original given triangle ABC. Since this is true for any parallelogram, such as FECD, moreover, the parallelogram may be chosen so that angle FEC is equal to the given angle.

Next, following I.43, it is proved that the complements of the parallelograms about the diameter are equal to one another. The proof begins by dividing any parallelogram by its diagonal into two equal triangles (I.34), and parallelograms EH and GF can be constructed, making parallelograms HF and EG its compliments, below left:

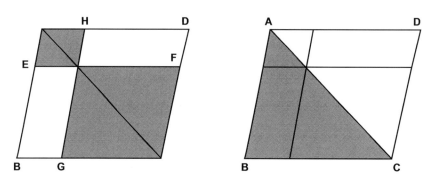

Figure 3.58.

Thus, the smaller two blue triangles are equal to one another, as are the two larger blue triangles—both are divided in half by the diagonal—and as we can see now, above right, the whole parallelogram is divided initially into two equal triangles, thus the compliments HF and EG must be equal to one another (above left): The task of I.44 is to begin with a triangle, C, in any specified angle, D, and then to construct a parallelogram, BL, on a given line, AB, equal in area to the given triangle.

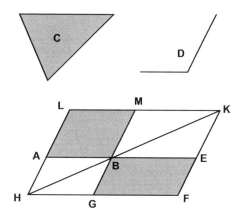

Figure 3.59.

To the given line is AB, parallelogram FB is produced equal to a given triangle, C (I.42), and then parallelogram BL, constructed on AB is shown to be equal in area to FB because the complements of the parallelograms about the diameter are equal to one another (I.43). What follows is the proof—by which I mean the line of mathematical intuitions—that secure the conclusion that might have been persuasive, even in a more rudimentary form, to the Greeks of the sixth century.

Having shown how a parallelogram can be constructed equal in area to a triangle (I.42) and that the complements of the parallelograms about the diameter are equal to one another (I.43), the proof of the "application of areas" in I.44 shows: The parallelogram FB is constructed equal in area to a given triangle in a specified angle, as a complement of parallelograms; the matching complement BL is constructed on the given line AB and is equal to it. Consequently, a parallelogram BL is constructed on a given straight line equal to a triangle in a specified angle. OED.[90]

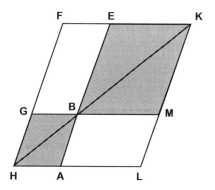

Figure 3.60.

192   The Metaphysics of the Pythagorean Theorem

Next, the task of I.45 is to extend the technique so that *any* rectilinear area, not only triangles, can be applied to a line in any given angle. The result is to *transform* any rectilinear figure into a parallelogram of whatever angle desired and with any one side of any specified length. This is what I mean by *transformational equivalences between appearances*. On a related point, I.45, below, answers the question "What is the area of this figure?"[91] And note carefully, *the area of any figure is determined by dissecting it into triangles*, just as I suggested in the Introduction was grasped by Egyptian land surveyors. Stated differently, all areas—all rectilinear figures—are built up out of triangles. And all triangles reduce ultimately to right triangles.

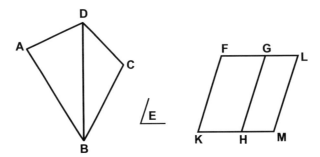

Figure 3.61.

The proof of I.45 begins with any rectilinear figure (below, left) in any specified angle and that figure is divided into triangles (below right).

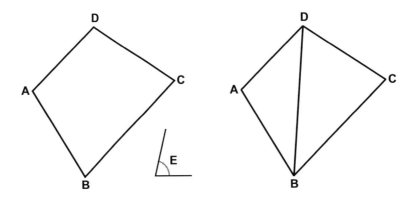

Figure 3.62.

Then, following I.44, parallelograms equal in area to the triangles are constructed, and those parallelograms are equal in area to the rectilinear figure. Thus, any rectilinear figure can be transformed into a parallelogram of whatever angle desired and with any one side of any specified length.

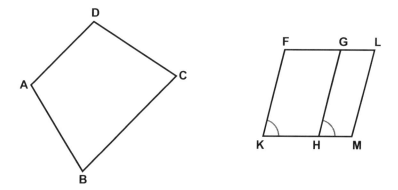

Figure 3.63.

I.2 The Application of Areas Theorems in Euclid VI.25, 28, 29 by Ratios and Proportions

Below, as Heath also conjectured, we have what seems to be the theorem explicitly attributed to Pythagoras by Plutarch, who records the following statement: "Two figures being given, to describe a third, which shall be equal to one and similar to the other. And it is reported that Pythagoras, upon the discovery of this problem, offered a sacrifice to the gods; for this is a much more exquisite theorem than that which lays down, that the square of the hypotenuse in a right-angled triangle is equal to the squares of the two sides."[92] The theorem is to construct a figure similar to one given rectilinear figure and equal to another.

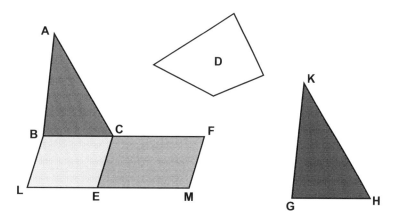

Figure 3.64.

Thus, the proof begins with a given triangle ABC in any angle and applies it to line BC to construct a parallelogram be equal in area to the given triangle.

194   The Metaphysics of the Pythagorean Theorem

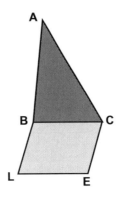

Figure 3.65.

Next, we take any rectilinear figure, D, divide it into triangles, and then construct a parallelogram equal to that rectilinear figure and apply it to line CE, and in a straight line with CF.

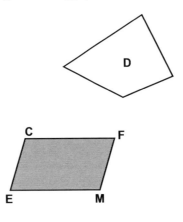

Figure 3.66.

Finally, we take a mean proportional GH—between BC, the base of the original given triangle, and CF, the side of the paralleogram whose area is equal to rectilinear figure D—and on it construct similar triangle KGH.

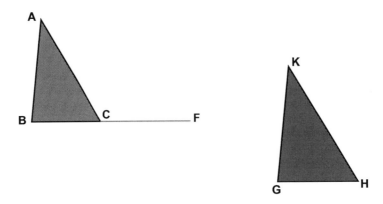

Figure 3.67.

Thus the proof is this: The given figure is triangle ABC and to it a similar figure KGH must be constructed whose area is equal to D, a rectilinear figure. Then, applied to BC is the parallelogram BE, which is equal to triangle ABC, and applied to CE is the parallelogram CM equal to D in the angle FCE, which equals the angle CBL. Since angles FCE and CBL are equal, BC is in a straight line with CF, and LE with EM. This much has already been established in the application of areas theorems that we already considered from Book I, 44 and 45.

Now, ratios and proportions are added; the mean proportional between BC and CF—GH—is taken (GH is the base of the triangle on the far right) and the triangle KGH is described as similar and similarly situated to ABC. Hence BC is to GH as GH is to CF, and if three straight lines are proportional, then the first is to the third as the figure on the first is to the figure on the second (*duplicate ratio*, Def. V.9). This means that the ratio of the line lengths of the first to the third (BC : CF) is the same ratio as the area of the figure on the first (ABC) to the similarly situated figure (area) described on the second (KGH). Therefore, BC is to CF as the triangle ABC is to triangle KGH. But BC is to CF as the parallelogram BE is to the parallelogram EF. Therefore, also the triangle ABC is to the triangle KGH as the parallelogram BE is to the parallelogram EF. And therefore, alternately, triangle ABC is to parallelogram BE as triangle KGH is to parallelogram EF.

But triangle ABC equals parallelogram BE, and therefore triangle KGH also equals parallelogram EF, and parallelogram EF equals D, therefore KGH also equals D.

And KGH is also similar to ABC. Therefore this figure KGH has been constructed similar to the given rectilinear figure ABC and equal to the other given figure D.   OED.

J

Pythagoras's *Other* Theorem in the Bigger Metaphysical Picture: Plato's *Timaeus* 53Cff

Let us now try to clarify how VI.25 fits within the metaphysics we have already discussed. The cosmos is filled with appearances, though there is some underlying unity such that there is an abiding permanence despite qualitative change in appearances. All differing appearances are to be accounted for by transformational equivalences; the deep intuition is that each appearance is related to every other by retaining equal area but in different rectilinear shapes; it seems to be a form of what we have come to refer to as the "principle of conservation of matter," though the essence of this underlying unity may be originally supersensible and capable of becoming visible and sensible.[93] Now, in the language of Plato's *Timaeus* 53Cff, those appearances all have volumes, and every volume (βάθος) has a surface (ἐπίπεδον).

> Πρῶτον μὲν δὴ πῦρ καὶ γῆ καὶ ὕδωρ καὶ ἀὴρ ὅτι σώματά
> ἐστι, δῆλόν που καὶ παντί. Τὸ δὲ τοῦ σώματος εἶδος πᾶν
> καὶ βάθος ἔχει. Τὸ δὲ βάθος αὖ πᾶσα ἀνάγκη τὴν ἐπίπεδον
> περιειληφέναι φύσιν, ἡ δὲ ὀρθὴ τῆς ἐπιπέδου βάσεως ἐκ
> τριγώνων συνέστηκε. Τὰ δὲ τρίγωνα πάντα ἐκ δυοῖν ἄρχεται
> τριγώνοιν, μίαν μὲν ὀρθὴν ἔχοντος ἑκατέρου γωνίαν,

196   The Metaphysics of the Pythagorean Theorem

> τὰς δὲ ὀξείας. ὧν τὸ μὲν ἕτερον ἑκατέρωθην ἔχει μέρος
> γωνίας ὀρθῆς πλευραῖς ἴσαις διῃρημένης, τὸ δὲ ἕτερον ἀνίσοις
> ἄνισα μέρη νενεμημένης. Ταύτην δὴ πυρὸς ἀρχὴν καὶ τῶν
> ἄλλων σωμάτων ὑποτιθέμεθα κατὰ τὸν μετ' ἀνάγκης εἰκότα
> λόγον πορευόμενοι. Τὰς δ' ἔτι τούτων ἀρχὰς ἄνωθεν θεὸς
> οἶδεν καὶ ἀνδρῶν ὅς ἐκείνῳ φίλος ᾖ.
>
> —Plato's *Timaeus* 53Cff.

In the first place, then, it is of course obvious to anyone that fire, earth, water, and air are bodies; and all bodies have volume (βάθος). Volume, moreover, must be bounded by surface (ἐπίπεδον), and every surface that is rectilinear (ὀρθή) is composed of triangles. Now all triangles derive their origin (ἄρχεται) from two triangles, each having one right angle and the other acute. Of these triangles, one has on either side half a right angle marked off by equal sides (the right-angled isosceles triangle); the other has the right angle divided into unequal parts by unequal unequal sides (the right-angled scalene triangle). This we assume as the first begininning and principles of fire and the other bodies, following the account that combines likelihood with necessity; the principles higher yet than these are known only to God and such men as God favors.

Thus, each surface is a plane—a flat surface that has been layered up or imagined as folded up to make the structure of that volume—and those flat surfaces are structured by straight lines and articulated by rectilinear figures. Each figure is reducible ultimately to triangles, and every triangle is reducible to a right triangle by dropping a perpendicular from the right angle to the hypotenuse, and further every right triangle divides ad infinitum into two similar right triangles—this last sequence is exactly the foundation of the proof of the "enlargement" of the Pythagorean theorem of Euclid VI.31. Then, the cosmos grows—literally unfolds areally from the right triangle—by certain increments describable by the mean proportional. To show this geometry, a triangle is transformed into a parallelogram, and to apply it to a given straight line (I.42, 44), some other rectilinear figure is transformed into a parallelogram (I.45) by constructing it out of triangles, and then (VI.25) a similar triangle is constructed on a line segment—a length—that is the mean proportional between the side of a parallelogram equal in area to the original triangle and another side of a parallelogram (in a straight connected line) equal in area to this other rectilinear figure. The triangle constructed on the mean proportional will be similar to the original triangle and equal in area to the second parallelogram, which was equal in area to the rectilinear figure. Thus Pythagoras's *Other* theorem—the application of areas, later formalized as Euclid VI.25—offers *the rules of transformational equivalence*—of structure, not process—from one appearance to another. Differences in appearances in the cosmos are not due to "change" but only to "alteration." How? The application of areas offers to explain how. The overall frame of just this project is described in Plato's *Timaeus* 54–56, the project of building all the primary, elemental bodies out of right triangles—isosceles and scalene.

Finally, I would like to end this section by a brief consideration of the application of areas theorems that Euclid places at VI.28 and 29. They are also generalizations on the problems in Book II, but they are part and parcel of the same, original metaphysical project. In Proclus's commentary on the first book of Euclid, the importance of the application of areas is discussed further with specific regard to what turns out to be theorems VI.28 and 29—the exceeding and falling short. For the metaphysical project, when things grow, they do not grow simply in quantum leaps—that is, in continuous proportions, though this I surmise was the underlying *rule* of growth—but rather incrementally, and so at any time might fall short or exceed a unit of growth that completes its transformational equivalence. That amount of excess or defect is addressed in VI. 28 and 29. Proclus writes as follows:[94]

> These things, says Eudemus (οἱ περὶ τὸν Εὔδημον), are ancient and are discoveries of the Muse of the Pythagoreans, I mean the *application of areas* (παραβολὴ τῶν χωρίων), their *exceeding* (ὑπερβολή) and their *falling short* (ἔλλειψις). It was from the Pythagoreans that later geometers [i.e., Apollonius] took the names, which again they transferred to the so-called conic lines, designating one of these *parabola* (application), another *hyperbola* (exceeding), and another *ellipse* (falling short), whereas those godlike men of old saw the things signified by these names in the construction, in a plane, of areas upon a finite straight line. For when you have a straight line set out and lay the given area exactly alongside the whole of the straight line, then they say you *apply* (παραβάλλειν) the said area; when however you make the length of the area greater than the straight line itself, it is said to *exceed* (ὑπερβάλλειν), and when you make it less, in which case, after the area has been drawn, there is some part of the straight line extending beyond it, it is said to *fall short* (ἐλλείπειν). Euclid, too, in the sixth book, speaks in this way both of *exceeding* and *falling short*; but in this place he needed the application simply, as he sought to apply to a given straight line an area equal to a given triangle in order that we might have it in our power, not only the construction (σύστασις) of a parallelogram equal to a given triangle, but also the *application* of it to a finite straight line.

In the case of VI.28: *To apply a parallelogram equal to a given rectilinear figure to a given straight line but falling short by a parallelogram similar to a given one; thus the given rectilinear figure must not be greater than the parallelogram described on the half of the straight line and similar to the given parallelogram.* "When this proposition is used, the given parallelogram D usually is a square. Then the problem is to cut the line AB at a point S so that the rectangle AS by SB equals the given rectilinear figure C. This special case can be proved with the help of the propositions in Book II."[95]

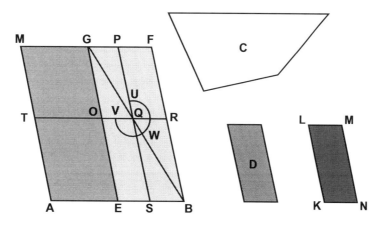

Figure 3.68.

198   The Metaphysics of the Pythagorean Theorem

In the case of VI.29: *To apply a parallelogram equal to a given rectilinear figure to a given straight line but exceeding it by a parallelogram similar to a given one.* This proposition is a generalization, again from the Pythagorean inspired Book II.6, where the figure presented is a square.

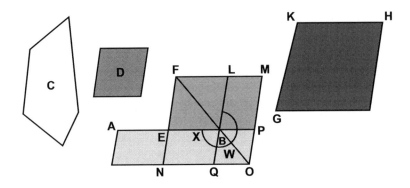

Figure 3.69.

K

Pythagoras and the Regular Solids: Building the Elements and the Cosmos out of Right Triangles

K.1 The Role of the Cosmic Figures in the Big Picture: Proclus's Insight into the Metaphysical Purpose of Euclid

These investigations of the application of areas, then, lead from the famous hypotenuse theorem to the the construction of regular solids, another project with which Proclus credits Pythagoras himself—he says that Pythagoras "put them together" (σύστασις).[96] This putting together of the regular solids, as Heath noted, is precisely what is described in Plato's *Timaeus*, and formalized in Euclid's last book, XIII, specifically in the "Remark" at the end of the last theorem, XIII.18. As Heath assessed the matter, he denies that Pythagoras or even the early Pythagoreans could have produced a complete theoretical construction such as we have in Euclid's Book XIII, but finds it perfectly plausible that that the σύστασις of the regular solids could have been effected by placing regular polygons around a point to form a solid angle, in a manner of trial and error. By such experimentation, it would be discovered that if four or more right angles were brought together around a point, they could not fold up into a solid angle, and thus could not fold up to create a regular solid.

The elements of the cosmos are constructed out of geometrical figures, each of which is built out of right triangles.[97] How does this take place? So far, we can see how triangles can be constructed as parallelograms with equal areas, rectilinear figures can be constructed as parallelograms and triangles, and parallelograms can be constructed as similar figures. The key to constructing a triangle similar to another triangle and equal in area to another rectilinear figure is an organic principle of growth—the geometric mean or mean proportional—that is expressed in ratios and proportions. The metaphysics of this geometry is revealed in the "Pythagorean theorem," VI.31. Timaeus refers to this in Plato's dialogue as a "continued proportion" (συνεχής ἀναλογία); the continued proportion "makes itself and the terms it connects a unity in the fullest sense" is described as "the most beautiful of bonds" (δεσμῶν κάλλιστος).[98] In Euclid XIII, the regular solids are born in the realm of proportion; the proportion determined by *the extreme and mean ratio*. This places the regular solids in a special place as a continued proportion (συνεχής ἀναλογία) whose three terms are the segments of a line and the whole line. And it is at least suggestive that Plato calls a continued proportion that "makes itself and the terms it connects a unity in the fullest sense," the "most beautiful of bonds" (δεσμῶν κάλλιστος).[99]

Proclus claims that the whole purpose of Euclid's *Elements* is to reach the construction of the regular solids. He does so no fewer than four times:[100]

> Euclid belonged to the persuasion of Plato and was at home in this philosophy; and this is why he thought the goal of the *Elements* as a whole to be the construction of the so-called Platonic figures.[101]

> If now anyone should ask what the aim of this treatise [i.e., Euclid's *Elements*] is. . . . Looking at its subject matter, we assert that the whole of the geometer's discourse is obviously concerned with the cosmic figures. It starts from the simple figures and ends with the complexities involved in the structure of the cosmic bodies, establishing each of the figures separately but showing for all of them how they are inscribed in a sphere and the ratios that they have with respect to one another. Hence some have thought it proper to interpret with reference to the cosmos the purposes of individual books and have inscribed above each of them the utility it has for a knowledge of the universe.[102]

> This then is its [i.e., Euclid's *Elements*] aim . . . to present the construction of several cosmic figures."[103]

> Its [i.e., Euclid's *Elements*] usefulness contributes to the study of the primary figures."[104]

Proclus's assessment of the purpose of Euclid's *Elements*, to construct the regular solids, the cosmic figures, has met with disapproval from mathematicians and historians of ancient mathematics. Fried points out that Proclus's view was rejected by distinguished mathematicians such as Simson, Playfair, and Legendre, and also by distinguished historians of mathematics such as Heath (and Morrow).[105] Fried identifies the kinds of objections that have been raised, starting with Heath's: (i) while it is true that the *Elements* ends with the construction of five regular solids in Book XIII, "the planimetrical portion has no direct relation to them, and the arithmetical no relation at all; the propositions about them are merely the conclusion of the stereometrical division of the work.[106] Fried adds, (ii) to accept Proclus's view, Book XIII must take precedence over Book X on incommensurables, Book V on the theory of proportion, and Books VII–IX on the theory of numbers. (iii) We must come to regard the substance of the learning of geometric techniques in Books I, III, IV, and VI as less central than that of Book XIII. And (iv), the deductive structure of Book XIII does not rely widely on propositions from all other parts of the *Elements*.[107] There can be no doubt that these are formidable objections to Proclus's position. Mueller, however, is less dismissive; he tries to mount a defense of Proclus's position that the construction of the regular solids is Euclid's goal (*telos*) in the *Elements*, but in the end even he does not share Proclus's professed insight.

> Proclus' remark is clearly due to his desire to associate Euclid with Plato, who used the regular solids in his *Timaeus*. However, although from the point of view of deductive structure[108] the remark is a gross exaggeration, one can see how book XIII might have led Proclus to make it. For in book XIII Euclid makes direct use of material from every other book except the arithmetic books and book XII; and XIII is ultimately dependent on the arithmetic books because book X is. In this sense the treatment of the regular solids does not constitute a synthesis of much of the *Elements*, a culmination of the Euclidean style in mathematics. However, the significance of the *Elements* lies less in its final destination than in the regions traveled through to reach it. To a greater extent than perhaps any other major work in the history of mathematics, the *Elements* are a mathematical world.[109]

It seems to me, however, that Proclus's assessment is right on the mark, and the tour de force that his comments represent suggest that he saw this underlying metaphysical narrative connected through the hypotenuse theorem of VI.31, the application of areas, and the construction of the regular solids. Those critical of Proclus's assessment object that his commitments to a certain view of Platonism colored his view of Euclid and impose a background in metaphysics where Euclid shows none. But if one looks to the *Timaeus* 53Cff—the construction of the cosmos out of the regular solids whose elements (στοιχεῖα) are ultimately reducible to right triangles—and comes to see the metaphysics of the ratios and proportions of the hypotenuse theorem of VI.31, and that the construction of regular solids are born in the realm of proportions of extreme and mean ratios that are central to it, Proclus's interpretation exhibits an insight missed by his critics. To interpret Euclid's *telos* as the construction of the regular solids is to grasp that Euclid's *Elements* can be read to reflect a metaphysics of sixth- and fifth-century origins; this is the metaphysical skeleton upon which Euclid fleshed out his system. Proclus's assessment is a result of reading sensitively the X-ray.

The lost narrative I am proposing to have recovered can be seen by starting with Plato's *Timaeus* and looking back through a train of thoughts—mathematical intuitions—that precede it. The "enlargement" of the hypotenuse theorem of VI.31, and the vision that the right triangle is the fundamental geometrical figure, stand at the beginnings of this development. In these earliest stages, the whole vista of the mathematical issues, interrelations, and concerns could hardly have been appreciated. Thales began with an intuition that there was a unity that underlies everything; all appearances of plurality are mere alterations of what is abidingly permanent. An account of cosmic reality includes an explanation of structural differences in appearances, and an explanation of the process by means of which these structural changes are effected. In the dim light that illuminates these earliest episodes, it seems to me that Thales's originating speculations were taken up by Anaximander and Anaximenes, on the one hand, and Pythagoras and his associates, on the other, though of course more were responding to these ideas. If we do not lose sight of this underlying metaphysical problem, then arguments bringing together both Thales and Pythagoras with the hypotenuse theorem merit a fresh hearing.

K.2 Did Pythagoras Discover the Cosmic Figures?

The launching point for the contemporary debate about connecting Pythagoras with the regular solids is an essay by Waterhouse. The old story had been to accept the testimony by Proclus that Pythagoras "discovered the structure of the cosmic figures" (i.e., the regular solids, κοσμικὰ σχήματα).[110] But the challenge has been to reconcile this testimony with a scholium to Euclid that reads "In this book, the thirteenth, are constructed the five figures called Platonic, which however do not belong to Plato. Three of these five figures, the cube, pyramid, and dodecahedron, belong to the Pythagoreans; while the octahedron and icosahedron belong to Theaetetus."[111] The problems presented by these two conflicting reports include that, if the Scholium is to be accepted, then the dodecahedron, a more complicated geometrical figure, was discovered before the octahedron; but the octahedron is only two pyramids put back-to-back with a square base. As Guthrie put it, the construction of the octahedron could have been accomplished based on principles known long before Theaetetus.[112] How shall we resolve this apparent inconsistency of placing the more complicated construction before the simpler one, and in the process sort out what belongs to Pythagoras and the Pythagoreans?

Waterhouse's proposal is that we must review the context; we must first focus on when the *idea* of the regular solids—as a group of geometrical figures—was first grasped. As Waterhouse points out, the cube and pyramid had been the subject of investigations for a long time; after all, geometry for the Greeks began with Thales's measurement of a pyramid. But the mention of the octahedron, icosahedron, and dodecahedron is invariably connected to discussions of the regular solids, and not elsewhere. The earliest definition of the "regular solids" occurs in Plato's *Timaeus* where the identification of the figures is connected with their relation to a circumscribed sphere.[113] Waterhouse urges us to see that the idea of a "regular solid"—figures whose faces are all the same and divide a sphere into equal and similar figures—is not something that can be generalized plausibly from any one of these figures. Through a process of generalization and abstraction, the idea must have dawned, and then our old friends started to look for them. It is Waterhouse's contention that the knowledge of the regular solids developed in two stages, the cube and pyramid being investigated first and then expanded on. Interest in the dodecahedron seems to have been a separate curiosity at first, and the existence of various minerals that exhibit dodecahedra and cubic crystals might well have sparked an interest in them by the Pythagoreans living in southern Italy.[114] Perhaps because icosahedral crystals are much less common, they were not considered earlier.

Waterhouse conjectures, then, that the octahedron should be placed in the second stage. He reasons, following Heath, that since it was a double pyramid sharing a square base, there was no particular compelling reason to investigate the octahedron more closely until the idea of a group of figures—regular solids—had been grasped through abstraction. The octahedron gets singled out only when someone goes looking for regular solids of these specific descriptions. And so Waterhouse argues his point, analogously, by referring to the discovery of the fifth perfect number: "Some Babylonian accountant may well have written down the number 33,550,336; but he did not observe the property that distinguishes this number from others. Similarly, the octahedron became an object of special mathematical study only when someone discovered a role for it to play."[115] Thus, Waterhouse's conclusion is to accept the Scholium and reject the report by Proclus. If we do so, what clarity does this bring to connecting, or disconnecting, Pythagoras himself with the regular solids?

Waterhouse's thoughtful thesis proposes to answer a question that by its focus misses something crucial that the approach here seeks to illuminate. The right *first* question to ask, accepting the thrust of Proclus's report, is why Pythagoras is exploring the regular solids at all? An analogous scholarly problem we considered already with Thales: what does it mean to place Thales "at the beginnings of Greek geometry?" If "Greek geometry" is demarcated by "formal, deductive proof," then we are looking for one story; if we see it as supplying a way to search for the fundamental structure that underlies permanently despite qualitative change—the fundamental geometrical figure—we are looking for a very different story, a very different context. Geometry was no less a search for the One over Many than the search for ὕδωρ, ἄπειρον, or ἀήρ—the One over Many in nature—because they are both part of the same project. The endless wrangling of whether, and to what extent, Thales "proves" the theorems with which he is connected, is wrongheaded anyway, as fascinating as the question is. I have already taken a stand, along the lines of van der Waerden, that Thales returned from his travels with various formulaic recipes to calculate heights and distances, lengths and areas, volumes and measures, and in reply to his querulous compatriots Thales sought to prove to himself, and to them, why these formulas and recipes were sound. At these early stages, some attempts at demonstration were empirical and some became more general and abstract; it makes sense to suppose that the more Thales sought to persuade his compatriots about the soundness of some formulas, the more aware he became of rules and principles exhibited in his replies. In this process of seeking to persuade himself and others of the soundness of these formulas, Thales and his retinue began to discover what were later called "elements" of geometry, more general principles and what followed from them. The debates about whether Thales learned geometry from the Egyptian priests, and what they may have taught him about deductive reasoning, again bypass the approach that offers us a chance for so much more insight. Probably, he did *not* learn deductive techniques from the Egyptians—this was his original innovation as he sought to persuade himself and his compatriots of these formulas—but from the surveyors and/or priests who recorded the surveyors' results, from whom he learned or confirmed that all of space could be imagined as flat surfaces ultimately reducible to triangles to calculate areas. And we have also explored the hypothesis that a comparable geometry in Egypt underlay the measurements of areas, volumes, and triangulation in their mathematical texts. What Thales did with the narrative that all triangles are reducible ultimately to right triangles may be uniquely his own, but substantive reasons encourage us to see Egypt supplying or confirming the primacy of triangles and their inextricable connection with rectangles and squares. As we already considered in the Introduction and the chapter on Thales, the solution to determining the area of a *triangular plot of land* (problem 51) was to find the *triangle's rectangle*; the Rhind Mathematical Papyrus urges us to see the visual solution to this calculation of area in the unusual phrase: "There is the triangle's rectangle." And so, we should not be lost in this abstract discussion of geometrical principles that land surveying and monumental architecture opened the way for extended communities in Egypt, and then in Greece, to reflect on geometrical principles. Indeed, in architecture, the right angle was perhaps the central most important principle for building. The very possibility of keeping walls erect, of holding up the roof, required and understanding of the right angle and the ability to secure it. The right angle, holding up the roof—the symbolic sky—literally held up the cosmic canopy. But, in any case, as Thales worked out the metaphysics of this geometrical insight, his project took on a character that seems purely a Greek invention and innovation.

Thus, we need to think again about Proclus's report that Pythagoras "put together" the cosmic figures in the context that he adopted this part of Thales's project and was trying to figure out how right triangles could be combined and repackaged to produce the world of appearances, literally built up and out of them. The question of whether he "proved" these solids—fascinating as is the question—misses the bigger story of *why* he is looking for them, and with that the plausibility of connecting Pythagoras with them. The case I propose to argue now is that in the sense of Proclus's term σύστασις, Pythagoras might certainly have "put together" the regular solids by folding them up from articulated rectilinear figures on flat surfaces. Waterhouse's conjecture is that the discovery of regular solids took place in two stages: in the first, the tetrahedron, cube, and dodecahedron were identified, and later the icosahedron and octahedron were identified only when the very idea of "regular solids" became conceptualized and so a project to investigate. Waterhouse is focused on the *proof* of these figures; I am arguing that Pythagoras is connected with their *discovery*. Pythagoras was trying to figure out how the right triangle—already grasped as the fundamental geometrical figure—unfolds into other figures, as atoms form molecules, how the fundamentals of the "little microscopic world" unfold into the "bigger macroscopic world." The project as it is related in Plato's *Timaeus*—building the elements of the cosmos out of right triangles—was the project that Pythagoras adopts (in my estimation inherited from Thales), and this is perhaps why Plato selects Timaeus—probably Timaeus of Locri, in southern Italy, in a dialogue that contains many Pythagorean themes—to narrate the εἰκὸς λόγος. This is how the project began to unfold; he was looking to see how the cosmos—imagined to be built out of flat spaces structured by straight lines and articulated at the most fundamental level by right triangles—could be unfolded into figures each of whose sides was the same uniform structure, creating a solid angle, and folded up and repackaged to form the building blocks of all other appearances. My opinion is that Proclus has it right, that Pythagoras "put together" the regular solids; it might have been Pythagoras's followers who proved three of the constructions and Theaetetus who proved the other two, but the "putting together" of them is plausibly Pythagoras's own doing. I now propose to show how this putting together may have been accomplished.

When we credit Pythagoras himself, as does Zhmud, with the theorems about odd and even numbers, and the musical intervals in the ratios of octave, fifth, and fourth, and the hypotenuse theorem, we see mathematical knowledge that brings together number, lines, areas, and a correlation between numbers and shapes—figurate numbers—in the μαθήματα he developed. Pythagoras's inspired vision was to see all things displaying numerical relations in terms of ratios and proportions; this was the most knowable underlying unity. And even though Zhmud regards the doctrine "All is number" to be a double reconstruction, there is still something fundamentally correct in attributing to Pythagoras a commitment to "number" being the ἀρχή of all things. His identification of numbers with shapes perhaps opened a way to see a numerical law-like rule underlying the sides of a right triangle. Pythagoras distinguished triangular numbers from square ones, and no doubt pentagonal numbers along with others. Becoming clear about relating numbers to shapes might well have provided the bridge between the sides of a right triangle and their assignment of numerical or metric values.

The sense of how Pythagoras *discovered the regular solids* needs to be clarified. Proclus writes that ὃς δὴ καὶ τὴν τῶν ἀναλόγων πραγματείαν καὶ τὴν τῶν κοσμικῶν σχημάτων σύστασιν ἀνεῦρεν, reading ἀναλόγων for ἀλόγων[116]—"[Pythagoras] discovered the theory of proportions and the putting together of the cosmic figures."[117] He discovered the theory of proportions—I am supposing this to mean that at least he grasped a theory behind proportional relations like those exhibited by Thales's measurement of the pyramid height and distance of a ship at sea, and probably the addition of proportions that is preserved at Euclid V.24, a key step in the enlargement of the Pythagorean theorem at VI.31—and the σύστασις of the cosmic figures, the regular solids. The cosmic figures are born in the realm of proportions, and so is the hypotenuse theorem in VI.31. Along the lines of thought that have been set out here, Pythagoras adopted Thales's program that placed the right triangle as the fundamental geometrical figure; by repackaging it, all compound appearances are produced. How does this happen? It is my conjecture that Pythagoras experimented with laying out triangles, arranging, rearranging, and folding them up from flat surfaces to form volumes. After some time, Pythagoras came to the mathematical intuition that there were intermediary elements out of which all appearances were produced. Consequently, he came to the project of looking for *elementary figures*—composed out of right triangles—from which the cosmos was made. To make this transition from flat surfaces to volumes, he focused on "solid angles"—στερεὰ γωνία. This means Pythagoras was experimenting by laying out combinations of triangles on flat surfaces, and folding them up, to see which would and which would not do so. In the process, what I believe he discovered was the general principle, later formulated by Euclid as XI.21, that to construct a *solid angle* around a point, the combined angles must always be *less than* four right angles. Below, is the proof as it later appears in Euclid XI.21:

Proposition 21

Any solid angle is contained by plane angles whose sum is less than four right angles.

Let the angle at A be a solid angle contained by the plane angles BAC, CAD, and DAB.

I say that the sum of the angles BAC, CAD, and DAB is less than four right angles.

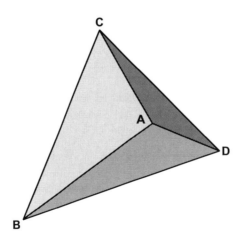

Figure 3.70.

Take points B, C, and D at random on the straight lines AB, AC, and AD respectively, and join BC, CD, and DB. **XI.20**

Now, since the solid angle at B is contained by the three plane angles CBA, ABD, and CBD, and the sum of any two is greater than the remaining one, therefore the sum of the angles CBA and ABD is greater than the angle CBD.

For the same reason the sum of the angles BCA and ACD is greater than the angle BCD, and the sum of the angles CDA and ADB is greater than the angle CDB. Therefore the sum of the six angles CBA, ABD, BCA, ACD, CDA, and ADB is greater than the sum of the three angles CBD, BCD, and CDB. **I.32**

But the sum of the three angles CBD, BDC, and BCD equals two right angles, therefore the sum of the six angles CBA, ABD, BCA, ACD, CDA, and ADB is greater than two right angles.

And, since the sum of the three angles of the triangles ABC, ACD, and ADB equals two right angles, therefore the sum of the nine angles of the three triangles, the angles CBA, ACB, BAC, ACD, CDA, CAD, ADB, DBA, and BAD equals six right angles. Of them the sum of the six angles ABC, BCA, ACD, CDA, ADB, and DBA are greater than two right angles, therefore the sum of the remaining three angles BAC, CAD, and DAB containing the solid angle is less than four right angles.

Therefore, *any solid angle is contained by plane angles whose sum is less than four right angles.*

Q.E.D. = Οπερ εδει δειξαι (= **thus it has been proved**)

It is on the principle of this theorem that "Any solid angle is contained by plane angles whose sum is less than four right angles" that Pythagoras and his retinue, through trial and error, plausibly discovered the regular solids. He was not at first looking for "regular solids"; *he was looking for intermediary figures compounded out of right triangles, to account subsequently for multifarious appearances.* His search for areal equivalences—different shaped figures with equal areas—led him to formulate the application-of-areas theorem, and that was also part of his broader project of explaining *transformational equivalences.* He latched onto the idea that the figures that seemed to serve this purpose were regular polyhedrons; they were solids with flat faces, each of which was the same size and shape, whether or not at this early time he imagined that they would divide a sphere into which they were inscribed into equal and similar figures. The deep intuition is driven by geometrical similarity: the little world shares the same structure as the big world. The little world dissects ultimately into right triangles, and packaging and repackaging of them constructs the elements of the big world. To explore the cosmos microscopically, or macroscopically, is to discover the same basic structure.

Now, the argument that follows is the last one in Euclid's *Elements*. In some ways, it should count as yet another theorem, and it has been suggested that it might well be treated as XIII.19. But in the text, it survives as a "Remark," thanks to later editors, but it follows the proof at XIII.18, "*To set out the sides of the five regular figures and compare them with one another.*" It is not set off, as are other theorems in Book XIII, as a porism. It simply follows as a remark or reflection. But it contains, I suggest, the basic thought of Pythagoras and the early Pythagoreans. It is for this reason that Proclus is able to report on some ancient authority, probably Eudemus, that Pythagoras "put together" the regular solids.

Remark

I say next that no other figure, besides the said five figures, can be constructed which is contained by equilateral and equiangular figures equal to one another.

For a solid angle cannot be constructed with two triangles, or indeed planes. **XI.21**

With three triangles the angle of the pyramid is constructed, with four the angle of the octahedron, and with five the angle of the icosahedron, but a solid angle cannot be formed by six equilateral and equiangular triangles placed together at one point, for, the angle of the equilateral triangle being two-thirds of a right angle, the six would be equal to four right angles, which is impossible, for any solid angle is contained by angles less than four right angles.

For the same reason, neither can a solid angle be constructed by more than six plane angles.

By three squares the angle of the cube is contained, but by four it is impossible for a solid angle to be contained, for they would again be four right angles.

By three equilateral and equiangular pentagons the angle of the dodecahedron is contained, but by four such it is impossible for any solid angle to be contained, for, the angle of the equilateral pentagon being a right angle and a fifth, the four angles would be greater than four right angles, which is impossible.

Neither again will a solid angle be contained by other polygonal figures by reason of the same absurdity.

Let us continue to explore further the regular solids to see how the σύστασις of them might have been achieved visually, and note that nothing is said about the suitability of any of these figures to be inscribed in a sphere so that it divides it into equal and similar figures. We now try to imagine the trials and errors by which the *elemental* figures—tetrahedron, hexahedron, octahedron, and icosahedron—may have been discovered. But now I draw the readers' speculative attention to the archaic period, at the very beginnings for the Greeks and not the end of the classical period when Euclid flourishes and publishes. In Euclid, while it is true that there is often a kind of reduction of solid problems to planar ones, with the exception of the dodecahedron, the solids are always conceived of as figures inscribed in a sphere. But I suspect that this is not how the geometry of the cosmos was imagined in the middle of the sixth century with Thales and Anaximander, and young Pythagoras after them, when geometrical diagrams and techniques first suggested themselves as a way to penetrate the fundamental structure of the cosmos. Already from Thales's pyramid measurements in Egypt, the pyramid was probably imagined as flat triangles folded up, and faceted stones and geodes may have confirmed this vision of an inner geometrical structure of all things, seen and unseen.

Let us now try to visualize how the regular solids might have been imagined as "folded up." As Thales sat full of wonder gazing on the pyramids, he might surely have imagined them as equilateral triangles each of whose sides is folded up to meet at the apex, below left.[118]

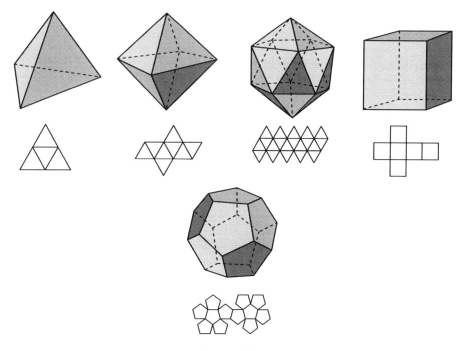

Figure 3.71.

It makes sense to suppose that Pythagoras first explored triangles and their compounds. A regular triangle is an equilateral triangle; it has three faces of the same shape and size, and three angles of the same size (we would say 60° each) that total two right angles. Of course Pythagoras knew that the sum of the angles of a triangle equals two right angles because, as I have argued, Thales before him knew this, and he learned many things about geometry from Thales and Anaximander or members of their school, though credit for the proof is explicitly given to the Pythagoreans by Eudemus. The point is that every triangle reduces to a right triangle, and any right triangle put together with its mirror image produces a rectangle. Every rectangle has four right angles, and every diagonal through the rectangle divides it back into two right triangles: there must be in half the figure two right angles if there were four in the whole figure. I argued in the introduction and chapter 2 on Thales that this is probably the meaning of Geminus's report that the "ancients" (= Thales) investigated how there were two right angles in each species of triangle. Thus, three of these triangles brought together around a single point to make a "solid angle" by gathering three of these equal angles together, would sum to two right angles. To see this, let us take a solid angle of this description and flatten it out, and it is obvious that they can fold up:

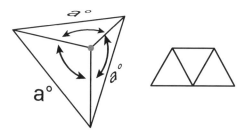

Figure 3.72.

Therefore, if there are two right angles in every triangle, then three of these equal angles placed together around a point would be equal to the angles in one triangle. And they can fold up into a solid figure.

If four such triangles are placed together, these four equal angles sum to two right angles plus a third of two right angles (we would say 180° plus 60°= 240°), that is, more than two but less than three right angles. And if we place these angles together around a single point making a solid angle and flatten it out, below, it also can be folded up.

Figure 3.73.

Finally, in the explorations of regular, equilateral triangles, five such equilateral triangles brought together around a single point make a solid angle of more than three right angles but less than four right angles (we would say 180° + 60° + 60° = 300°). And these angles put together around a single point make a solid, flattened out, below, that also can be folded up.

Figure 3.74.

But this is the end of possibilities for the equilateral triangle, for six of them placed together around a single point cannot be folded up; at the limit of four right angles, the shape, flattened out, has no "gaps" for it to fold up. The sum of the angles of an equilateral triangle gathered around a single point must be less than four right angles if they are to fold up. So, here are three of the four elemental solids, the regular solids. When three equilateral triangles meet we have the tetrahedron, with four we have the octahedron, with five we have the icosahedron.

When three squares meet we have a solid angle around a single point that contains three right angles.

Figure 3.75.

Two of these right angles will not fold up to contain a figure, and four right angles produce a solid angle that will have no gap or space to allow the folding up of them, and Pythagoras noticed again that four such angles would be equal to four right angles; the possibility of a solid figure with equal faced sides must be less than four right angles. And, of course, each face of the square dissects to two isosceles right triangles.

The dodecahedron is constructed of regular pentagons, each of whose angles is slightly more than a right angle (= 108°). Two of these angles placed around a single point will not contain a closed figure, but three will; the sum of these angles is less than four right angles (=108° + 108° + 108° = 324°), and they will fold.

Figure 3.76.

More than three of these angles gathered together around a single point will not fold up, and by reflection we can see that they would exceed four angles.[119]

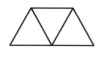    

Figure 3.77.

Thus, I propose that the *discovery* of these five figures—that later came to be known as "regular solids" or "Platonic solids"—came about in an empirical way by trial and error, starting with the intuition that appearances in this cosmos were ultimately reducible to right triangles and by compounds of triangles, and the squares and pentagons into which these figures dissect; solid angles could be formed and consequently figures folded up from them. All these figures have areas that reduce to triangles, and every triangle reduces ultimately to a right triangle. It may very well be that Pythagoras's intuition did not include the awareness that these five unusual figures can be inscribed in a sphere so that each of their points touches the sphere and divides it into equal and similar figures, and there is no mention of this in Proclus's report. The scholium to Euclid claims that the tetrahedron, cube, and dodecahedron were discovered by the Pythagoreans while the octahedron and icosahedron were discovered by Theaetetus. In

210   The Metaphysics of the Pythagorean Theorem

light of our presentation here, I think something else is going on. Pythagoras "put together" the regular solids in the manner I have described, through trial and error, and recognized the general principle that the limits of any solid angle must be less than four right angles. That insight is preserved later in Euclid XI.21, and the Remark at XIII.18. On the other hand, the discovery of the *geometrical construction* of them—the lines of thought that comprise the mathematical proof—may very well trace their lineage to the Pythagorean school, and the *geometrical construction* of the octahedron and icosahedron, together with the *systematic exposition and formal proofs*, may trace their lineage to Theaetetus.[120]

At each Vertex:	Angles at Vertex (Less than 360°)	Number of Faces	Number of Vertices	Solid	Shape
3 triangles meet	180° two right angles	4	4	tetrahedron	
4 triangles meet	240° more than two right angles but less than three	8	6	octahedron	
5 triangles meet	300° more than three right angles but less than four	20	12	icosahedron	
3 squares meet	270° three right angles	6	8	cube	
3 pentagons meet	324° more than three right angles but less than four	12	20	dodecahedron	

Table 3.1.

K.3 Pythagoras's Regular Solids and Plato's Timaeus: The Reduction of the Elements to Right Triangles, the Construction of the Cosmos out of Right Triangles

> There is nothing in all this that would be beyond Pythagoras or the Pythagoreans provided that the construction of the regular pentagon was known to them; the method of the formation of the solids agrees with the known facts that the Pythagoreans put angles of certain regular figures together round a point and showed that only three of such angles would fill up the space in one plane round the point.
>
> —Heath[121]

Plato there shows how to construct the regular solids in the elementary sense of putting triangles and pentagons together to form their faces. He forms a square from four isosceles right-angled triangles, and an equilateral triangle from three pairs of triangles which are halves of equal equilateral triangles cut in two by bisecting one of the angles. Then he forms solid angles by putting together (1) squares three by three, (2) equilateral triangles three by three, four by four, and five by five respectively; the first figure so formed is a cube with eight solid angles, the next are the tetrahedron, octahedron, and icosahedron, respectively. The fifth figure, the dodecahedron, has pentagonal faces, and Plato forms solid angles by putting together equal pentagons three by three; the result is a regular solid with twelve faces and twenty solid angles.

—Heath[122]

The story of Pythagoras and the five regular solids is the story of the importance of triangles in the structuring of space. We have a report that Pythagoras discovered all five of those solids, and we have a conflicting report that credits the discovery of the octahedron and icosahedron to Theaetetus. It is not my argument that Pythagoras *proved* them, if by "proved" we mean something along the lines of Euclid's Book XIII. The most important point for my case is to make clear *why* Pythagoras was looking for them at all. Having begun with the supposition that the right triangle was the fundamental geometrical figure, he was looking for composites of triangles—since all triangles can be reduced to right triangles—that would constitute the elements out of which were constructed the myriad of appearances that make up our world. Like Thales, he came to imagine space as articulated by rectilinear figures, all of which were reducible ultimately to triangles.

The tetrahedron, octahedron, and icosahedron are composed of equilateral triangles, and are identified in Plato's *Timaeus* with the elements of fire, air, and water, respectively. The hexahedron or cube is identified with the element earth, and is composed of isosceles triangles, and the dodecahedron is identified sometimes with aether, and sometimes with the "all"—space—in which the elements are contained;[123] it is a figure that consists in twelve pentagonal sides. In imagining the elements out of which our world is composed, the geometrical construction dissects to triangles, and right triangles.

It is not my ambition in this study to analyze Plato's *Timaeus* to show similarities and difference with this narrative. My concern here is only to direct the reader's attention to the *Timaeus* as an echo and a refinement of the intuitions of Thales and Pythagoras and those who followed their lead and contributed to their project. Many developments took place, of course, throughout the fifth and fourth centuries, but I have tried to isolate the metaphysical kernel that points back to the initial intuition in the sixth century. As Plato's narrative unfolds, two kinds of right triangles are central: (i) isosceles right triangles have two equal sides and, as we considered already in terms of one of the theorems with which Thales is credited, the only way there can be two sides of equal length is if the angles opposite those sides are equal. Thus in the isosceles right triangles, in addition to the right angle, there are two angles each equal to 1/2 of a right angle, or 45°. The isosceles right triangle is the building block of a special type of rectangle, an equilateral rectangle or square. The other species of right triangle is called "scalene," which means that all three sides are of unequal length; there are limitless varieties of them. In short, then, we can say that every rectangle can be divided into two right triangles, or that two right triangles equal to one another—sharing the same angles and line

212   The Metaphysics of the Pythagorean Theorem

lengths—can be placed together in such a way to form a rectangle. The only case in which we construct an equilateral rectangle is when two isosceles right triangles are so joined. These points may seem both obvious and, shall we say, elementary, to many readers. But the import of these claims takes a while to settle in.

In Plato's *Timaeus*, the most perfect scalene triangle is half an equilateral triangle, and thus one side has half the length of its hypotenuse, whose corresponding angles are 90°, 60°, and 30°—1/2 of a right angle, 2/3 of a right angle, and 1/3 of a right angle. The construction of the four elements consists in each of these two kinds of right triangles—isosceles and scalene—uniting with its pair; these are the atoms[124] out of which all other things are composed. They come together to make conglomerates, collide with each other and break off into triangles, and recombine and realign to produce the myriad appearances of our world. The scalene triangles are the ultimate building block of the tetrahedron, octahedron, and icosahedron, while the isosceles triangles are the ultimate building block of the hexahedron or cube. But let us be clear, *the cosmos is constructed not only of triangles but is all built out of <u>right triangles</u>—and every triangle can be reduced to right triangles by dropping a perpendicular from the vertex of an angle to the side opposite*. This is the program started by Thales and followed by Pythagoras and his followers. The technique in the *Timaeus*, however it may have been refined after much discussion and reflection, nevertheless points back to archaic times.

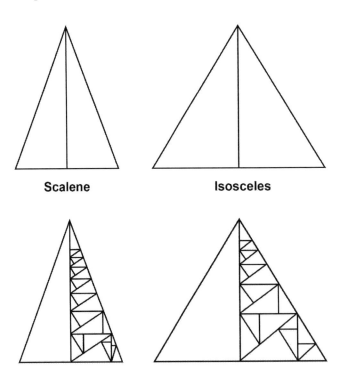

Figure 3.78.

Chapter 4

Epilogue

From the Pythagorean Theorem to the Construction of the Cosmos Out of Right Triangles

It is plausible that Thales knew an areal interpretation of the so-called "Pythagorean theorem." He discovered it looking for the fundamental geometrical figure into which all other rectilinear figures dissect. Having imagined the world of diverse appearances to be expressions of a single underlying unity—ὕδωρ—that altered without changing, he sought the fundamental geometrical figure that transformed into, or was the building block of, cosmic diversity. That figure is the right triangle, and this is central to what the famous theorem plausibly meant in the sixth century BCE as it was "discovered" in the context of exploring ratios, proportions, and similar triangles; by recombining and repackaging right triangles, isosceles and scalene, all alterations are accounted for structurally. This is the lost narrative, echoed in Plato's *Timaeus* 53Cff, in a dialogue containing Pythagorean themes; the project points directly to Pythagoras and his school, but Thales plausibly inspired the original project.

Thales's geometrical vision imagined space as flat surfaces, structured by innumerable straight lines and articulated by countless rectilinear figures—that is, polygons; volumes were built up by layers, or folded up. He reached the conclusion that all polygons are built out of triangles, that every polygonal figure can be expressed as the summation of triangles into which it can be divided—he learned or confirmed this in Egypt—and that inside every triangle are two right triangles, and so what is true of right triangles is true also of all triangles, and a fortiori, of all polygons. We have the testimony of Geminus, who reports that while the Pythagoreans *proved* that there were two right angles in every triangle,[1] the "ancients" (who could be none other than Thales and his school) *theorized* about how there were two right angles in each species of triangle—equilateral, isosceles, and scalene. There are several ways this may have been investigated, though no more details are supplied by our ancient sources; nevertheless, all of them could have been revealed to have two right angles by placing each triangle in its rectangle, as below, and as I have argued, this may very well have been the way that the Egyptian Rhind Mathematical Papyrus problems 51 and 52 (= Moscow Mathematical Papyrus problems 4, 7, and 17) were solved *visually*—the problem of calculating a triangular plot of

214   The Metaphysics of the Pythagorean Theorem

land, and a truncated triangular plot of land (i.e., a trapezium). By dropping a perpendicular from the vertex to the base, the rectangle can be completed: This is the triangle's rectangle.

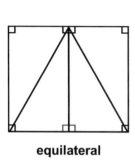 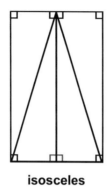 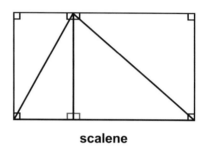

equilateral   **isosceles**   **scalene**

Figure 4.1.

The *triangle's rectangle* is a rectangle equal in area to a triangle (constructed on the same base and between the same two straight lines). Whereas we ordinarily regard rectangles as having twice the area of triangles—between the same straight lines and sharing the same base—the triangle's rectangle is the figure that has the *same area* as the triangle. Now, each rectangle contains four right angles and is divided in half by the side of the triangle—the diagonal—that creates its own rectangle; the two triangles that result are shown to have four right angles—each half of the total angles in each rectangle. By subtracting the two right angles where the perpendicular meets the base (as indicated in all three diagrams), each species of triangle is revealed to contain two right angles. At the same time that the triangle's rectangle reveals that each species of triangle contains two right angles, it reveals also that *every triangle has within it two right triangles*. If the division continues by dropping a perpendicular from the right angle, the right triangles divide ad infinitum into right triangles.

Thales's conception was not atomistic because there was no smallest right triangle but rather a continuous and endless divide; to adapt the quip attributed variously to Bertrand Russell or William James, for Thales it was "right triangles"—not "turtles"—all the way down. Not only did Thales plausibly conclude that the right triangle was the fundamental geometric figure, but he noticed that it collapsed, and expanded, according to a pattern. That pattern came to be called a "mean proportional" or geometric mean; the right triangle collapsed and expanded *in continuous proportions*. The proof of this—what plausibly persuaded Thales and his retinue—was the areal equivalences created by similar and similarly drawn figures on the sides of a right triangle. When the rectangle made from the extremes has the same area as the square on the "mean," then the three lengths stand in continuous proportions. *This is what the discovery of the Pythagorean theorem by ratios, proportions, and similar triangles plausibly meant in the middle of the sixth century BCE. Its discovery was part of a resolution to a metaphysical problem: How does a single underlying unity appear so diversely by altering without changing?*

The train of thoughts that led to this discovery by Thales became formalized and refined later in Euclid's Book VI.31, the so-called "enlargement" of the Pythagorean theorem. This narrative has been lost but is echoed in Plato's *Timaeus* at 53Cff—the construction of the cosmos out of right triangles. Thus the endless wrangling about whether the tertiary testimonies secure or fail to secure the claim that Pythagoras discovered the theorem or had anything to do with it at all (or even mathematics in general) is now illuminated in a new light. First answer the question: What does anyone know when they know the Pythagorean theorem at Euclid VI.31? What could it have meant to Thales and Pythagoras in Archaic Greece had they discovered it when geometry was in its infancy in Greece? Follow the diagrams and see that geometry was part of a metaphysical inquiry to show how a single underlying unity could alter without changing. Let us, in summary, do just that.

Let us focus again on the "enlargement" of the Pythagorean theorem at VI.31:

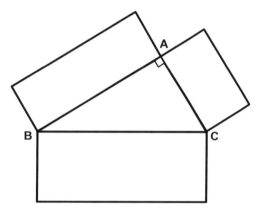

Figure 4.2.

As we have already considered, the rectangles, not squares, were the work of Heiberg, whose edition of Euclid in the 1880s was adopted by Heath, and thus by recent generations of educated people who were guided by his book. There is an edition from the sixteenth century CE by Commandino that also displays rectangles, but the medieval manuscript tradition's schematic diagrams favor what seems to be an isosceles right triangle with the right angle at A,[2] which may very well go back to Thales himself and the traditions that followed these earliest forays in geometry. Despite the schematic nature of the diagram, below, the theorem of VI.31 applies to *all* right triangles, and moreover *all* figures similar and similarly drawn on its sides—this is why it is regarded as the "enlargement": *In right-angled triangles the figure on the side opposite the right angle equals the sum of the similar and similarly described figures on the sides containing the right angle*. Thus, the diagram from the medieval sources—Codices B, V, and b, below—is regarded as schematic, in part, because it applies to *all* right triangles, and there are no figures of any kind presented in the diagram.

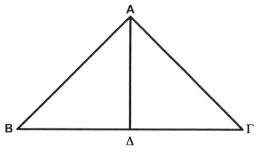

Figure 4.3.

216   The Metaphysics of the Pythagorean Theorem

From the time Thales sought to measure the height of a pyramid at a time of day when the shadow was equal to its height, the isosceles right triangle played a pivotal role in his reflections because areal equivalences are immediately obvious (below, left) in a way they are not with the famous "Pythagorean triples" such as the 3, 4, 5 triangle (below, right).

Areal Interpretation

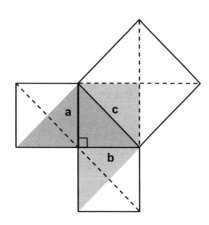 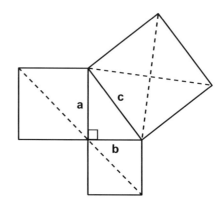

Isoceles Right Triangle
Triangles a,b,c are all the same size
(areal equivalence is obvious)

3,4,5 Scalene Right Triangle
Triangles a,b,c are all different sizes

Figure 4.4.

In the latter case of the 3, 4, 5 right triangle, the whole-number relations jump out but not the areal equivalences, and in so many cases there are no whole-number relations among all three sides of a right triangle. Because the hypotenuse of an isosceles right triangle *never* has a whole number to measure it when the length of its two legs are expressed in whole numbers, it seems likely that Thales could not find a metric or numeric interpretation of the relation between the sides of a right triangle—though he was undoubtedly looking for it; instead he focused on the areal equivalences. He started with the more obvious isosceles right triangle and saw the connection to *all* right triangles as he investigated further, as he generalized from the obvious case.

In Euclid, the steps of the proof bring together earlier theorems in Book VI, a book devoted to *similar figures*, especially VI.8, 13, 17, 19, and 20.

> VI.8—Every right triangle is divided by a perpendicular from the right angle to the base into two similar right triangles; also the corollary, that the perpendicular is the mean proportional between the two parts into which the hypotenuse is divided.

> VI.13—We construct a mean proportional by dividing a circle into two equal parts by its diameter (a theorem attributed to Thales), and then in the semi-

circle, every triangle that connects the ends of the diameter with the vertex touching the semicircle is right-angled (III.31, in some form it is another one of Thales's theorems). The perpendicular that extends from that point that touches the semicircle to the hypotenuse is a mean proportional between the two lengths that is now divided by it on the hypotenuse.

VI.17—The relation between the three lengths—two segments into which the hypotenuse has been divided and the perpendicular that is the mean proportional between them—forms a continuous proportion; the first is to the second as the second is to the third. This relation is expressed by showing that the rectangle made by the first and third is equal to the square on the second.

VI.19—Two similar triangles are to one another in the duplicate ratio of the corresponding sides. As side length increases, the area of the figure increases in the duplicate ratio of the corresponding sides. This is how the ancient Greeks expressed what our modern language calls "squaring."

VI.20—Similar polygons can be dissected into similar triangles, and consequently, since similar triangles are to one another in the duplicate ratio of their corresponding sides, similar polygons are to one another in the duplicate ratio of their corresponding sides.

Let us now review them in an order that proposes to connect the intuitions of a discovery in the middle of the sixth century.

The theorem at VI.20 is that similar polygons dissect into similar triangles; all rectilinear figures dissect into triangles and the little figures and bigger figures—polygon to polygon—have a ratio duplicate of that which the corresponding side has to the corresponding side. The cosmos is imagined as countless rectilinear figures—polygons—and the little worlds and the bigger worlds are structurally interrelated—*similar*—by a pattern revealed by their dissection into similar triangles. Because every polygon reduces to triangles, what holds for every triangle holds also for every polygon.

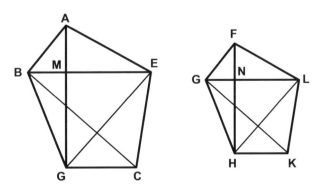

Figure 4.5.

218   The Metaphysics of the Pythagorean Theorem

Thus, at VI.19 the theorem is "Similar triangles are to one another in the duplicate ratio of their corresponding sides." What holds for the world of all similar polygons is a consequence of the fact that it holds for all similar triangles, since every polygon can be expressed as the summation of triangles into which it can be divided.

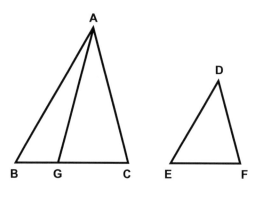

Figure 4.6.

Now every polygon can be reduced to triangles, and every triangle can be divided into two right-angled triangles by dropping a perpendicular from the vertex to the side opposite. And each of these right triangles can be further dissected into pairs of similar right triangles, each of which is similar to the other and similar to the ever-larger or ever-smaller right triangles, shown below:

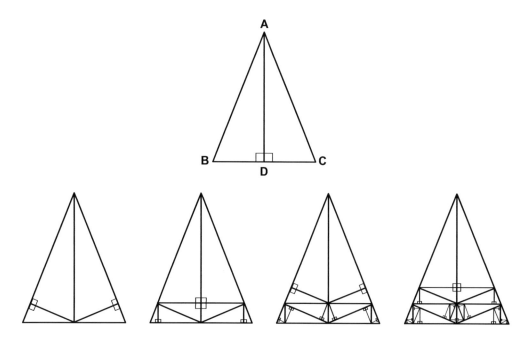

**All triangles divide into two right triangles and
all right triangles divide indefinitely into right triangles.**

Figure 4.7.

In Book VI, Euclid first established that VI.8: If in a right-angled triangle a perpendicular is drawn from the right angle to the base, then the triangles adjoining the perpendicular are similar both to the whole and to one another. In the diagram below, an isosceles right triangle is used and the areal equivalences are obvious because the right triangle is divided equally into two right triangles, each of which shares the same angles, and hence the same shape; the two right triangles together make the whole triangle. It would not matter, of course, where the perpendicular was placed—so long as it came from the right angle—because the result will always be two right triangles similar to each other (= same angles and thus same shape) that together make up the larger right triangle.

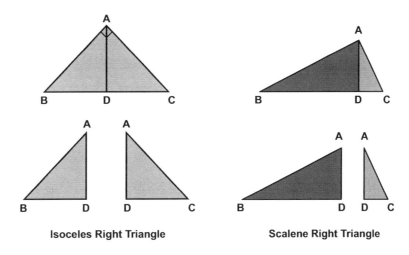

Figure 4.8.

And the corollary is: From this it is clear that, if in a right-angled triangle a perpendicular is drawn from the right angle to the base, then the straight line so drawn (in this case AD) is a *mean proportional* between the segments of the base.

Now Euclid shows this next step and its interconnection at VI.13. The guarantee that the figure is right-angled at A is the inscription of the triangle in the semicircle, the very theorem with which Thales is credited by Pamphile. First, the circle is bisected by its diameter, another theorem credited to Thales, and then every triangle in the (semi-)circle is right. When the perpendicular is dropped from the right angle at A, the line AD is the mean proportional between the two segments into which the hypotenuse BC is divided, namely BD the larger and CD the smaller (below, right), though of course they could be equal (below, left).

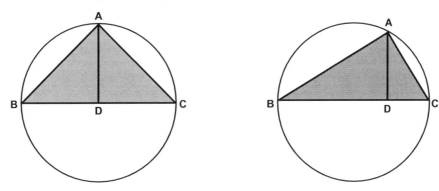

Figure 4.9.

220   The Metaphysics of the Pythagorean Theorem

To identify the perpendicular AD as the mean proportional is to identify the pattern with which the right triangle collapses when divided, or conversely, expands when enlarged. The pattern is revealed in Euclid VI.17, that is, when three straight lines BD, AD, and CD (in the diagram above) are proportional, then the rectangle formed by the first and third is equal in area to the rectangle (= square) made by the second.

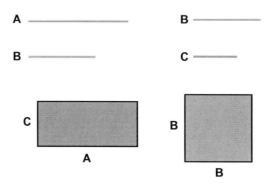

Figure 4.10.

And this insight revealed, on closer inspection of the right triangle, that not only was the hypotenuse divided into two parts so that the rectangle made by the first and third was equal to the square made on the second, but also that the ratio of the lengths of the first to the third was the *same ratio* as the figure on the first was to the figure on the second. This revealed a way not only to express but to later prove geometrically the pattern of continuous proportions in which the right triangle collapses upon reduction and expands in growth. What was required was a clear understanding of the areal relations of squares and rectangles constructed on the sides of a right triangle, and that the perpendicular divided it into two similar right triangles.

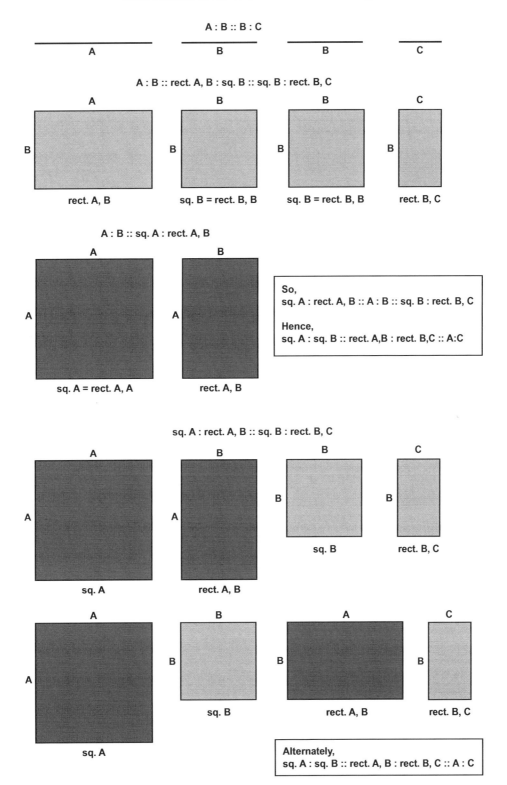

Figure 4.11.

222   The Metaphysics of the Pythagorean Theorem

These diagrams in red and blue, above, show how the figures corresponding to line lengths A, B, and C in continuous proportion stand in relation to each other. The ratio of line lengths A to C is the same ratio as the figure (= square) on A to the figure (= square) on B, namely in duplicate ratio. As one continues to make these diagrams, the relations of line lengths and figures becomes clearer. For example, if we start with the isosceles triangle in the semicircle (below, left), and then make a square on the hypotenuse BC across from the right angle at A, and a square on hypotenuse AC across from the right angle of the similar triangle at D, and then continue to divide both figures into triangles by diagonals—each of these triangles is now the same size—we can see immediately and obviously that the square on AC contains four right triangles and the rectangle on CD also contains four right triangles of the *same size*. Thus the square on AC is equal to the rectangle on half the hypotenuse CD. The same areal relations can be shown to hold for the scalene right triangle (below, right), and by completing the other half of the right triangle at A (below, left and right) we have an areal interpretation of the Pythagorean theorem. The plausibility that Thales discovered the hypotenuse theorem in this way rests on grasping that he was looking for the fundamental geometrical figure, that he concluded it was the right triangle, and that he consequently invested his sharpest focus on the properties of the right triangle and the relation among its sides.

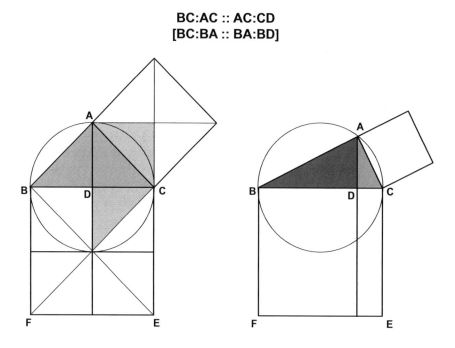

Figure 4.12.

The visualization, though not the proof, that led Thales to this discovery can be expressed by showing—placing them next to one another—the areal relations between squares and rectangles constructed on the sides of the right triangle, and its perpendicular.

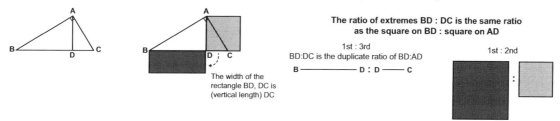

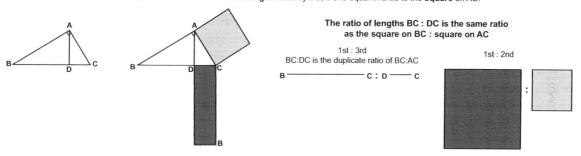

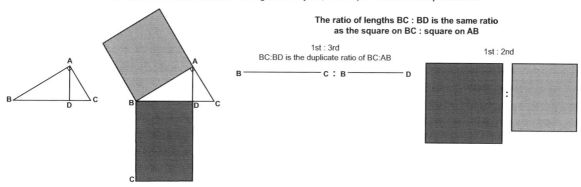

Figure 4.13.

Moreover, the route to this discovery was paved by his exploration of the triangle in the (semi-)circle, a theorem credited to both Thales and Pythagoras, and I have argued for the plausibility that it was discovered by Thales in the context of this metaphysical problem, and that Pythagoras's progress plausibly followed the same path. The report crediting them both was not an error, nor a dittography, but rather testifies that both worked through these same

224   The Metaphysics of the Pythagorean Theorem

lines of thought. First, Thales showed that every circle is bisected by its diameter (one of the theorems credited to him), and then he inscribed triangles in the divided circle, below:

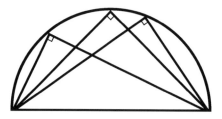

Figure 4.14.

Thales plausibly came to the realization that every triangle inscribed was right by applying the isosceles theorem with which he is also credited: When a triangle has two sides equal in length, there must be equal base angles opposite those sides. Thus, below, AD is equal to BD in triangle ADB, and AD is equal to CD in triangle ADC—AD, BD, and CD, BC are both radii of the circle—and thus their respective angles opposite them are equal.

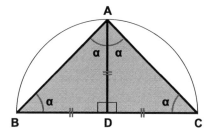 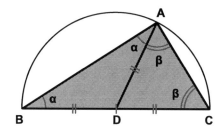

Figure 4.15.

Starting with the diagram on the left, the isosceles right triangle in the semicircle, the right triangle is divided into two similar right triangles, and in this case the angles marked α and β are the same. Thus each triangle contains a right angle at D and two equal angles, 1/2 a right angle each. Thus there are two right angles in each of the triangles, plus 2 α + 2 β; and therefore 2 α + 2 β plus two right angles (at D) = 4 right angles. By subtracting the 2 right angles at D, 2 α + 2 β = 2 right angles, and hence α + β = 1 right angle at A. Here was Thales's proof—his lines of reasoning that he and his compatriots could regard as persuasive. The same lines of reasoning apply to the scalene triangle inscribed in the semicircle on the right; although α and β are different, they sum to a right angle at A.

As Thales continued to gaze at his diagram he realized also that every possible right triangle was contained in the semicircle, and if the diagram was continued for the whole circle, every rectangle was also contained. Moreover, it is plausible, *ceteris paribus*, that he realized that the vertices of every right triangle in the (semi-)circle form a circle—the concept of a geometric locus—that the right triangle is even the basic geometrical figure of the circle: In every circle is every right triangle, and by plotting all the right angles, the vertices form a circle. In a sense, then, the circle, too, grows out of right triangles, or collapses into them.

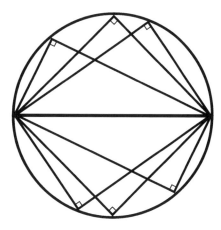

Figure 4.16.

Let us review one more time the discovery of the hypotenuse theorem by ratios and proportions. Consider the diagrams below:

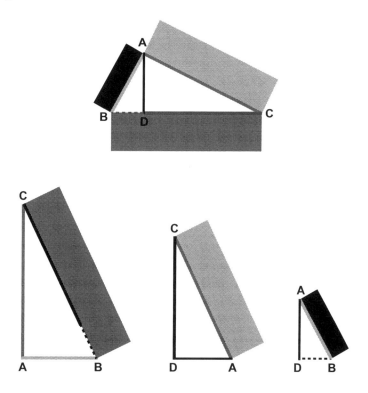

Figure 4.17.

226 The Metaphysics of the Pythagorean Theorem

The simple intuition of areal equivalences is this: Since the large right triangle divides into two similar right triangles, the two triangles into which the large triangle has been divided sum to that one large triangle. Thus, the figures on the hypotenuses of the two smaller right triangles sum to the similar and similarly drawn figure on the hypotenuse of the large right triangle because they are the parts that constitute it.

The proof sequence is as follows, by ratios and proportions:[3]

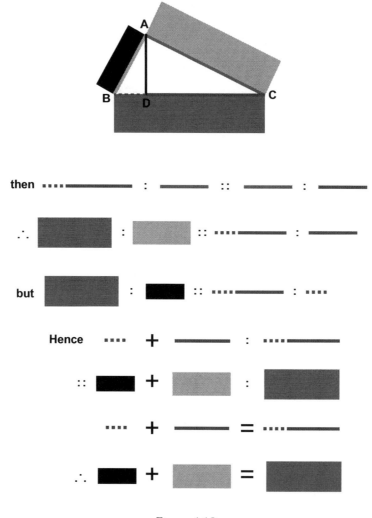

Figure 4.18.

As the hypotenuse of the biggest triangle is to the hypotenuse of the middle-sized triangle, which is also the longest side of the big triangle, so the hypotenuse of the middle-sized triangle is to its own longest side, which is the longer part of the hypotenuse into which the perpendicular has divided it. Stated differently, these three line lengths stand in continuous proportion, and thus the middle one (———) is the mean proportional between the two extremes:

Figure 4.19.

And therefore, since the three straight lines are in continuous proportion, the ratio of the figure on the first to the similar and similarly drawn figure on the second is the same ratio as the line length of the first to the length of the third:

Figure 4.20.

And this same relation holds in the other, smaller but similar right triangle into which the larger right triangle has been divided by the perpendicular from the right angle. So, the figure on the first is to the figure on the second in the same ratio as the length of the first is to the length of the third:

Figure 4.21.

And so the hypotenuse of the largest triangle is divided into two parts such that the smaller + the larger parts are to the whole hypotenuse as the respective similar figures on the two similar triangles' hypotenuses are to the similar and similarly drawn figure on the whole hypotenuse:

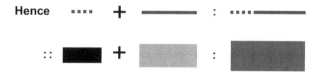

Figure 4.22.

And since the two segment lengths into which the hypotenuse on the largest triangle is divided sum to the whole hypotenuse, thus the two figures drawn on the hypotenuses of the two similar triangles into which the largest has been divided sum to the similar figure on the largest hypotenuse:

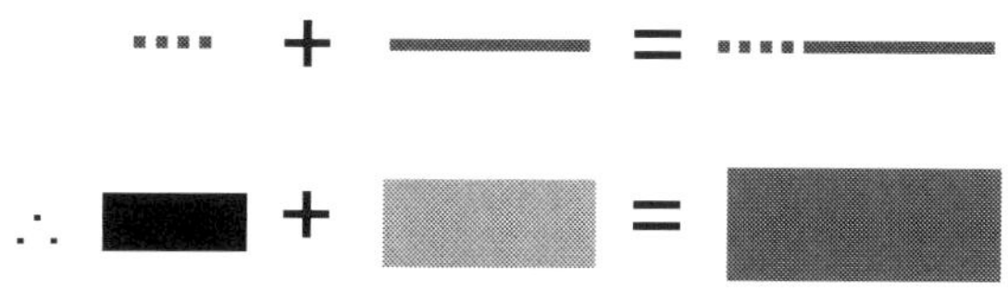

Figure 4.23.

Thales had already become fully absorbed in reflecting on similar triangles when measuring the height of a pyramid when the shadow length was equal to its height. He could only have known the exact time to make the measurement by setting up a gnomon at the tip of the pyramid's shadow or even nearby; when the gnomon's shadow was equal to its height, that was precisely the time to make the measurement:

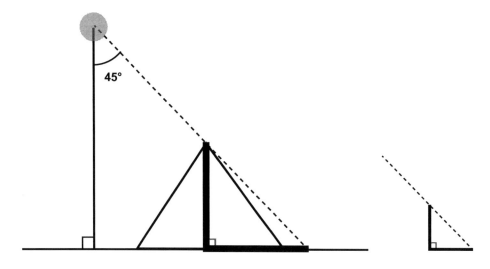

Figure 4.24.

And Thales is also credited with making the measurement when the shadow length was unequal but proportional; this also shows a command of the idea of similar triangles:

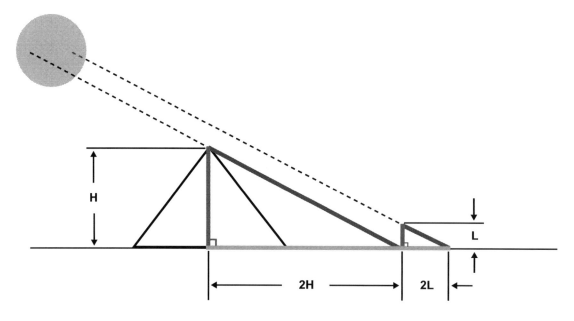

Figure 4.25.

And he measured the distance of a ship at sea, probably more than one way, including equal and similar triangles. The technique directly below requires an understanding of ASA triangle equality (I.26), with which he is also credited in ancient testimony, though by inference:

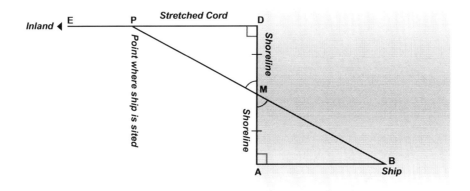

Figure 4.26.

Had the measurement been made from an elevated tower, many possibilities present themselves. Once he grasped the idea of similar triangles, which the argument supposes he indeed did, Thales may have attempted several different measurements to confirm his understanding:

Figure 4.27.

And since we now understand that Eupalinos had made contemporaneously a scaled-measured diagram for the tunnel in Samos, Thales also may well have measured the distance of a ship at sea by constructing a simple scale model, as Anaximander is claimed to have done for making a *sphairos* of the heavenly wheels. The scale model, of course, is an exercise in similar figures.

230   The Metaphysics of the Pythagorean Theorem

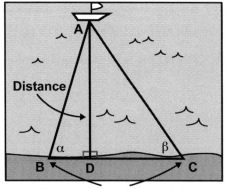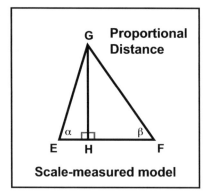

The actual distance is calculated by proportionality. That is:
BC : AD :: EF : GH

Figure 4.28.

Thus, connecting the dots:

- Geminus's testimony that Thales investigated two right angles in each species of triangle revealed also that there were two right triangles in every triangle.

- The measurement of the height of the pyramid and distance of a ship at sea, both of which required an understanding of ratios, proportions, and *similarity*, showed that the little world and big world shared the same structure.

- The theorems credited by Proclus on the authority of Eudemus (the circle is bisected by its diameter, two straight lines that intersect make the vertical and opposite angles equal, the base angles of an isosceles triangle are equal, and two triangles that share two angles and a side in common are equal—ASA equivalence) and the discovery of the right triangle in the (semi-)circle for which he celebrated by sacrificing an ox—it was the discovery that the right triangle was the fundamental geometric figure (i.e., the lost narrative) show the core of what Thales knew.

- The metaphysical search for a unity that appears divergently, altering without changing—collectively, these points constitute a plausible case that Thales knew an areal interpretation of the hypotenuse theorem.

Had Thales connected the series of thoughts as set out here, then the knowledge of the hypotenuse theorem was current by the middle of the sixth century, when Pythagoras was only a young man, roughly the age of our undergraduates, the age of a Greek *kouros*. Given the contemporary project of digging the tunnel by Eupalinos in Pythagoras's backyard, and the proximity of Miletus to Samos where we have accounts, although late, that place Pythagoras meeting with both Thales and Anaximander, there is every reason to think that Pythagoras, along with other Ionians, would have known an areal interpretation of the hypotenuse theorem, had they sought geometrical knowledge. If we grant to Pythagoras himself some engagement with the theorems concerning odd and even numbers, later formalized into Euclid's Book VII for which there is broad scholarly consensus, we have evidence for his interest in numbers; had he believed that numbers held the hidden nature of things, the successful digging of the tunnel by measures, lengths, and numbers may well have provided special inspiration in fostering or furthering such a vision. Had Pythagoras sought to confirm Eupalinos's hypotheses of north and south tunnel lengths by imagining the project as two right-angled triangles sharing the theoretical line connecting the crest of the hill with the planned meeting point directly under it, perhaps he extended Thales's project to include the metric interpretation of the theorem. Had Pythagoras and his school taken up Thales's project as conjectured here, connecting Pythagoras also with his *other* theorem—the *application of areas*—fits seamlessly within the project of building the cosmos out of triangles.

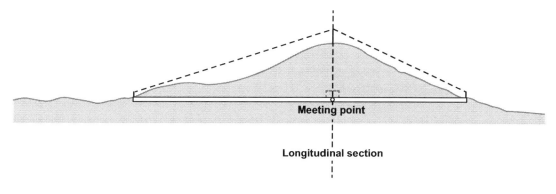

Figure 4.29.

Keeping in mind that every polygon reduces to triangles—the area of every polygon is the sum of the triangles into which it divides—and every triangle reduces to two right triangles, *the building of the cosmos is ultimately made out of right triangles.*

Thus, Pythagoras and his compatriots sought to show how triangles in any angle could be transformed into parallelograms. Parallelogram FC (or AE—they are both equal figures) has twice the area of triangle AEC, and because triangle ABC has been divided into two equal triangles by AE, now the whole triangle ABC is equal in area to parallelogram FC (or AE), below. Stated differently, and in the context of our discussion of the Egyptian RMP 51, FC is triangle ABC's "rectangle" (i.e., parallelogram).

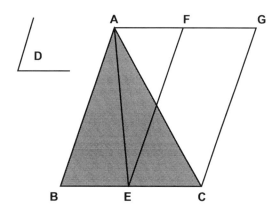

Figure 4.30.

232 | The Metaphysics of the Pythagorean Theorem

The next step is to show that complements (green) of every parallelogram are equal to one another, below; HF is equal to EG because the complements of the parallelograms about the diameter are equal to one another:

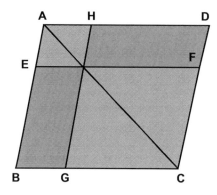

Figure 4.31.

The next step is to construct a parallelogram, BF, below, equal to a given triangle, C, in any specified angle, D, and apply it to a given straight line, AB. By connecting it to a given straight line we have the project of expressing transformational equivalences of shapes. The figure LB is applied to AB. In Euclid, the proof of the "application of areas" is in I.44, below; having shown how a parallelogram can be constructed equal in area to a triangle (I.42) and that the complements of the parallelograms about the diameter are equal to one another (I.43), the parallelogram FB is constructed equal in area to a given triangle in a specified angle, as a complement of parallelograms. The matching complement BL is constructed on the given line AB and is equal to it. Consequently, a parallelogram BL is constructed on a given straight line equal to a triangle in a specified angle. OED.[4]

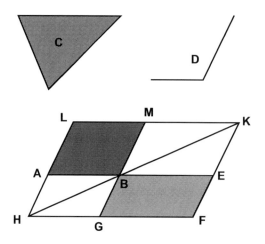

Figure 4.32.

The proof offers an insight into the structure of *transformational equivalences*, how from one shape it can be transformed into another while still retaining *the same area*. The metaphysical point is that it is as if the basic triangle transforms into another shape with the same area, and grows, expands, or diminishes into similar figures larger and smaller.

Next, the task of I.45 is to extend the technique so that *any* rectilinear area, not only triangles, can be applied to a line in any given angle. The result is to transform any rectilinear figure into a parallelogram of whatever angle desired and with any one side of any specified length. Stated differently, I.45, below, answers the question "What is the area of this figure?"[5] And note carefully, *the area of any figure is determined by dissecting it into triangles*, and then summing them up, or expressing the figure as a transformational equivalent—a figure of different shape but equal in area. This is the same principle as finding the triangle's rectangle but now extended to find the "rectangle" (= parallelogram) of every rectilinear figure regardless of its shape. This, we argued, was a key point behind ancient Egyptian surveying and appears in RMP 51 to determine the area of a triangular plot of land. Stated differently, all areas—all rectilinear figures—are built up of triangles. And all triangles reduce ultimately to right triangles.

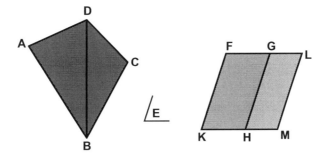

Figure 4.33.

And finally we move to the theorem that Heath conjectured seems to match the description that Plutarch gives to Pythagoras's achievement in the application of areas—Pythagoras's *Other* theorem. First, a parallelogram is constructed equal to a given triangle—parallelogram BE is equal to triangle ABC:

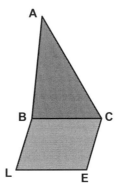

Figure 4.34.

234   The Metaphysics of the Pythagorean Theorem

Next, any rectilinear figure, D, is divided into triangles and then a parallelogram is constructed equal to that rectilinear figure and applied to a given line CE, in a straight line with CF, and thus parallelogram CM is equal to rectilinear figure D.

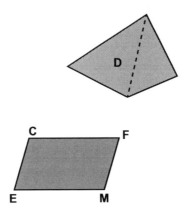

Figure 4.35.

And finally, we take a mean proportional GH—between BC, the base of the original given triangle, and CF, the side of the paralleogram whose area is equal to rectilinear figure D—and on it construct similar triangle KGH.

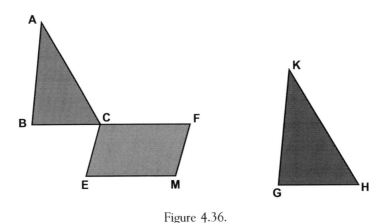

Figure 4.36.

Thus the result is this proof:

Pythagoras's Other Theorem

παραβολὴ τῶν χωρίων

The Application of Areas, VI.25

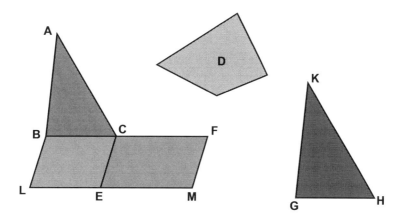

Figure 4.37.

The given figure is triangle ABC and to it a similar figure KGH must be constructed whose area is equal to D, a rectilinear figure. Then, applied to BC is the parallelogram BE, which is equal to triangle ABC, and applied to CE is the parallelogram CM equal to D in the angle FCE, which equals the angle CBL. Since angles FCE and CBL are equal, BC is in a straight line with CF, and LE with EM. This much has already been established in the application of areas theorems that we already considered from Book I, 44 and 45.

Now, ratios and proportions are added; the mean proportional between BC and CF—GH—is taken (GH is the base of the triangle on the far right), and the triangle KGH is described as similar and similarly situated to ABC. Hence BC is to GH as GH is to CF, and if three straight lines are proportional, then the first is to the third as the figure on the first is to the figure on the second (*duplicate ratio*, Def. V.9). This means that the ratio of the line lengths of the first to the third (BC : CF) is the same ratio as the area of the figure on the first (ABC) to the similarly situated figure (area) described on the second (KGH). Therefore, BC is to CF as the triangle ABC is to triangle KGH. But BC is to CF as the parallelogram BE is to the parallelogram EF. Therefore, also the triangle ABC is to the triangle KGH as parallelogram BE is to parallelogram EF. And therefore, alternately, triangle ABC is to parallelogram BE as triangle KGH is to paralleogram EF.

But triangle ABC equals parallelogram BE, and therefore triangle KGH also equals parallelogram EF, and parallelogram EF equals D, therefore KGH also equals D.

And KGH is also similar to ABC. Therefore this figure KGH has been constructed similar to the given rectilinear figure ABC and equal to the other given figure D. QED.

Let us now recall how VI.25 fits within the metaphysics we have already discussed. The cosmos is filled with appearances, though there is some underlying unity such that there is no change, only alteration. All differing appearances are to be accounted for by transformational equivalences; the deep intuition is that each appearance is related to every other by retaining equal area, but in different rectilinear shapes—it seems to be a form of what we have come to refer to as the "principle of conservation of matter," though the essence of this underlying unity may be originally supersensible and capable of becoming visible and sensible, like the

236   The Metaphysics of the Pythagorean Theorem

Greek gods themselves. In the language of Plato's *Timaeus* 53Cff, those appearances all have volumes, and every volume (βάθος) has a surface (ἐπίπεδον). This is where Thales's original project of constructing the cosmos out of right triangles, taken up by Pythagoras, is preserved.

To think again about the passage in Plato's *Timaeus*, we might ask *why* build the cosmos out of right triangles? What understanding propels the vision to construct the cosmos out of them? The answer seems to be clear, although it may not have been routinely emphasized: If we imagine the cosmos as flat surfaces structured by straight lines and articulated by rectilinear figures, all polygons reduce to triangles and all triangles reduce to right triangles. The project of building the cosmos out of right triangles is grounded in the understanding that the right triangle is the fundamental geometrical figure.

Now the world of experience does not appear as flat surfaces but rather three-dimensional objects. Having imagined space as flat surfaces, the *Timaeus* points to an early conception that Proclus attributes to Pythagoras himself, of the σύστασις of the regular solids, "putting them together." The elements, and the sphere in which they are inscribed, are layered up or folded up into volumes. To produce this result Pythagoras had to understand that: Any solid angle is contained by plane angles whose sum is less than four right angles.

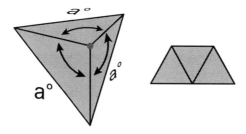

Figure 4.38.

And if we continue to experiment by placing regular polyhedra around a single point, we not only confirm that they cannot fold up if they are equal to or exceed four right angles, but that the successful possibilities include only three equilateral triangles, four equilateral triangles, five equilateral triangles, three squares, and three pentagons:[6]

Figure 4.39.

When these polygons are folded up, the five regular solids appear. The four elements are built explicitly from triangles, one of which must be right and the other scalene:[7]

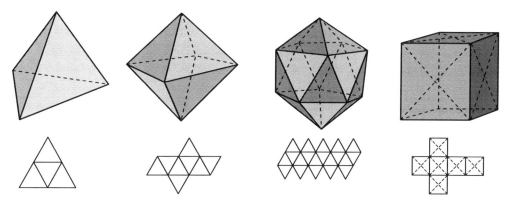

Figure 4.40.

The elements are all placed in the whole, represented as the dodecahedron and constructed out of regular pentagons. Again, like all polygons, the dodecahedron reduces to triangles, and

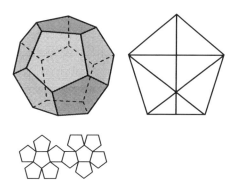

Figure 4.41.

every triangle has contained within it right triangles. Consequently, all the regular solids are built out of right triangles, both the ones corresponding to the four elements and the one containing the whole.

238   The Metaphysics of the Pythagorean Theorem

In Plato's *Timaeus*, all the constructions of the regular solids—the four elements and the whole—are expressed as right triangles—internally dissected into right triangles, below:

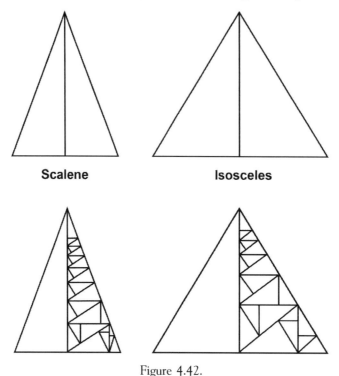

Figure 4.42.

Even the dodecahedron, constructed out of regular pentagons, divides into triangles, and every triangle divides into right triangles.

Finally, let us recall that doubts about connecting Pythagoras with the famous theorem can be traced back as early as Cicero. Van der Waerden, and Burkert who followed him in this respect, echoed the same doubts; the report that upon his discovery Pythagoras sacrificed an ox seemed hopelessly at odds with "Pythagoras the vegetarian" and "Pythagoras the believer in reincarnation" where human souls can be reborn in animal bodies.[8] The doubts, then, about connecting Pythagoras with the theorem emerge not from any ancient testimony to the contrary nor by appeal to the internal history of geometry, but from the apparent incongruity of animal sacrifice by a person who seemed unable to sanction the killing of animals. So these doubts about Pythagoras and the theorem are really doubts about the reported animal sacrifice; the baby was thrown out with the bath water.

But now let us wonder again just *when* Pythagoras's discovery and celebration might have taken place. If we imagine these events to have transpired—if they transpired at all—in southern Italy, when Pythagoras was an elder statesman and sage committed to vegetarianism and reincarnation, then the objections cannot be easily discounted. But, had the discovery of an areal interpretation been common knowledge among Thales and his school by the middle of the sixth century, and had Pythagoras taken up Thales's project and further perfected or refined a proof that he celebrated while still a young man in Samos, a splendid sacrifice for a divine inspiration on the gigantic altar of the temple of Hera appears with a new plausibility. For a twenty-something young man in Samos, not yet a vegetarian or a believer in reincarnation, a splendid sacrifice of oxen would have been fitting and understood in the context of the culture of the Samian Heraion. Thus, to dismiss the doxographical reports that connect Pythagoras and the theorem on the grounds that the sacrifice is inconsistent with his vegetarianism and transmigratory beliefs is to fail to consider—without any good reason—that his achievement may well have happened while he was still a young man, a *kouros*, in Samos.

Notes

Introduction

1. Aristotle, *Metaphysics* 983b21ff.
2. Cf. Patricia Curd, "Presocratic Philosophy," in the *Stanford Encyclopedia of Philosophy*. Cf. Aristotle, *Metaphysics* 983b2–4: καὶ τοὺς πρότερον ἡμῶν εἰς ἐπίσκεψιν τῶν ὄντων ἐλθόντας καὶ φιλοσοφήσαντες περὶ τῆς ἀληθείας [speaking of the earliest philosophers] "of those who before us approached the inquiry into what is truly real and philosophized about Truth." Cf. also *Part of Animals* 640b4ff: Οἱ μὲν οὖν ἀρχαῖοι καὶ πρῶτοι φιλοσοφήσαντες περὶ φύσεως περὶ τῆς ὑλικῆς ἀρχῆς καὶ τῆς τοιαύτης αἰτίας ἐσκόπουν, τίς καὶ ποία τις, καὶ πῶς ἐκ ταύτης γίνεται τὸ ὅλον. (Now the earliest men who first philosophized about nature, devoted themselves to try to discover what was the material principle or material cause, and what is was like; they tried to find out how the whole cosmos was formed out of it.)
3. Aristotle, *Metaphysics* 983B7.
4. Cf. Nails 2002, p. 293. There have been some doubts whether Timaeus is a Pythagorean. While he comes from Locri in southern Italy, a region populated by Pythagoreans, some views in the dialogue are Empedoclean and others Parmenidean, though others are clearly Pythagorean. Archer-Hind regarded Timaeus as Plato's mouthpiece for Pythagorean-inspired doctrine: (when Timaeus is described in the dialogue at 20A) "he [Plato] must have set a high estimate on the Pythagorean's philosophical capacity [as] he has proved by making him the mouthpiece of his own profoundest speculations" (p. 63). On the other hand, Cornford (1937, p. 3) pointed out that "Plato nowhere says that Timaeus is a Pythagorean. . . . [but] much of the doctrine is Pythagorean." When A. E. Taylor discusses Pythagorean elements in the dialogue, he claims "This is exactly what we shall find Timaeus trying to do in his famous geometrical construction of body." (*Timaeus* 53C–55C) (1928, p. 18). For my argument, it is not pivotal that Timaeus be revealed as a Pythagorean but rather that the views he expresses of the construction of the cosmos out of right triangles (*Timaeus* 53Cff) are plausibly Pythagorean, for this is the lead that points back to Pythagoras himself.
5. There is also a distinction discussed by philosophers referred to as *priority monism*. This is the view that only one thing is ontologically basic or prior to everything else, and sometimes applied to Neoplatonic adherents. From this ontological priority, things that *differ* from it come forth; this is not Thales's view as I see it.
6. Graham 2006 is an exception; see also Zhmud (2006), who doubts Aristotle's report and the identification that Thales posited an ἀρχή.
7. Burkert 1962/1972.
8. Since Neugebauer's translation of the cuneiform tablets (see 1957, chapter 2), we know that the hypotenuse theorem was known in some forms more than a millennium earlier, and probably known earlier elsewhere. My focus is on when the Greeks became aware of it in such a way that it found a place

within their mathematical reflections, and my emphasis is that from the middle of the sixth century, its place was within the context of metaphysical speculations.

9. Netz 1999, 2004.
10. Netz 1999, p. 61.
11. Netz 2004.
12. Netz 1999, p. 60.
13. Netz 1999, p. 61.
14. Netz 1999, p. 60.
15. Von Fritz 1945, p. 259.
16. Here I am following the clear outline in Kastanis and Thomaidis 1991.
17. Neugebauer 1957: In his chapter "Babylonian Mathematics," he writes "All these problems were probably never sharply separated from methods which we today call 'algebraic.' In the center of this group lies the solution of quadratic equations for two unknowns" (p. 40). Cf. also Rudman 2011, p. 70: "Such was Neugebauer's stature that his interpretation of that OB scribes were doing quadratic algebra to solve this and other problem texts became the gospel spread by his disciples [esp. van der Waerden and Aaboe]."
18. Tannery 1912, pp. 254–80.
19. Zeuthen 1885, pp. 1–38; and 1896, pp. 32–64. It was Zeuthen who noticed that there were propositions in Euclid II and VI that might normally be written out as algebraic formulas.
20. Cf. Heath 1908, vol. I, p. 351: "practically the whole doctrine of Book II is Pythagorean." In discussing Plutarch's attribution of both the theorem at I.47 and also the application of areas to Pythagoras: "The theorem [sc. at I.47] is closely connected with the whole of the matter of Euclid Book II, in which one of the most prominent features is the use of a gnomon. Now the gnomon was a well-understood term with the Pythagoreans (cf. the fragment of Philolaus.) Aristotle also (*Physics* III, 4, 203a 10–15) clearly attributes to the Pythagoreans the placing of odd numbers as gnomons round successive squares beginning with 1, thereby forming new squares, while in other places (*Categories* 14, 15a30) the word *gnomon* occurs in the same (obviously familiar) sense: "e.g. a square, when a gnomon is placed around it, is increased in size but is not altered in form." As a point of clarification, II.14, the last proposition in this shortest book of *Elements*, finishes the quadrature of rectilinear figures. (The narrow meaning of the word "quadrature" is to find a square with the same area of a given figure, also called "squaring" the figure. In a broader sense, quadrature means finding the area of a given figure.) Besides Heath and van der Waerden, and it seems Artmann, Pythagorean origins of Euclid Book II are presumed (Artmann 1999, pp. 324–25), though Fowler 1987 thinks otherwise.
21. Neugebauer 1936, pp. 245–59. The reason the Greeks "geometrized algebra" is, Neugebauer conjectures, that after the discovery of irrational quantities, they demanded that the validity of mathematics be secured by switching from the domain of rational numbers to the domain of ratios of arbitrary quantities; consequently, it then became necessary to reformulate the results of pre-Greek "algebraic" algebra in "geometric" algebra. Cf. also Neugebauer 1957, pp. 147ff.
22. Neugebauer 1957, p. 147.
23. Van der Waerden 1954, pp. 118–124. Cf. especially "thus we conclude that all the Babylonian normalized equations have, without exception, left their trace in the arithmetic and geometry of the Pythagoreans. It is out of the question to attribute this to mere chance. What could be surmised before, has now become a certainty, namely that the Babylonian tradition supplied the material with which the Greeks, the Pythagoreans in particular, used in constructing their mathematics" (p. 124).
24. Szabó 1969.
25. Unguru 1975, p. 171.
26. Van der Waerden 1975, pp. 203 and 209. Italics supplied for emphasis. Cf. also Kastanis and Thomaidis 1991. Italics supplied for emphasis of i.
27. Fowler 1980, pp. 5–36.

28. Mueller 1981, pp. 41–52.

29. Berggren 1984, pp. 394–410.

30. Knorr 1986, p. 203 n. 94.

31. Rudman 2010, p. 18. I do not share the same enthusiasm with which Rudman identifies this visualization as "geometric algebra," especially because he thinks the phrase is usefully applied to the Greeks. But what I do share, and why I have quoted him, is that visualization of geometric problems through diagrams was a familiar practice from the Old Babylonian period onward.

32. Zhmud 2012.

33. Zhmud 2012, pp. 242–46.

34. This story is told by Herodotus II.154 and concerns how the Saite Pharaoh Psammetichus I (Psamtik) (c. 664–610) of the Twenty-Sixth Dynasty of Egypt, overthrown and desperate, seeks the advice of the Oracle of Leto at Buto, who answered that "he should have vengeance when he saw men of bronze coming from the sea." But when a short time after, it was reported to Psamtik that Ionian and Carian pirates were foraging nearby and wearing bronze armor, Psamtik quickly made friends with them, and seeing in their presence the fulfillment of the oracle, he promised them great rewards if they would join him in his campaign to regain the throne. Upon the success of this endeavor, Psmatik kept his promises and bestowed on the mercenaries two parcels of land (or "camps" στρατόπεδα) opposite each other on either side of the Nile near to his capital in Sais.

35. Herodotus II.154, 6–8.

36. Herodotus II.154, 9–11.

37. Lloyd 1975, pp. 52–53, voiced just this view of Egyptian influence on what the Greeks could have imported from Egypt.

38. Isler 2001, pp. 135ff.

39. Robins 2001, p. 60, emphasizes cases where paint splatter can still be detected.

40. The figure on the right, after Robins 1994, p. 161. The photograph on the left taken by the author.

41. Robins 1994, p. 162.

42. Bryan 2001, p. 64.

43. Senenmut, perhaps the lover of Queen Hatshepsut, belongs to the Eighteenth Dynasty (early New Kingdom), and his tomb chapel (SAE 71) is near Hatshepsut's. The ostrakon was found in a dump heap just below Senenmut's tomb chapel, Eighteenth Dynasty.

44. After an artifact in the British Museum; here we have a carefully outlined figure of Tuthmosis III. Robins speculates that it might have been a prototype for seated figures of a king in monumental scenes, Robins 2001, p. 60 and plate 15.1.

45. This useful illustration after www.pyramidofman.com/proportions.htm and also http://library.thinkquest.org/23492/data/. One can see that the standing figure is divided into 18 equal units or squares starting from the soles of the feet to the hairline. The bottom of the feet to the knee is between 5 and 6 squares. The elbow line at 2/3 height appears at square 12, and so on. The seated figure occupies 14 squares from the soles of the feet to the hairline, etc. See also Robins 1994, p. 166ff.

46. Cf. Hahn 2003, p. 126; Iverson 1975, plate 23, and Robins 1994, p. 161.

47. Robins 2001, p. 60; cf. also Robins 1994, p. 182.

48. Robins 2001, p. 61.

49. Robins 1994, p. 63; also pp. 177–81.

50. Peet 1926, p. 420.

51. Robins and Shute 1987. Cf. the chapter entitled "Rectangles, Triangles, and Pyramids."

52. Peet 1923, p. 91.

53. The royal cubit consisted of 7 palms, not 6, and is described as the distance between a man's elbow and the tip of his extended middle finger. Its usual equivalence is 52.4 centimeters. One hundred royal cubits was called a *khet*. The *aroura* was equal to a square of 1 *khet* by 1 khet, or 10,000

cubits squared. Herodotus tells us that each man received a square parcel of land, but while we know them as *arouras* (= rectangles), the expression of these plots in terms of squares 1 *khet* by 1 *khet* might explain this discrepancy.

54. Imhausen 2003, pp. 250–51.
55. Clagett, vol. 3, 1999, p. 163.
56. Peet 1935, p. 432.
57. Peet 1935, p. 432.
58. Here I am following the lead of Brunes 1967, pp. 218–20, but without getting entangled with the broader theory of the origins of geometrical thinking upon which he speculates. I am following him by placing the geometric problems on a square grid, as he does, but he never acknowledges or suggests the grid technique as a familiar process of making large-scale paintings for tombs and temples. I am also following some of his written descriptions of how the problems were reckoned.
59. Robins 1994, p. 57, where, discussing Egyptian statues, she emphasizes the importance of the grid system used by Egyptian sculptors.
60. Peet 1923, p. 94.
61. Clagett 1999, vol. 3, p. 164.
62. Peet 1923, p. 98.
63. Peet 1923, p. 97.
64. Cf. Brunes 1967, pp. 223–25.
65. Again, I am following Brunes 1967, pp. 226–27, without arguing for his general, speculative theories. I place the visual problem on the square grid following the common technique by tomb painters.
66. Proclus 250.22 (Friedlein 1873, and Morrow 1970, p. 195).
67. Proclus 157.11 (Friedlein 1873, and Morrow 1970, pp. 124ff).
68. Proclus 299.3 (Friedlein, 1873, and Morrow 1970, p. 233ff).
69. Burkert 1972, pp. 416–17.
70. Proclus 352,14–18 (Friedlein 1873, and Morrow 1970, p. 275).
71. And to show that angles match up to angles (even when the sides are unequal but proportional, when Thales measured at the time of day when the shadow cast was not equal but proportional) to show similarity.
72. Von Fritz, *ABG* 1955, 76ff, *ABG* 1959 *passim*.
73. Becker, 1957, p. 37ff.
74. Szabó 1958, pp. 106–31.
75. Diogenes Laertius (D.L.) I.27; cf. also Heath 1921, I. 129, who regards this report as the original of the three.
76. Pliny *N.H.* XXXVI.12 (17).
77. D.L. II, 1–2 (= DK12A1) and also in the *Suda* (cf. K-R 1971, p. 99, #97).
78. Aristotle, *Metaphysics* 983b9–11.
79. Aristotle, *Metaphysics* 983b10–11.
80. Aristotle, *Metaphysics* 983b13–14.
81. Proclus 379, 2–5 (Friedlein 1873, and Morrow 1970, p. 298).
82. Allman 1889, pp. 11–13. Cf. also Heath 1921, I, pp. 133–36.
83. Allman 1889, pp. 11–12, who refers to them as οἱ παλαιοί. Geminus is quoted in Apollonius's *Conica* (Cf. Fried and Unguru, 2001); cf. also Heath 1908/1956, vol. II, pp. 318ff.
84. Cf. Fried and Unguru 2001, referring to the passage (Heib.) 170-4-10-3.
85. Here I am using the translation by Michael Fried.
86. Gow 1923, p. 143. Italics supplied for emphasis.
87. D.L. I.24–25.

Introduction   245

88. Heath 1921, p. 143.
89. Heisel, *Antike Bauzeichnungen*, 1993.
90. Schädler 2004.
91. I acknowledge gratefully some of the phrasing here concerning these roof tiles by Ulrich Schädler, who clarified some of these points in correspondence with me.
92. The technical German expression is *Kreissegment*.
93. Schädler 2004, which he identifies as Abb.7: *Projektion der Reconstruktion auf das Blattstabfragment*.
94. Kienast 2005, p. 37.
95. Herodotus 3.60.
96. Kienast 1995, p. 172; cf. also pp. 150–55. "Es kann kein Zweifel bestehen, daß vor der Inangriffnahme dieser Korrektur eine genau Bauaufnahme der beiden Stollen erarbeitet worden war, und es kann auch kein Zweifel bestehen, daß anhand dieser Bestandspläne der weitere Vortrieb überlegt wurde, auf daß mit möglichst geringem Aufwand mit möglichst hoher Sicherheit der Durchstoß gelänge." I am including here a series of quotations from his report as he argued that Eupalinos must have made a scaled-measured plan:

p. 150
Das Besondere dieses Systems liegt darin, dass die gemessene Einheit klar benannt ist und vor allem natürlich auch darin, dass es mit seiner Zählung von außen nach innen den Vortrieb des Tunnels dokumentiert.

"The most special feature about this system is that the units of measurement are clearly defined and that it documents the digging of the tunnel from the outside to the inside."

p. 166
Da das Dreieck durch die oberirdisch gemachten Beobachtungen nur in groben Zügen festgelegt war, die genaue Größe aber ausschließlich im Ermessen des Baumeisters lag, kann diese Koinzidenz kein Zufall sein—sie zeigt vielmehr, dass die gewählte Form des Dreiecks mit Maßstab und Zirkel am Zeichentisch erarbeitet wurde.

"This cannot be a mere coincidence. The triangle was only defined broadly according to the observation above ground. The exact size, however, could only be defined at the discretion of the architect. Thus, it indicates that the chosen form of the triangle was constructed with a scale and a compass at the drawing table."

p. 167
Zurückzuführen lässt sich dieser Fehler am ehesten auf die Eupalinos zur Verfügung stehenden Zeichengeräte und vor allem auf den starken Verkleinerungsmaßstab, mit dem er seinen Tunnelplan gezeichnet haben wird.

"This mistake is most likely attributable to the drawing devices which were available for Eupalinos and especially to the immense reduction of the scale with which Eupalinos had apparently drawn his tunnel plan."

p. 165
Die beiden Stollen werden nicht als gleichwertig pendelnde Größen behandelt, sondern einer wird zum Fangstollen, auf den der andere auftrifft, und die Längen der T-Balken sind eine klare Vorgabe für die einkalkulierte Fehlerquote; sie werden nur—und erst

246   Notes to Introduction

dann—verlängert, wenn der Durchschlag nach Ablauf der Sollzeit immer noch nicht erfolgt ist.

"The two tunnels were not considered equal but one of them became the 'Fangstollen' that the other one was supposed to meet. The lengths of the I-beams were a clear guideline for the calculated error rate; they were only lengthened if the meeting of both tunnels would not occur after the passing of the calculated required time."

p. 170
Wie die tatsächlichen Fehler gemessen und verifiziert wurden, wie die Summe der Abweichungen in Länge und Richtung in der Praxis ermittelt wurde, lässt sich nicht mehr feststellen. Erzielen lässt sich dergleichen am ehesten mit einer regelrechten Bauaufnahme, bei der die Messergebnisse vor Ort und maßstäblich auf einem Plan aufgetragen werden.

"How the actual errors were measured and verified, and how the amount of the deviations of the length and direction were calculated, cannot be determined anymore. This was most probably achieved by making an architectural survey for which the measurement results were added on a plan on-site and true to scale."

p. 170
"When analyzing the original and the new measurement it became obvious that the Λ in one measuring system has shifted by 3 m against the K of the other system. Thus, Eupalinos had assumed the additional digging length resulting from the triangle detour to be 17.60 m. Nevertheless, the additional length that resulted from a triangle with a ratio of 2 to 5 and a total length of both legs being 270 m was 19.20 m. The difference of 1.60 m between these two numbers amounts to 8 percent and is therefore relatively high. However, at this distance, this was scarcely of any importance. This mistake is most likely attributable to the drawing devices that were available for Eupalinos and especially to the immense reduction of the scale with which Eupalinos had apparently drawn his tunnel plan."

97. Hahn 2003, pp. 105–21.
98. Kienast 2005, p. 49.
99. Vitruvius III, 3.7.
100. This is a central thesis of all three studies—Hahn 2001, 2003, and 2010.
101. Hahn 2001, chapter 2; 2003, pp. 105–9.
102. Cf. Kienast 2005, p. 47; on the eastern wall there are some 400 different measure marks, painted in red, belonging to other marking systems that Kienast identifies with controlling the depth of the water channel.
103. The inscription is bordered by two vertical lines that give a distance of 17.10 m. The difference to be measured at the starting point counts 17.59 m and the distance that can be earned at the two letters Λ (original system) and K (shifted system) is 239.80–236.80, that is, 3 m. If we subtract 3 m from the average distance of the system, that is, 20.60, the outcome is 17.60 m. That means that we have two times the same distance of 17.60 m—and the distance at the inscription differs from this ideal measure some 50 cm. But note also, there are those such as Wesenberg who have suggested, instead, that PARADEGMA is intended to indicate the ideal of a strengthening wall, since it is written on one.
104. Kienast 2005, p. 54.

105. Wesenberg 2007 had a very different conjecture about PARADEGMA. He speculated that PARADEGMA painted on the strengthening wall might have indicated that this "wall" was an example of how a strengthening wall should be made.

106. Netz 1999, p. 60.

107. Hahn 2001, 2003, 2010.

108. Cf. Coulton 1977, pp. 49–52. Cf. also Hahn 2003.

109. Suda s.v. Cf. also K-R p. 99, and Hahn 2001, chapters 2 and 3; and 2003, pp. 105–25.

110. Netz 2004, p. 248.

Chapter 1

1. Cf. my discussion of "geometric algebra" in the Introduction, section C. Some caution and consideration should be mentioned here. Both Unguru and van der Waerden had a running debate over whether and to what degree Euclid and geometrical knowledge was amenable to algebraic form. It has been customary, even in Heath, to treat the contents of Euclid Book II as "geometric algebra," that is, as covering how simple algebraic equations would be solved by geometric methods alone. Cf. also Victor J. Katz and Karen Hunger Parshall, Taming the Unknown: A History of Algebra from Antiquity to the Early Twentieth Century, Princeton University Press 2014. I side with Unguru and against van der Waerden on this question. The Greeks did not have algebra.

2. While the term intuitions is used by mathematicians in specific ways, I am using it in a general sense to suggest having an insight into how things are without giving (or, perhaps, being able to give) clearly articulated reasoning. When I discuss the intuitions that are presupposed or must be connected to understand this theorem or any other, I mean the ideas or principles that must be connected.

3. Though some have doubted that Thales knew this theorem, on the grounds that Proclus infers that Thales must have known it, since he regarded it as a presupposition of measuring the distance of a ship at sea.

4. Cf. Szabó 1958 and 1978.

5. Cf. Saito 1985, p. 59: "The necessity and significance of these propositions [in Book II of the Elements] lies in the distinction of 'visible' and 'invisible' figures in Euclid. The propositions II 1–10 are those concerning invisible figures, and they must be proved by reducing invisible figures to visible ones, for one can apply to the latter the geometric intuition which is fundamental in Greek geometric arguments."

6. Proclus, 376–77 (Friedlein 1873, and Morrow 1970, pp. 295–96).

7. Proclus, 403–4 (Friedlein 1873, and Morrow 1970, pp. 318–19).

8. The term ἀναγράφομαι is an interesting choice of words; it literally means to inscribe, as if to put down in a definite way or place, not to be moved. It also became a technical term by which architects specified the details of a building, like the one built in Peireus—the anagrapheus supplied the contract in writing, inscribed on a stele.

9. I owe this clarification to Ken Saito, who added that the Latin translation of 1572 should have the same diagrams.

10. Cf. Saito's website for the diagrams.

11. Von Fritz 1945.

12. The strategy of V.24 has nothing to do with triangles or compounding (with one reservation): the important tools that Euclid has developed in Book V are designed with an eye to ratios and proportions of all kinds of magnitudes, even if he uses lengths as his standard representation (which he

also uses for numbers). Book V and Book X are unique in this regard, that is, in this kind of generality. The three important moves he employs, besides the implicit general definition of proportion, are componendo (sunthesis), separando (diairesis), and ex aequali (di' isou): none of these is really an operation; rather, they say that if you have one proportion of a set of proportions, you will have another as well. Componendo is not compounding: if A : B :: C : D then one can say, COMPONEDO that (A + B) : B :: (C + D) : D. Ex aequali says that if A : B :: C : D and B : E :: D :: F then, EX AEQUALI, A : E :: C : F. As for the reservation, there are times in Apollonius, but not in Euclid, as far as I recall, that Apollonius will say this is by compounding, namely, the ratio A : E is compounded of A : B and B : E. But this, as we have discussed in the past, is as strange as it is common (it probably comes from music theory). Still, note that ex aequali is defined for any number of ratios.

13. Madden (James), "Ratio and Proportion in Euclid." Unpublished manuscript, 2009. www.lamath.org/journal/vol5no2.

14. Cf. Artmann 1999, p. 136.

15. Just a side comment: A modern would be tempted to say that equality is a kind of proportion, that is, A = B can just as easily be written A : B :: 1 : 1. This may well be a subtle error because numbers are magnitudes, but 1 is a very problematic number in Greek mathematics and philosophy, as the extensive secondary literature shows.

16. Let me note here that while big and little share the same structure in similar figures, some consideration of subtleties is in order. When we enlarge a figure to produce a similar figure, the sides increase at a different rate than the areas, and for that matter, the volumes (which increase as the cube of the ratio of the sides); in fact, this is what it means that the areas are in the duplicate ratio of the sides! Thus the subtlety here is that we cannot have an elephant-sized ant—its volume, and therefore its weight, would increase faster than the cross-section of the legs and it would not be able to support itself. Thus, scaling is very subtle.

17. Cf. the Introduction, section D.

18. I owe this phrasing to my colleague Michael Fried, who was kind enough to clarify this point for me.

19. Again, I owe this clarification and phrasing to Michael Fried. To get even clearer about this point that duplicate ratio is a state of affairs and not a procedure or operation, we should become attentive to Euclid's use of the perfect-passive-imperative Εἰλήφθω, "Let BG have been taken." How? By whom? According to Marinus Taisbak 2003, it is taken by the ever-ready "helping hand"! In other words, this is not an operation, but a state of affairs prepared for us so that we can proceed.

20. Usually the shorthand U : W = comp(V : W, U : W) is used.

21. Giventhal, https://math.berkeley.edu/~giventh/.

22. Of course it should be pointed out that historians of mathematicians such as Heath disagree. It is hard to accept Proclus's claim and still explain Books VII–IX, for instance. A familiar objection is that Proclus's assertion of a metaphysical underpinning reflects his Neoplatonic motives. He understands things in a Platonic light, as he interprets it, and so expressing such a metaphysical underpinning should come as no surprise.

23. Proclus, 65.20 (Friedlein 1873, and Morrow 1970, pp. 52ff.).

24. Netz 1999, p. 255.

25. Gow 1923: "Thales might be held to have known the first six books of Euclid." P. 143.

26. Cf. Hahn 2001, chapter 2.

27. For a very nice presentation of a proof, cf. "The Pythagorean Theorem Is Equivalent to the Parallel Postulate" at www.cut-the-knot.org.

28. For a fuller presentation of the proofs that the Pythagorean theorem is equivalent to the parallel postulate, cf. Scott E. Broadie's discussion in Bogomolny 2016.

29. Neugebauer 1957, pp. 208–26. Cf. my discussion of geometric algebra in the Introduction, section C.

30. I owe the phrasing here to Michael Fried, who kindly helped me to express this point.

31. In fact, if you think of the Pythagorean theorem as a fundamental expression of the distance between points, then a sense of distance, or, more technically, a metric, in these geometries, Euclidean and non-Euclidean, is very different from how we commonly think of distances in everyday experience.

32. In non-Euclidean geometry—both elliptic (Riemannian) and hyperbolic (Lobachevskian)—the angle sum of a triangle is a function of its size. We know this in retrospect, but of course, Thales, Pythagoras, Plato, and Euclid could hardly have appreciated it. Now if two triangles are similar they must at least have their angle sums be the same, so that in non-Euclidean geometry they would have to be also the same size, that is, congruent. Accordingly, one cannot speak about similar figures (for example, squares) on the sides of a right triangle in these non-Euclidean geometries, only in Euclidean geometry.

33. I include semicircles here but with the caveat that Euclid had to wait until Book XII to show that circles are to one another as the squares on their diameters, and thus the situation with circles and semicircles is different from that with rectilinear figures. But the situation is true nonetheless.

34. I owe this reflection and the phrasing in the next two paragraphs to mathematician Travis Schedler.

35. A *tessellation* of a flat surface is the tiling of a plane using one or more geometric shapes.

36. I owe the phrasing of this point to mathematician Jerzy Kocik.

37. Couprie 2003, and more fully in 2011.

Chapter 2

1. Heath 1931, pp. 87–88.

2. Heath 1931, pp. 87–88. I have added in italics "*the base angles of an isosceles triangle are equal,*" which is Euclid I.5, and its converse "*and if the base angles are equal, the sides subtended by the equal angles are also equal.*"

3. Neugebauer 1957, "It seems to me evident, however, that the traditional stories of discoveries made by Thales or Pythagoras must be discarded as totally unhistorical" (p. 148).

4. Dicks 1970, p. 43, and Dicks *CQ*, 1959.

5. van der Waerden 1954, p. 89.

6. van der Waerden 1954, p. 89.

7. Burkert 1972, p. 416.

8. Aristophanes, *Birds*, 995ff., 1009.

9. Becker, *MD* 1957, pp. 37ff.

10. von Fritz, *ABG* 1955.

11. Burkert 1972, p. 417, referring to Szabó 1958.

12. Burkert 1972, p. 417, referring to the assessment by Szabó 1958.

13. Even in Pappus, fourth century CE, we have reference to books containing not theorems but diagrams. Heath and others tend to put the word "proposition" or "demonstration" in actual or implied brackets. The reader may benefit by reviewing Netz 1999, esp. chapter 1, "The Lettered Diagram."

14. Callimachus. *Iambi* I. 55–58. While Callimachus recites this observation to suggest that it was a geometrical diagram discovered by Phrygian Euphorbus, an earlier incarnation of Pythagoras, Thales is believably imagined making geometrical sketches in the sand: ἐν τοῦ Διδυμέος τὸν γέροντα κωνήῳ ξύοντα τὴν καὶ γράφοντα τὸ σχῆμα.

15. Thus, proof begins for Thales as a response to his doubting compatriots; if this supposition seems troubling, I urge the reader to make a trip to Greece and meet the Greek people. Make any claim about anything—anything at all—and the natural character of the people appears. They want to know how you know whatever it is you claim. Even today, they still demand—in a way remarkable to their character—the proof of one's assertions, an explanation or justification of the things someone asserts!

16. van der Waerden, of course, is an exception because he is providing a systematic study of ancient mathematical knowledge.

17. Zeller 1866.

18. Burnet 1892, 1930.

19. Kirk-Raven 1957, Kirk-Raven-Schofield 1983.

20. Guthrie 1962.

21. Barnes 1982, pp. 38–57.

22. Allen 1991.

23. Taylor 1997, the essay by Malcolm Schofield is pp. 47–87.

24. Roochnik 2004.

25. McKirahan 1994.

26. O'Grady 2002. Cf. chapters 2 and 10.

27. Derrida 1978.

28. Heath 1931, p. 87.

29. Von Fritz 1945, p. 259.

30. D.L. I. 27. Heath (1921) translation I. 129. Heath regards this version to be the original over the other two, including Plutarch's description of a proportional technique when the shadow is unequal to its height.

31. Pliny, *Natural History* XXXVI. 12 (17). Heath (1921) translation I. 129.

32. Robins and Shute, 1987, pp. 47–49.

33. Peet 1923, p. 3: "The Rhind Papyrus is dated in the 33rd year of the Hyksos King Aauserre Apophis who must have ruled sometime between 1788 and 1580 BC. The papyrus contains mention by the scribe that it was a copy of an older document written in the time of King Nemare, Amenemmes III of the twelfth dynasty who was on the throne from 1849 to 1801 B.C." Thus the scribe Ahmose lived during the time of the Hyksos and the second intermediate period; he tells us his papyrus is a copy of another dating to the reign of Amenenhat III; and it is likely that this still points to a much older time, reaching into the early part of the Old Kingdom when pyramid building was in its heyday and the calculations would have been immediately relevant to building programs.

34. In Peet's translation of the RMP (1923), he refers to the *seked* as the "batter," p. 97–102, and the specific reflections on "batter" appear on p. 98.

35. It is of course possible that he made the measurement in "measured steps"—pacing out the length of the shadow and adding to it the number of paces across half of any one of the sides of the pyramid's ground-level square. But the availability of the royal cubit cord seems more likely.

36. Peet 1923, RMP problem 57. Note that the diagram of the pyramid is too steep to be a drawing accurate to the problem, and so the illustration seems clearly intended to be suggestive of the problem, that is, schematic.

37. Robins and Shute 1987, p. 47, state the problem slightly differently: "This is obtained by dividing 7 by twice the *seked* to get 1/3 palms, which is multiplied by 140 to give the height of 93 + 3 parts cubits."

38. DK 11A21, which is from Plutarch, commonly referred to by the Latin title *Convivium septem sapientim* 2, which translates the Greek Συμπόσιον τῶν ἑπτὰ σοφῶν. Italics supplied for emphasis. For the arguments for its authenticity as a work of Plutarch, see Betz 1978.

39. Because of the precession of the equinoxes, the window of opportunity 2,700years ago would have involved a different assignment of dates. These dates reflect the 2012 calendar.

40. Diogenes Laertius II. 1–2 (DK 12A1); cf. also my chapter 6 in AOP (2010) where I propose a reconstruction of what the seasonal sundial might have looked like, give the fact that the earliest surviving Greek sundials date to Hellenistic times, some three centuries later.

41. Proclus on the authority of Eudemus. 379 (Friedlein 1873, and Morrow 1970, p. 298).

42. Clearly, while it seems obvious that every isosceles triangle contains two right angles, the proof that *all* triangles contain two right angles needs a different approach, and this is by way of parallel theorems that I shall get to shortly.

43. The solar calculator consulted http://www.esrl.noaa.gov/gmd/grad/solcalc/.

44. The solar calculator consulted was http://www.esrl.noaa.gov/gmd/grad/solcalc/.

45. I do acknowledge, however, that he could also have noticed more simply that in early February, the shadow of the sun at noon crept from day to day nearer toward the length of the gnomon, and at the right day have made his measurements. And similarly, mutatis mutandis, for the east or west shadow times. Thus some might say in objection: the simpler the method, the better the chance that it was used. But the reports that connect him, and Anaximander, with marking both solstices and equinox suggest a more reflective knowledge.

46. Burkert 1972, pp. 416ff., argued that a work by this title found its way into Proclus's hands and accounts for his reports on Thales, whether or not the Milesian is the author of that book. But while the book may not have been authored by Thales, the reports form a picture that is hard to discard.

47. Diggins 1965, pp. 59–68.

48. Proclus, 427 (Friedlein 1873, and Morrow 1970, p. 337).

49. My argument in no way turns on the awareness of the irrational. At this stage, I am making no claim that Thales was aware that there was no number that corresponded to the length of the hypotenuse, but rather that he must have supposed it was some fraction above the nearest whole number; nor am I supposing that Thales was aware that there would be no fundamental length in terms of which all lengths could ultimately be reduced.

50. Callimachus, *Iambi*, 4: the poem relates how Thales was making a diagram in the dust in front of the temple of Apollo at Didyma. But in a good-natured way, Callimachus suggests that Thales learned this from the soldier Euphorbus, who was an earlier incarnation of Pythagoras.

51. The measurement by comparison with the height of his own shadow conveys the idea but lends itself to great imprecision. Setting up a gnomon is far more precise, but is also not so easy to construct. He would need a long flat stone, and by means of the *diabêtês* ("level with a plumb line") could be certain that the stone was level and the gnomon was at right angle to it.

52. I wish to acknowledge some caution about these remarks. It may be pointed out that this sort of measurement is an *analogia* in an ordinary sense, what Aristotle refers to as "likeness," which of course is related to "similarity." But this kind of nonmathematical way of thinking about these things is very different from the precise way developed by Eudoxus. I wish to make it clear that I accept such a point, in broad stroke, but to see the little world and big world sharing the same structure does not require such precision. Even without the fine, rigorous argument, the general interconnection can nevertheless be seen.

53. Cf. von Fritz 1945, p. 261.

54. Von Fritz 1945, p. 250.

55. In this report, Thales set up a stake or gnomon at the end of the pyramid's shadow and then reckoned that as the length of the shadow cast by the stake was to its shadow, so also was pyramid's shadow was to its height. They both were in the same ratio.

56. DK11A21, which reports *Convivium septem sapientiae*. 2, p. 147 A, translated by Heath 1921, p. 129.

57. There is a likelihood that this is just how projects were carried out at the building sites under the supervision of the architects. Exemplars—*paradeigmata*—of architectural elements would have been set up at the building sites and workers would have replicated these elements, producing products as close to identical to the exemplars as possible.

58. Naturally, the measurement from a height above the sea adds another difficulty to the measurement, and for these concerns, cf. Heath, 1921, I, pp. 131–33.

59. Dantzig 1954, pp. 97ff.

60. Proclus, 250.23–251.1 (Friedlein 1873, and Morrow 1970, p. 195).

61. Proclus (Morrow 1970), p. 52 (Prologue).

62. Peet 1923, p. 92. See my Introduction to trace the proposed diagrams.

63. Herodotus II.109. The term is τετράγωνον at II.109.4.

64. Cf. Hahn, "Heraclitus, Milesian Monism, and the Felting of Wool," 2016, where I discuss the "process" by which structural change is effected in Thales's picture.

65. D.L. I.24–25.

66. Heath 1931, p. 82.

67. Heath 1908, Book I, definition 17.

68. Hahn 2010, chapter 3, pp. 53–86.

69. Ibid. p. 84.

70. Most recently, cf. Zhmud 2012, pp. 252–53.

71. Aristotle, *Prior Analytics*, 41b13–22.

72. There is a proof in Pappus (Cf. It is reported by Proclus [Friedlein], pp. 249–50, Morrow, pp. 194–95.) that is so obvious that presenting it could be conceived as simply noticing; since AB = AC and angle A is common, triangles BAC and CAB are equal. Therefore, angles ABC = ACB (being the corresponding angles).

73. Cf. P. Lafitte, vol. II, p. 291; Finger 1831, p. 20. For the general overview of "naming" Thales's theorems, cf. Patsopoulos and Patronis 2006.

74. From Ken Saito's website, 2011.

75. Heath 1931, p. 87.

76. Heath 1931, p. 87. I have changed only the assignment of letters to match up to my other figures.

77. D.L. I. 24–25.

78. In light of the issues that Heath raises, and trying to be exceedingly cautious, he also proposes how Thales might have proved the theorem attributed to him by Pamphile, *without* assuming that the sum of the angles of a right triangle sum to two right angles, that is, by appeal to the knowledge of the principle that the base angles of a triangle with sides of equal length must be equal—one of the theorems with which Thales is credited. Heath constructs the figure with rectangles, as shown in 3.1, but it amounts to the same figure to place an equilateral rectangle that divides by a diameter into two isosceles right triangles, in the semicircle with angle relations remaining the same, below:

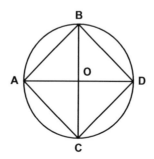

Figure N.2.1.

Heath's claim is that on the basis of Thales's knowledge of isosceles triangles (namely, that a triangle is isosceles if and only if its base angles are equal), he might have shown that any rectangle *could* be inscribed in a circle such that its diagonals will be diameters. And so it might well be that in his *Manuel* (1931) Heath interprets Pamphile's remark to suggest a kind of indirect argument, though his exact words appear at the opening of this chapter. In effect, Heath argues that if ABC is a right triangle with its right angle at A, then ABC can be inscribed in a circle whose diameter is BC. But, in his edition of *Euclid's Elements* (1908), when discussing I.32, Heath states explicitly of Pamphile's testimony that "in other words, he [sc. Thales] discovered that the angle in a semi-circle is a right angle."

79. Proclus (Morrow 1970), p. 298.

80. Cf. also Proclus, 379 (Friedlein 1873, and Morrow 1970, p. 298), who cites Eudemus in crediting the discovery that there were two right angles in every triangle to the Pythagoreans. Proclus supplies the following diagram:

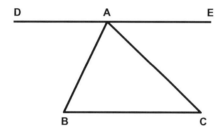

Figure N.2.2.

Thus, alternating angles DAB and ABC are equal, and angles EAC and ACB are equal, and BAC is common. And so if DAB equals ABC, and EAC equals ACB, then if angle BAC is added to each—since DAB + BAC + EAC equal a straight angle (= straight line) or two right angles, there must be two right angles in every triangle. Cf. also Allman 1889, pp. 11–12, referring to *Apolloni Conica*, ed. Halleius, p. 9 Oxon 1710. Cf. Heath 1908, I. p. 318), who credits Allman (1889, p. 12).

81. Dantzig 1954, p. 30.

82. Dantzig 1954, p. 97.

83. Cf. Gregory 2013, p. 52, just for the most recent mention and discussion of the Milesians as hylozoists, though Gregory prefers the term panpsychists. But the point is that that scholarship has recognized for a very long time that the Milesians regarded the whole cosmos as alive.

84. Dantzig 1954, p. 95.

85. Dantzig 1954, p. 96: "the Pythagorean relation between the sides of a right triangle was equivalent to the Euclidean *postulate of parallels*."

Chapter 3

1. Zhmud 1989, pp. 249–68.
2. D.L. VIII, 12.
3. Zhmud 2012, p. 59.
4. Vitruvius, Book IX, 6–8.
5. Who originally learned it from Apollo, that is, from the god.
6. Zhmud 2012, pp. 256–57.
7. I note here that trying to connect the dots backward, as it were, from Hippocrates to Hippasus and earlier still, follows the general lines that lead to the hypotenuse theorem of I.47 and not VI.31.

254   Notes to Chapter 3

In this line of thought, the requirement that the triangle be right is relaxed—it is still squares built on the sides, but in any case the lines of thought through Hippocrates are not by ratios and proportions.

8. Cf. Fried 2001. Eutocius, in his commentary on Apollonius, for example, says explicitly that he made some of the proofs clearer. And if we compare Eudemus to what mathematicians do today, he could very well have clarified what Hippocrates's argument "really" was.

9. Heath 1921, I, p. 193.

10. Heath 1921, I, p. 195.

11. Here we have a specific generalization. Hippocrates, according to Eudemus, was able to show that circles are to one another as the square on their diameters (Euclid's proof of this is given in Book XII of the Elements), from which, apparently, Hippocrates inferred that also segments of circles, like AGB and AFC related to one another as the squares of their bases, that is, AGB : AFC :: sq.AB : sq.AC.

12. Euclid X.1: Two unequal magnitudes being set out, if from the greater there is subtracted a magnitude greater than its half, and from that which is left a magnitude greater than its half, and if this process is repeated continually, then there will be left some magnitude less than the lesser magnitude set out.

13. Heath 1949, pp. 23–24, though the diagram here follows Zhmud 2012, p. 253, rather than Heath. Zhmud proposes a diagram that begins with the diameter of the circle, while Heath, claiming a close similarity to that of Pacius, does not.

14. Cf. Patsopoulos and Patronis 2006, esp. the account of "Thales's theorem" on isosceles triangles in French textbooks, p. 61.

15. Aristotle, *Prior Analytics*, 23, 41a23–27, and again at 44, 50a35–38.

16. The mathematics lesson in Plato's *Meno* shows how to find the side of a square which is twice a given square. It does not rely on the Pythagorean theorem. It also does not show that the side which is found is incommensurable with the side of the original square (although that fact is undoubtedly in the background—for example when Socrates tells the slave to point if he cannot say).

17. The proof that the diagonal of a square is incommensurable with its side (Euclid Book X, proposition 117). Heiberg Appendix 27.

Προκείσθω ἡμῖν δεῖξαι, ὅτι ἐπὶ τῶν τετραγώνων
σχημάτων ἀσύμμετρός ἐστιν ἡ διάμετρος
τῇ πλευρᾷ μήκει.

Ἔστω τετράγωνον τὸ ΑΒΓΔ, διάμετρος δὲ αὐτοῦ
ἡ ΑΓ λέγω, ὅτι ἡ ΓΑ ἀσύμμετρός ἐστι τῇ ΑΒ μήκει.
Εἰ γὰρ δυνατόν, ἔστω σύμμετρος λέγω ὅτι συμβήσεται
τὸν αὐτὸν ἀριθμὸν ἄρτιον εἶναι καὶ περισσόν.
φανερὸν μὲν οὖν, ὅτι τὸ ἀπὸ τῆς ΑΓ διπλάσιον τοῦ
ἀπὸ τῆς ΑΒ. Καὶ ἐπεὶ σύμμετρός ἐστιν ἡ ΓΑ τῇ ΑΒ,
ἡ ΓΑ ἡ ἄρα πρὸς τὴν ΑΒ λόγον ἔχει, ὃν ἀριθμὸς πρὸς
ἀριθμόν. ἐχέτω, ὃν ὁ ΕΖ πρὸς Η, καὶ ἔστωσαν οἱ
ΕΖ, Η ἐλάχιστοι τῶν τὸν αὐτὸν λόγον ἐχόντων αὐτοῖς.
οὐκ ἄρα μονάς ἐστιν ὁ ΕΖ. εἰ γὰρ ἔσται μονὰς ὁ
ΕΖ, ἔχει δὲ λόγον πρὸς τὸν Η, ὃν ἔχει ἡ ΑΓ πρὸς
τὴν ΑΒ, καὶ μείζων ἡ ΑΓ τῆς ΑΒ, μείζων ἄρα καὶ
ἡ ΕΖ τοῦ Η ἀριθμοῦ, ὅπερ ἄτοπον. οὐκ ἄρα μονάς
ἐστιν ὁ ΕΖ, ἀριθμὸς ἄρα. καὶ ἐπεί ἐστιν ὡς ἡ ΓΑ
πρὸς τὴν ΑΒ, οὕτως ὁ ΕΖ πρὸς τὸν Η, καὶ ὡς ἄρα
τὸ ἀπὸ τῆς ΓΑ πρὸς τὸ ἀπὸ τῆς ΑΒ, οὕτως ὁ ἀπὸ
τοῦ ΕΖ πρὸς τὸν ἀπὸ τοῦ Η. διπλάσιον δὲ τὸ ἀπὸ

τῆς ΓΑ τοῦ ἀπὸ τῆς ΑΒ, διπλασίων ἄρα καὶ ὁ ἀπὸ
τοῦ ΕΖ τοῦ ἀπὸ τοῦ Η. ἄρτιος ἄρα ἐστὶν ὁ ἀπὸ τοῦ
ΕΖ, ὥστε καὶ αὐτὸς ὁ ΕΖ ἄρτιός ἐστιν. εἰ γὰρ ἦν
περισσός, καὶ ὁ ἀπ᾽ αὐτοῦ τετράγωνος περισσὸς ἦν,
ἐπειδήπερ, ἐὰν περισσοὶ ἀριθμοὶ ὁποσοιοῦν συντεθῶσιν,
τὸ δὲ πλῆθος αὐτῶν περισσὸν ᾖ, ὁ ὅλος περισσός ἐστιν,
ὁ ΕΖ ἄρα ἄρτιός ἐστιν. Τετμήσθω δίχα κατὰ τὸ Θ,
καὶ ἐπεὶ οἱ ΕΖ, Η ἐλάχιστοί εἰσι τῶν τὸν αὐτὸν λόγον
ἐχόντων [αὐτοῖς], πρῶτοι πρὸς ἀλλήλους εἰσίν. καὶ
ὁ ΕΖ ἄρτιος, περισσὸς ἄρα ἐστὶν ὁ Η. εἰ γὰρ ἦν
ἄρτιος, τοὺς ΕΖ, Η δυὰς ἐμέτρει. πᾶς γὰρ ἄρτιος
ἔχει μέρος ἥμισυ, πρώτους ὄντας πρὸς ἀλλήλους, ὅπερ
ἐστὶν ἀδύνατον. οὐκ ἄρα ἄρτιός ἐστιν ὁ Η περισσὸς
ἄρα, καὶ ἐπεὶ διπλάσιος ὁ ΕΖ τοῦ ἀπὸ ΕΘ τετραπλάσιος
ἄρα ὁ ἀπὸ ΕΖ τοῦ ἀπὸ ΕΘ. διπλάσιος δὲ ὁ ἀπὸ τοῦ
ΕΖ τοῦ ἀπὸ τοῦ Η. διπλάσιος ἄρα ὁ ἀπὸ τοῦ Η τοῦ
ἀπὸ ΕΘ. ἄρτιος ἄρα ἐστὶν ὁ ἀπὸ τοῦ Η. ἄρτιος ἄρα
διὰ τὰ εἰρημένα ὁ Η. ἀλλὰ καὶ περισσός. ὅπερ ἐστὶν
ἀδύνατον. οὐκ ἄρα σύμμετρός ἐστιν ἡ ΓΑ τῇ ΑΒ
μήκει. ὅπερ ἔδει δεῖξαι.

18. Zhmud 2012, p. 272; Becker 1934, pp. 533–53, and 1966, pp. 44ff.; Reidemeister 1949, pp. 31ff.; von Fritz 1963, pp. 202ff.; Knorr 1975, pp. 135ff.; van der Waerden 1954, pp. 108ff.; Waschkies 1989, pp. 29ff., 269ff.; Szabó 1978, pp. 246ff.; Burkert 1972, pp. 434ff.

19. I gratefully acknowledge the clarification offered by Michael Fried in clarifying the proof, and for helpful phrases, that I follow here.

20. Why Heiberg removed "X.117" from the rest of Book X and saw it as a later addition is probably because it is somewhat different in character from the general direction of the book. But to say that it is a later addition is not necessarily to say that the provenance of the proposition is also late. Indeed, the famous passage in Aristotle's *Prior Analytics* where he says that the incommensurability of the side and diagonal of a square is obtained by assuming they are commensurable and then showing that "an odd number is equal to an even number" hints at a proof like that in X.117.

21. But it might be objected that there are, in fact, an infinite number of Pythagorean triples—and this can be shown with pebbles, as I suggest later in this chapter—so perhaps the problem was not the rarity of Pythagorean triples, as we tend to say today.

22. In this paragraph, I am indebted to Dantzig 1954, chapter 8: "The Hypotenuse Theorem," pp. 95–107.

23. Dantzig 1954, p. 106.

24. Even if it be objected that it was the Greeks who marveled at Thales's technique and result—not the Egyptians at all—it still would not have been memorable had there not been an awareness of an Egyptian result that was preciously close.

25. Cf. DK 12A1 (D.L. II, 1–2).

26. Cf. Hahn 2001, p. 62.

27. There are no such semicircles in the Oxyrhynchus papyrus for example.

28. Heath 1949, p. 20.

29. The familiar phrase is "the harmony of the spheres" but the expression, traceable to Plato in *Republic* X, is to "circles" not "spheres."

30. Riedweg 2005, 80.

31. Zhmud 2012, pp. 394ff.
32. Zhmud 2012, p. 268.
33. Zhmud 2012, p. 286.
34. Zhmud 2012, p. 290.
35. Nicomachus *Harm.* 6. Cf. also Zhmud 2012, p. 291.
36. Creese 2010, pp. 87ff., who also refers to Barker 1989, p. 495 n.4.
37. Barker 1991, p. 52.
38. Barker 1991, p. 52.
39. Cf. also Barker 1991, p. 53. I would like to gratefully acknowledge musicologist Jay Matthews, who reviewed my discussion of Barker and made helpful comments.
40. Szabó 1978.
41. Creese 2010 dates the monochord to the fourth century BCE, pp. 11, 47, 67–68; and see also p. 87 for reflections on what Pythagoras may have known about this.
42. DK 47 B2.
43. Cf. Luigi Borzacchini 2007.
44. Cf. Aristotle, *Pseudepigraphus*, Rose, fragment 43.
45. The music of the spheres is mentioned by many authors, but Aristotle comments, *De Caelo* 290B12ff., that while the notion of a music of the spheres is pretty and ingenious, it's not true. He reports the (Pythagorean?) explanation that we do not hear it because we have been accustomed to it from birth. For a discussion of the Pythagorean theory of the harmony of the spheres see Heath 1913, pp. 105–13, where this passage is both translated and discussed.
46. Plato. *Republic*, 617c.
47. Aristotle. *De Caelo* 290B12–29.
48. Proclus, 65.16–18 (Friedlein 1873, and Morrow 1970, p. 52).
49. Young 1966.
50. Aristotle, *Metaphysics* 985b24–986b7.
51. Herodotus III.60, though Herodotus mentions the tunnel first, the sea-mole second, and the temple third. Since Theodorus is identified by Vitruvius as the architect of the Samian Heraion, this suggests he is connected first with Dipteros I (begun shortly after 575 BCE and taken down around 550). Since Herodotus mentions a temple in Samos as the greatest he had ever seen, it must be Dipteros II (begun sometime soon after 550 when Dipteros I began to list under its own immense weight and the new temple was begun some 43 meters to the west), and thus this temple is identified with the architect Rhoikos, the younger contemporary of Theodorus. Cf. these important resources for details and argument, the most important of which is Hermann Kienast's *Die Wasserleitung des Eupalinos auf Samos* (1995), *and the short version in English,* The Aqueduct of Eupalinos on Samos (2005). Cf. also Van der Waerden 1968, "Eupalinos and His Tunnel"; Goodfield, Toulmin, and Toulmin (1965), "How Was the Tunnel of Eupalinus Aligned?"; Evans 1999, "Review of Hermann Kienast, *Die Wasserleitung des Eupalinos auf Samos*"; Burns 1971, "The Tunnel of Eupalinus and the Tunnel Problem of Hero of Alexandria"; Apostol 2004, "The Tunnel of Samos"; Olson 2012, "How Eupalinos Navigated His Way Through the Mountain—An Empirical Approach to the Geometry of Eupalinos."
52. Cf. Hahn 2001, chapter 2.
53. The double peristyle was *inside* the temple at Luxor, not *outside*, as evidenced in the Samian Heraion, Ephesian Artemision, and Didymaion just outside of Miletus.
54. Kienast 1992, pp. 29–42.
55. Coulton 1977, p. 25, italics supplied for emphasis.
56. Hahn 2003, pp. 105ff.
57. Vitruvius III.3.7.

58. Herodotus III.60.3.

59. Cf. the abstract by Panaghiota Avgerinou: "The Water Management in Ancient Megara." "Although the investigation has just begun the first data show that it was constructed in the sixth century BC. It consists of tunnels with the help of which the water of the underground aquifer was conveyed to the city following the natural inclination. The tunnels vertical air shafts were equally distributed on the surface. Both the tunnels and the vertical air shafts reach depths of 4.30 m to 8 m under the surface. A total of 70 air shafts can be seen today arranged in four branches of length 1100 m and width of 560 m. The chronology of technical characteristics and the whole concept allow us to ascribe this work to Eupalinos, the famous Megarian engineer, who constructed the most amazing of the ancient Greek water supply lines, the great tunnel at Samos, which carried water more than a kilometer through a mountain." Thus, it has been is speculated that perhaps Eupalinos gained fame and distinction by engineering the underground water channels in mainland Megara, to explain why he might have been selected for this project in Samos.

60. Olsen 2012 disagrees.

61. Kienast 2005, p. 52.

62. Kienast 2005, p. 53: "The increase in the length of the tunnel as a result of a triangular detour was planned simply by using a compass and counting backwards. By displacing the new numbering, which took account of the increase in length as a result of the triangular detour, it proved possible to continue the measurements in the straight sections with the original numbering."

63. Kienast 2005, p. 54.

64. Kienast 2005, p. 54.

65. Vitruvius, III.3.7. Cf. my discussion in Hahn 2003, pp. 105ff.

66. Cf. Hahn 2010, chapters 1, 2, and 12.

67. I wish to emphasize here, and repeat again, that the four elements that correspond to the tetrahedron, hexahedron, octahedron, and icosahedron are all constructed explicitly out of right triangles. The dodecahedron is constructed out of right triangles too, but only derivatively, since each pentagon resolves to triangles and every triangle resolves to right triangles.

68. Proclus 379 (Friedlein 1873, and Morrow, p. 298). Cf. Eudemus fragment 136. In Wehrli 1955, this triangle is equilateral.

69. While the reader might insist that we clarify that the two horizontal lines *must* be parallel, I am presuming that Thales, Pythagoras, and the architects knew this but might not have stipulated that they were any more than "straight lines" evenly horizontal to each other.

70. I gratefully acknowledge Michael Fried, who clarified my understanding of duplicate ratio and geometric mean by these two sets of diagrams that he prepared.

71. Heath 1921, I. pp. 147ff.

72. Cf. Zhmud 2012, p. 265.

73. Zhmud 2012, pp. 265–66.

74. Cf. Hahn 2016, where the Milesian "process" by means of which the structure is "compressed" or "rarefied" is explored through photographing modern–day felters working in traditional methods. Cf. Hahn 2016.

75. Though it should be noted that while right triangles compose the faces of the solids in a clear way, there is no real nexus between that and assembling the faces into solids. This requires a different approach to "putting together" (σύστασις).

76. Heath 1931, p. 100.

77. Heath 1931, p. 100. Heath's point seems to be something like this: If you are dealing with commensurable magnitudes, then proportion becomes only a matter of counting since the common measure of the magnitudes can be taken as a unit. Thus 3 : 5 :: 6 : 10 because 5 groups of 3 units is

the same as 3 groups of 5 just as 5 groups of 6 units is the same as 3 groups of 10. What he refers to as the Pythagorean theory of proportion (if there truly was a well-worked-out theory), accordingly, is one in which all ratios of magnitudes are the same as a ratio of whole numbers (collections of units), and the definition of proportion will be something like that in Book VII (def. 20) of the *Elements*. But you cannot use the theory when you have incommensurable magnitudes—for example, you cannot say what the ratio is of the side of the square to the diagonal, you cannot write BA : BC in your left figure. If you could, you would have to say that a certain multiple of BA contains the same number of units as a certain multiple of BC, which is impossible.

Consider VI.1, triangles under the same height are to one another as their bases. If all lines were commensurate, then BC could be divided into m units (e.g., Bx, xy, yC) of length k and CD could be divided into n units (e.g. Cu, uv, vw, wD) also of length k. Hence, by *Elem.* I.38, the triangles BAx, xAy, . . . wAD are all equal—call them unit triangles. Hence, BAC contains m unit triangles and CAD contains n unit triangles. Therefore BAC : CAD :: m : n :: BC : CD. But if it is possible that BC and CD are not commensurable, that is, where you cannot count unit segments and triangles, then this proof would not hold.

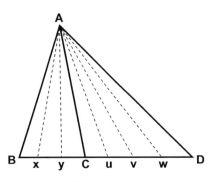

Figure N.3.1.

One more consideration. It may appear that Eudoxus's new definition of proportion is only a modification of the Pythagorean definition (~*Elem.* VII, def. 20); however, it is only superficially similar. The claim that equimultiples ALWAYS exceed or fall short demands an entirely different way of thinking: one cannot merely count.

78. I owe an explanation of this problem in Heath to my colleague Michael Fried: "The theory of ratio and proportion has two appearances in the *Elements*. One is of course in Book V and the other in the three arithmetic books, Books VII–IX, one for general magnitudes and the other for numbers (meaning whole numbers)—one concerning measuring and the other concerning counting. The two would be the same if all magnitudes were commensurable, so that the ratio between two magnitudes would be the same as a ratio between two numbers. The discovery of incommensurables made it clear that that was not always the case, and, accordingly, a more subtle theory of ratio and proportion was needed. That was the theory developed, apparently, by Eudoxus in the time of Plato. That latter theory is absolutely necessary for any discussion of similar figures such as are discussed in Book VI. In fact, if you look at VI.1, 'triangles and parallelograms under the same height are to one another as their bases,' you can really see how this would be pretty easy to prove, without going beyond Book I, IF every two lines were multiples of a common unit (you would simply use I.36, that parallelograms in the same parallels and on equal bases are equal, over and over—and, in practical situations, that is probably what WAS

done): but if the bases of the triangles were, for example, as the side of a square to the diagonal, then such a proof would no longer work—you could not say the base of the one is M times a unit length and the other N times. In the case of VI.31, the ratios of the sides of the similar figures, except where the length of the sides of the right triangle are Pythagorean triples (or similar to such a triangle), are precisely ratios which are NOT those of a number to a number. So, although a proof of VI.31 or even ONE SINGLE case TOGETHER WITH A THEORY OF RATIO that allows one to speak generally about similar rectilinear figures would make the Pythagorean theorem almost trivial, that theory of ratio is itself NOT trivial. Thus the problem."

79. Cf. Hahn 2016, article on "felting."

80. For a detailed consideration of the incommensurability, cf. Knorr 1975.

81. Cf. Herda 2013, p. 96 and p. 97 fn 168. Herda details relevant testimonies that I note here: Hipponax fr. 4, 63, 123 (West). Alcaeus fr. 448 Lobel/Page (= Himerios, *Orationes* 28.7 Colonna) may already refer to the story of the seven sages. Xenophanes (DK 21 B 19) stresses Thales's astronomic achievements as Heraclitus did (DK 22 B 38): Classen 1965, p. 931. According to Diogenes Laertius 1.22, quoting Demetrios of Phaleron (c. 350–280 BCE) in his List of Archons, Thales was the first to be called "sage" in the year of the Athenian Archon Damasias (Olympiade 49.3 = 582 BCE), exactly the year of the introduction of the Panhellenic ἀγὼν στεφανίτης in Delphi: Marmor Parium, IG XII 5, 444 ep. 38 (= *FGrHist* 239, 38); Eusebius *Chronicle*, p. 125. Kirk et al. 2001, 84 n. 1 assume that Damasias was the first who "canonized" the seven sages.

82. DK 12A40: πολυμαθίη νόον ἔχειν οὐ διδάσκει.

83. Burkert dismissed the argument for Pythagoras's vegetarianism.

84. I benefitted considerably by consulting Abraham 2003.

85. Saito 1985, p. 59ff.

86. Plutarch, *Quaestiones Convivales*. Book VIII. 720a.

87. Knoor 1975, pp. 192ff, and 1986, pp. 66ff.

88. The first "Elements" are credited to Hippocrates of Chios, circa 440 BCE.

89. It might well be that parallelograms appeared in Greek geometry after the sixth century, not before, since by some accounts what turned out to be the first book of Euclid was rewritten in the fourth century. At this earlier stage, perhaps simply the transformation of triangles into rectangles, and vice versa, different shapes yet both sharing the same area, provided the pivotal insight into transformational equivalences. And this might well be the inspiration behind what came to be Euclid II, the Pythagorean proof that a rectangle can be shown to be equal to a square, II.14.

90. Probably the right phrase here is ὅπερ ἔδει ποιῆσαι—what was to be done, at the end of a completed construction.

91. I am here paraphrasing aleph0.clarku.edu/~djoyce/java/elements/bookI/propI45.html.

92. Plutarch, *Quaestiones Convivales*. Book VIII, 720a.

93. As Cornford 1913, pp. 136ff. argued for Thales—that "water" originally had a "supersensible" meaning.

94. Proclus, 419.11ff. (Friedlein 1873, and Morrow 1970, pp. 332ff.).

95. Cf. http://aleph0.clarku.edu/~djoyce/java/elements/bookVI/propVI28.html.

96. Proclus, 65.20 (Friedlein 1873, and Morrow 1970, p. 52ff.). (ὃς δὴ καὶ τὴν τῶν ἀναλόγων πραγματείαν καὶ τὴν τῶν κοσμικῶν σκημάτων σύστασιν ἀνεῦρεν, 65, 16, reading ἀναλόγων for ἀλόγων, cf. Friedlein, and Heath, 1921, I.84f).

97. It must be acknowledged that the construction of the dodecahedron, while regular, is not composed of right triangles, and Plato seems keenly aware of this. But the pentagon can be dissected into triangles, and every triangle can be reduced further to right triangles.

98. Plato. *Timaeus* 31c.

99. Michael Fried, 2012.
100. Cf. Michael Fried 2012, whose line of thought I follow here.
101. Proclus, 68.20–25 (Friedlein 1873, and Morrow 1970, p. 58).
102. Proclus, 70.19–71.5 (Friedlein 1873, and Morrow 1970, p. 58),
103. Proclus, 71.22–24 (Friedlein 1873, and Morrow 1970, p. 59).
104. Proclus, 74.11–13 (Friedlein 1873, and Morrow 1970, p. 61).
105. Fried 2012, p. 609, who also identifies the four places in which Proclus emphasizes his reading of Euclid. And Fried also emphasizes Proclus's assessment that the second purpose of the book is to guide students, an important point that nevertheless is not part of my argument.
106. Heath 1921, I. p. 2.
107. Fried 2012, p. 609.
108. NB, Mueller 1991 measures claims about the *Elements* according to its deductive structure. This is a modern view. It is the reason, for example, that Euclid is condemned for introducing definitions in Book I that he never uses. Fried pointed out in a note to me that he, for instance, does not deny the importance of deductive structure, which is why he did not reject completely Heath's observations. But he insists that there are other measures of what was important to Euclid, and what was not, what he found primary and what he found secondary, and so Fried rejects the final conclusion of Heath and others regarding Book XIII.
109. Mueller 1991, pp. 302–3.
110. Proclus, 65.20 (Friedlein 1873, and Morrow 1970, p. 52ff.). For the controversies, cf. Heath 1921, I, 158–62. The whole sentence is ὃς δὴ καὶ τὴν τῶν ἀνάλογων πραγματείαν καὶ τὴν τῶν κοσμικῶν σκημάτων σύστασιν ἀνεῦρεν. I follow Friedlein, and Heath (I.84ff.) in emending ἀνάλογων for ἀλόγων.
111. Cf. Waterhouse 1972, p. 212; cf. also Euclid XIII, Scholium 1 (ed. Heiberg, vol. 5, p. 654). In the *Suda*, there is an entry that Theaetetus "first constructed the so-called five solids." The Greek term here is γράφειν; in Proclus the term is σύστασις.
112. Guthrie 1962, vol. I, p. 268.
113. Plato, *Timaeus* 55a.
114. Waterhouse 1972, p. 213. "The Discovery of the Regular Solids," *Archive for History of Exact Sciences* 9, no. 3 (30, XII, 1972), pp. 212–21.
115. Waterhouse 1972, p. 216.
116. Though he might have grasped incommensurability as well. Heath 1921, I.84–5, notes that Friedlein thought the discovery was of "irrationals." Fabricius records the variant ἀνάλογων that is also noted by August, and followed by Mullach. Heath's assessment is that ἀνάλογων is not the correct form of the word, which should be τῶν ἀναλογιῶν ("proportions"). Diels emends τῶν ἀνὰ λόγον, and Heath finally declares that this reading is most probably correct. The consensus formed was that ἀλόγων is wrong and the theory that Proclus is attributing to Pythagoras is more likely "proportions" or "proportionals."
117. Proclus, 65.20 (Friedlein 1873, and Morrow 1970, p. 52ff). Cf. Friedlein, and Heath I.84f.
118. This image, below, was inspired by the website: http://mathworld.wolfram.com/PlatonicSolid.html.
119. The images below, after Casselman 2011.
120. It is impossible to have more than five platonic solids, because any other possibility would violate simple rules about the number of edges, corners, and faces you can have together. The idea that the solids could be characterized by a single relation involving their edges, corners, and faces was intimated by Descartes already in 1639, but fully developed by Euler in his 1758 paper "Elementa Doctrinae Solidorum." "The key observation is that the interior angles of the polygons meeting at a vertex of a polyhedron **add to less than 360 degrees.** To see this, note that if such polygons met in a plane, the interior angles of all the polygons meeting at a vertex would add to exactly 360 degrees. Now cut an

angle out of paper, and fold another piece of paper to that angle along a line. The first piece will fit into the second piece when it is perpendicular to the fold. Think of the fold as a line coming out of our polyhedron. The faces of the polyhedron meet at the fold at angles less than 90 degrees. How can this be possible? Try wiggling your first piece of paper within the second. To be able to incline it with respect to the fold you have to decrease the angle of the first piece, or increase the angle of the second."

121. Heath 1931, p. 107.

122. Heath 1931, pp. 106–10.

123. In Plato's *Phaedo*, 110B, the dodecahedron is identified with the shape of the Earth.

124. When I use the expression "atom" I am thinking in terms of our contemporary analogy with "atoms" as the basic building blocks of all things. But, in the ancient Greek sense, "atom" is "*a-toma*," literally "cannot be cut in half," and this is certainly not the "a-tomic" sense of the elements that Timaeus is describing.

Chapter 4

1. Cf. also Proclus (Morrow) 379, who cites Eudemus in crediting the discovery that there were two right angles in every triangle to the Pythagoreans. Proclus supplies the following diagram:

Fig. N.4.1

Thus, alternating angles DAB and ABC are equal, and angles EAC and ACB are equal, and BAC is common. And so if DAB equals ABC, and EAC equals ACB, then if angle BAC is added to each—since DAB + BAC + EAC equal a straight angle (= straight line) or two right angles, there must be two right angles in every triangle.

2. Cf. Saito 2011, p. 195.

3. Here I am following Oliver Bryne's 1847 colored diagrams, p. 259. I have made some small changes to them to emphasize points in our narrative.

4. Probably the right phrase here is ὅπερ ἔδει ποιῆσαι—what was to be done, at the end of a completed construction.

5. I am here paraphrasing aleph0.clarku.edu/~djoyce/java/elements/bookI/propI45.html.

6. These images, directly below, after Casselman.

7. This image, below, was inspired by the website http://mathworld.wolfram.com/PlatonicSolid.html.

8. After all, we know of the account by Xenophanes, a contemporary, who claimed that when Pythagoras heard the yelp of a puppy he recognized in its sound an old friend.

Bibliography

Aaboe, Asger. Episodes from the Early History of Mathematics. New York: Random House, 1964.

Abraham, Ralph H. "The Visual Elements of Euclid." Visual Math Institute, Inc. Restored September 18, 2003. http://www.visual-euclid.org/elements/.

Allen, R. E. Greek Philosophy: Thales to Aristotle. New York: Free Press, 1991.

Allman, George Johnston. Greek Geometry from Thales to Euclid. Dublin: Hodges, Figgis, & Co., 1889.

Apostol, Tom M. "The Tunnel of Samos." Engineering and Science 1 (2004): 30–40.

Archer-Hind, R. D. The Timaeus of Plato. London: MacMillan, 1888. Reprint, New York: Arno Press, 1973.

Aristophanes. Birds. Translated by Jeffrey Henderson. Vol. 3. Cambridge, MA: Harvard University Press, 2000.

Aristotle. Categories. Translated by Harold P. Cooke. Cambridge, MA: Harvard University Press, 1967.

———. De Caelo. Translated by W. K. C. Guthrie. Cambridge, MA: Harvard University Press, 1939.

———. Metaphysics. 2 vols. Translated by Hugh Tredennick. Cambridge, MA: Harvard University Press, 1933.

———. Physics. Translated by Philip Wicksteed and Francis M. Cornford. Cambridge, MA: Harvard University Press, 1970.

———. Prior Analytics. Translated by Hugh Tredennick. Cambridge, MA: Harvard University Press, 1938.

———. Pseudepigraphus. Edited by Valentini Rose. Leipzig: B. G. Teubner, 1863.

Arnold, Dieter. Building in Egypt: Pharaonic Stone Masonry. Oxford, UK: Oxford University Press, 1991.

Artmann, Benno. Euclid: The Creation of Mathematics. New York: Springer-Verlag, 1999.

Averinou, Panagiota. "Water Management in Ancient Megara." Academia. www.academia.edu/6046045/water_management.

Barker, Andrew. Greek Musical Writings. Vol. 2. Harmonic and Acoustic Theory. Oxford, UK: Oxford University Press, 1989.

———. "Three Approaches to Canonic Division." In PERI TÔN MATHÊMATÔN. APEIRON, Vol. XXIV (1991): 49–83.

———. The Science of Harmonics in Classical Greece. Cambridge:, UK: Cambridge University Press, 2007.

Barnes, Jonathan. The Presocratic Philosophers. London: Routledge, 1982. First paperback edition, revised from original printing in 1979.

Becker, Oskar. "Die Lehre von Geraden und Ungeraden in IX. Buch der Euklidischen Elemente." Quellen und Studien zur Geschichte der Mathematik, Astronomie und Physik. Abteilung, B3 (1934): 533–53.

———. Das mathematische Denken der Antike. Göttingen, Germany: Vandenhoeck & Ruprecht (MD), 1957.

Berggren, J. L. "History of Greek Mathematics: A Survey of Recent Research." Historia Mathematica 11 (1984): 394–410.

Betz, Hans Dieter. *Studia Corpus Hellenisticum Novi Testamenti*. Leiden: E. J. Brill, 1978.

Borzacchini, Luigi. "Incommensurability, Music and Continuum: A Cognitive Approach." *Archive for History of Exact Sciences* 61, no. 3 (2007): 273–302. http://dx.doi.org/10.1007/s00407-007-0125-0.

Bridges, Marilyn. *Egypt: Antiquities from Above*. Boston: Little Brown, 1996.

Bogomolny, A. "The Pythagorean Theorem Is Equivalent to the Parallel Postulate." From *Interactive Mathematics Miscellany and Puzzles*. Accessed April 14, 2016. http://www.cut-the knot.org/triangle/pythpar/PTimpliesPP.shtml.

Broadie, Scott. "The Pythagorean Theorem Is Equivalent to the Parallel Postulate." In A. Bogomolny. "The Pythagorean Theorem Is Equivalent to the Parallel Postulate." From *Interactive Mathematics Miscellany and Puzzles*. Accessed April 14, 2016. http://www.cut-the knot.org/triangle/pythpar/PTimpliesPP.shtml.

———. "The Pythagorean Theorem Is Equivalent to the Parallel Postulate. www.cut-the-knot.org/triangle/pythpar/PTimpliesPP.shtml.

Brunes, Tons. *The Secrets of Ancient Geometry and Its Use*. Copenhagen: Rhodos Publ., 1967.

Bryan, Betsy M. "Painting Techniques and Artisan Organization in the Tomb of Suemniwer." In *Colour and Painting in Ancient Egypt*, edited by W. V. Davies, 63–72. London: British Museum Press, 2001.

Brynes, Oliver. *The First Six Books of the Elements of Euclid (in which Coloured Diagrams and Symbols are used instead of Letters for the Greater Ease of Learners)*. London: William Pickering, 1847.

Burkert, Walter. *Lore and Science in Ancient Pythagoreanism*. Translated by E. L. Minar. Cambridge, MA: Harvard University Press, 1972. First published in 1962 in German, *Weisheit und Wissenschaft: Studien zu Pythagoras, Philolaus un Platon*. Nuremberg, Germany: Hans Carl, 1962.

Burnet, John. *Early Greek Philosophy*. London: 1892; 4th ed. reprinted and expanded, 1930.

Burns, Alfred. "The Tunnel of Eupalinus and the Tunnel Problem of Hero of Alexandria." *Isis* 62, no. 2 (1971): 172–85.

Callimachus. *Iambi*. Translated by C. A. Trypanis. Cambridge, MA: Harvard University Press, 1958.

Casselman, Bill. "Why Only 5 Regular Solids?" *American Mathematical Society*. February 2011. http://www.ams.org/samplings/feature-column/fcarc-five-polyhedra.

Clagett, Marshall. *Ancient Egyptian Science: A Source Book*. 3 vols. Philadelphia: American Philosophical Society. 1999.

Cornford, Francis M. From *Religion to Philosophy: A Study in the Origins of Western Speculation*. 1913, reprint New York: Harper and Row, 1957.

———. *Principium Sapientiae: A Study of the Origins of Greek Philosophical Thought*. Cambridge, UK: Cambridge University Press, 1952.

———. *Plato's Cosmology: The Timaeus of Plato*. London: Routledge and Kegan Paul, 1937.

Coulton, J. J. *Ancient Greek Architects at Work*. Ithaca, NY: Cornell University Press, 1977.

Couprie, Dirk. *Heaven and Earth in Ancient Greek Cosmology: From Thales to Heraclides Ponticus*. This work includes three monographs: Couprie: "The Discovery of Space: Anaximander's Astronomy"; Robert Hahn: "Proportions and Numbers in Anaximander and Early Greek Thought"; Gerard Naddaf: "Anthropogony and Politogony in Anaximander of Miletus." Astrophysics and Space Science Library 374. New York: Springer, 2011.

Couprie, Dirk, Robert Hahn, and Gerard Naddaf. *Anaximander in Context: New Studies in the Origins of Greek Philosophy*. Ancient Philosophy series. Albany: State University of New York Press, 2003.

Creese, David. *The Monochord in Ancient Greek Harmonic Science*. Cambridge, UK: Cambridge University Press, 2010.

Curd, Patricia. "Presocratic Philosophy." *Stanford Encyclopedia of Philosophy*. Summer 2016 edition. http://plato.stanford.edu/archives/sum2016/entries/presocratics/.

Dantzig, Tobias. *Mathematics in Ancient Greece*. New York: Dover Books, 1954.

Derrida, Jacques. *Edmund Husserl's Origin of Geometry: An Introduction*. Translated by J. P. Leavey, Jr. Boulder, CO: Nicolas Hayes, 1978.

Dicks, D. R. "Thales." *Classical Quarterly* 53 (1959): 294–309.

———. *Early Greek Astronomy to Aristotle*. Ithaca, NY: Cornell University Press, 1970.

Diels, Hermann, and Walter Kranz. *Die fragmente der Vorsokratiker* (DK). 3 vols. Dublin: Weidmann, 1972 (first edition 1903).

Diggins, Julie E. *String, Straight-Edge, and Shadow: The Story of Geometry*. New York: Viking Press, 1965.

Diogenes Laertius (D.L.). *Lives of the Philosophers*. 2 vols. Translated by R. D. Hicks. Cambridge, MA: Harvard University Press, 1925.

Euler, Leonhard, *Elementa doctrinae solidorum*, Novi Commentarii Academiae Scientiarum Petropolitanae 4, 1752/3 (1758): 109–40; reprinted in *Opera Omnia*, series prima, vol. 26: 71–93. Zürich: Orell Füssli, 1953.

Evans, Harry B. "Review of Hermann Kienast, *Die Wasserleitung des Eupalinos auf Samos*." *American Journal of Archaeology* 103, no. 1 (1999): 149–50.

Finger, F. A. *De Primordiis Geometriae apud Graecos*. Heidelbergae 1831. http://reader.digitale-sammlungen.de/de/fs1/object/display/bsb10847231_00001.html.

Fowler, David. "Book II of Euclid's Elements and a Pre-Eudoxian Theory of Ratio." In *Archive for the History of Exact Sciences*, vol. 22 (1980): 5–36.

———. *The Mathematics of Plato's Academy: A New Reconstruction*. Oxford, UK: Oxford University Press, 1987.

Fried, Michael, and Sabetai Unguru. *Apollonius of Perga's Conica, Mnemosyne*. Supplemental vol. 222. Leiden: Brill, 2001.

———. "Book XIII of the *Elements*: Its Role in the World's Most Famous Mathematics Textbooks." In *History and Pedagogy of Mathematics* 2012, 16–20 July 2012, DCC, Daejeon, Korea.

Furley, David, and R. E. Allen, eds. *Studies in Presocratic Philosophy*, i. London: Routledge and Kegan Paul, 1970.

Gillings, Richard J. *Mathematics in the Time of the Pharaohs*. New York: Dover Publications, 1972.

Giventhal, Alexander. "The Pythagorean Theorem: What Is It About?" Department of Mathematics, University of California Berkeley. https://math.berkeley.edu/~giventh/.

Goodfield, June, and Stephen Toulmin. "How Was the Tunnel of Eupalinus Aligned?" *Isis* 56, no. 1 (1965): 46–55.

Gow, James. *A Short History of Greek Mathematics*. New York: G. E. Stechert & Co., reprint 1923.

Graham, Daniel W. *Explaining the Cosmos: The Ionian Tradition of Scientific Philosophy*. Princeton, NJ: Princeton University Press, 2006.

Gregory, Andrew. *The Presocratics and the Supernatural: Magic, Philosophy and Science in Early Greece*. London: Bloomsbury T&T Clark, 2013.

Guthrie, W. K. C. *A History of Greek Philosophy*. Vol. I. Cambridge, UK: Cambridge University Press, 1962.

Hahn, Robert. *Anaximander and the Architects: The Contributions of Egyptian and Greek Architectural Technologies to the Origins of Greek Philosophy*, Ancient Philosophy series. Albany: State University of New York Press, 2001.

———. *Archaeology and the Origins of Philosophy*. Ancient Philosophy series. Albany: State University of New York Press, 2010.

———. "Heidegger, Anaximander, and the Greek Temple," http://www.tu-cottbus.de/BTU/Fak2/TheoArch/Wolke/eng/Subjects/071/Hahn/Hahn.htm, 2007.

———. "Heraclitus, Monism, and the Felting of Wool." In *Heraklit im Kontext*. Studia Praesocratica. Berlin: De Gruyter, 2016.

Hahn, Robert, Dirk Couprie, and Garard Naddaf. *Anaximander in Context: New Studies in the Origins of Greek Philosophy*. Ancient Philosophy series. Albany: State University of New York Press, 2003.

Hartshorne, Robin. *Geometry: Euclid and Beyond*. New York: Springer, 2000.

Heath, Thomas L. *Aristarchus of Samos, the Ancient Copernicus*. Oxford, UK: Clarendon Press, 1913.

———. *A History of Greek Mathematics*. Oxford, UK: Clarendon Press, 1921. Dover reprint, 2 vols., 1981.

———. *A Manuel of Greek Mathematics*. Oxford, UK: Clarendon Press, 1931. Dover reprint, 1963.

———. *The Thirteen Books of Euclid's Elements*. Cambridge, UK: Cambridge University Press, 1908. Dover reprint, 3 vols., 1956.

———. *Mathematics in Aristotle*. Oxford, UK: Clarendon Press, 1949.

Heiberg, Johan Ludvig, and H. Menge. *Euclidis opera omnia*. 8 vol. & supplement, in Greek. Teubner, Leipzig, 1883–1916. Edited by J. L. Heiberg and H. Menge.1883–1888.

Heisel, Joachim P. *Antike Bauzeichnungen*, Darmstadt: Wissenschaftliche Buchgesellschaft, 1993.

Herda, Alexander. "Burying a Sage: The Heroon of Thales on the Agora of Miletos. With Remarks on Some Other Excavated Heroa and on Cults and Graves of the Mythical Founders of the City." In O. Henry (ed.), Les mort dans la ville. Pratiques, contextes et impacts de inhumation intro-muros en Anatolie, de debut de l'Âge du Bronze à l'époque romaine, 2èmes recontres d'archéologie d'IFÉA, Istanbul, 2013.

Herodotus. *Histories*. 4 vols. Translated by A. D. Godley. Cambridge, MA: Harvard University Press, 1981.

Høyrup, J. *Measure, Number, and Weight: Studies in Mathematics and Culture*. Albany: State University of New York Press, 1994.

Iamblichus. *The Life of Pythagoras*. Translated by Thomas Taylor. Reprint Rochester, VT: Inner Traditions International, 1986.

Imhausen, Anette. *Ägyptische Algorithmen. Eine Untersuchung zu den mittelägyptischen mathematischen Aufgabentexten*. Wiesbaden, 2003.

Isler, Martin. *Sticks, Stones, & Shadows: Building the Egyptian Pyramids*. Norman: Oklahoma University Press, 2001.

Iverson, E. *Canon and Proportion in Egyptian Art*. Wiltshire, UK: Aris and Phillips, 1975.

Ivins, William M., Jr. *Art and Geometry: A Study of Space Intuitions*. New York: Dover Publications, 1946.

Joyce, David E. "Euclid's Elements." Department of Mathematics and Computer Science, Clark University. Last modified 1999. http://aleph0.clarku.edu/~djoyce/java/elements/elements.html.

Kahn, Charles. *Pythagoras and the Pythagoreans: A Brief History*. Bloomington: Indiana University Press, 2001.

Kastanis, Nikos, and Thomaidis Yannis. "The term 'Geometrical Algebra,' target of a contemporary epistemological debate." 1991. *Aristotle University of Thessaloniki*. http://users.auth.gr/~nioka/history_of_mathematics.htm.

Katz, Victor J., and Karen Hunger Parshall, *Taming the Unknown: A History of Algebra from Antiquity to the Early Twentieth Century*. Princeton, NJ: Princeton University Press. 2014.

Kienast, Hermann. *Die Wasserleitung des Eupalinos auf Samos*. Deutsches Archäologisches Institut, Samos, Vol. XIX. Bonn: Dr. Rudoplh Habelt GMBH, 1995.

———.*The Aqueduct of Eupalinos on Samos*. Athens: Ministry of Culture, Archaeological Receipts Fund, 2005.

———. "Fundamentieren in schwierigem Gelände." *Mitteilungen des deutschen Archäologischen Instituts, Athenische Abteilung* 107 (1992): 29–42.

———. "Geometrical knowledge in designing the tunnel of Eupalinos." German Archaeological Institute, Athens, Greece. Proceedings, Progress in Tunneling after 2000, Milano, June 10–13, 2001.

Kirk, Geoffrey, and John Raven. *The Presocratic Philosophers*. Cambridge, UK: Cambridge University Press, 1957.

Kirk, Geoffrey, John Raven, and Malcolm Schofield. *The Presocratic Philosophers*. Cambridge, UK: Cambridge University Press, 1983.

Klein, Jacob. *Greek Mathematical Thought and the Origins of Algebra*. Cambridge, MA: Harvard University Press, 1968.

Knorr, Wilbur. *The Evolution of the Euclidean Elements*. Dordrecht, the Netherlands: D. Reidel, 1975.

———. *The Ancient Tradition of Geometric Problems*. New York: Dover Publications. 1986.

Laffitte, P. "Les Grands Types de l'Humanite." Vol. II in *Appreciation de la Science Antique*. Paris: Ernest Leroux, 1876.

Landels, John G. *Engineering in the Ancient World*. Berkeley: University of California Press, 1978.

Lawlor, Robert. *Sacred Geometry*. London: Thames and Hudson, 1982.

Lewis, M. J. T. *Surveying Instruments of Greece and Rome*. Cambridge, UK: Cambridge University Press, 2001.

Lloyd, Alan B. *Herodotus Book II, Introduction*. Leiden: Brill, 1975.

Lloyd, G. E. R. *Polarity and Analogy*. Cambridge, UK: Cambridge University Press, 1966.

Lockhart, Paul. *Measurement*. Cambridge, MA: Harvard University Press, 2012.

Madden, James. "Ratio and Proportion in Euclid." *LATM (Louisiana Association of Teachers of Mathematics) Journal* 5, no. 3 (2009). http://www.lamath.org/journal/vol5no2/vol5no2.htm.

McKirahan, Richard D. *Philosophy Before Socrates*. Indianapolis, IN: Hackett Publishing Co., 1994.

Mueller, Ian. *Philosophy of Mathematics and Deductive Structure in Euclid's Elements*. New York: Dover Publications. 1981.

———, ed. PERI TÔN MATHÊMATÔN. *APEIRON*, Vol. XXIV, 1991.

Muss, Ulrike, ed. *Griechische Geometrie im Artemision von Ephesos*. Vienna: University of Wien, 2001.

Nails, Debra. *The People of Plato: A Prosopography of Plato and Other Socratics*. Indianapolis, IN: Hackett Publishing Co., 2002.

Netz, Reviel. *The Shaping of Deduction in Greek Mathematics*. Cambridge, UK: Cambridge University Press, 1999.

———. "Eudemus of Rhodes, Hippocrates of Chios and the Earliest Form of a Greek Mathematical Text." *Centaurus* 46 (2004): 243–86.

Neugebauer, Otto. *Exact Sciences in Antiquity*. 2nd ed. Providence, RI: Brown University Press, 1957.

———. "Zur geometrischen Algebra. Studien zuer Geschichte der antiken Algebra III." *Quellen und Studien zur Geschichte der Mathematik, Astronomie und Physik*. Abteilung B, Studien, 3, 1936.

Nicomachus. *The Manuel of Harmonics of Nicomachus the Pythagorean*. Translated and with commentary by Flora R. Levin. Michigan: Phanes Press, 1994.

NOAA Solar Calculator. NOAA Earth System Research Laboratory Global Monitoring Division. http://www.esrl.noaa.gov/gmd/grad/solcalc/.

O'Grady, Patrica. *Thales of Miletus: The Beginnings of Western Science and Philosophy*. Aldershot, UK: Ashgate, 2002.

Olson, Åke. "How Eupalinos navigated his way through the mountain? An empirical approach to the geometry of Eupalinos." *Anatolia Antiqua, Institut Français d'Études Anatoliennes* XX (2012): 25–34.

Patsopoulos, Dimitris, and Tasos Patronis. "The Theorem of Thales: A Study of the Naming of Theorems in School Geometry Textbooks." *The International Journal for the History of Mathematics*, 2006.

Peet, Thomas Eric. *The Rhind Mathematical Papyrus*. London: University of Liverpool Press, 1923.

———. *Mathematics in Ancient Egypt*. Manchester, UK: Manchester University Press, 1931.

Plato. *Phaedo*. Translated by W. R. M. Lamb. Cambridge, MA: Harvard University Press, 1938.

———. *Republic*. 2 vols. Translated by Paul Shorey. Cambridge, MA: Harvard University Press, 1969.

———. *Timaeus*. Translated by R. G. Bury. Cambridge, MA: Harvard University Press, 1929.

Pliny the Elder. *Natural History*. Edited and translated by H. Rackham. Cambridge, MA: Harvard University Press, 1968.

Plutarch. *Moralia*. 16 vols. Cambridge, MA: Harvard University Press. Vol. II: *Septem sapientium convivium* (Dinner of the Seven Wise Men). Translated by F. C. Babbitt, 1928. Vol. 9: *Quaestiones Convivales* (Dinner Talk 7–9). Translated by E. L. Minar, F. H. Sandbach, and W. C. Helmbold, 1961.

Posamentier, Alfred S. *The Pythagorean Theorem: The Story of Its Power and Beauty*. New York: Prometheus Books, 2010.

Proclus. *A Commentary on the First Book of Euclid's Elements*. Translated and edited by Glenn Morrow. Princeton, NJ: Princeton University Press, 1970.

———. *Procli Diadochi. In primum Euclidis elementorum librum comentarii*. Greek edition, Gottfried Friedlein. Leipzig: B. G. Teubner, 1873.

Rhodes, P. J. *Aristotle: The Athenian Constitution*. London: Penguin Books, 1984. Reidemeister, K. *Das exakte Denken der Griechen*. Leipzig: Classen and Goverts, 1949.

Riedweg, Charles. *Pythagoras: His Life, Teaching, and Influences*. Ithaca, NY: Cornell University Press, 2005.

Robins, Gay. *Proportion and Style in Ancient Egyptian Art*. London: Thames and Hudson, 1994.

———. *Egyptian Statues*. Buckinghamshire, UK: Shire Publications, 2001.

———. "Use of the square grid as a technical aid for artists in Eighteenth Dynasty painted Theban tombs." In *Colour and Painting in Ancient Egypt*, edited by W. V. Davies: 60–62. London: British Museum Press, 2001.

Robins, Gay, and Charles Shute, *The Rhind Mathematical Papyrus: An Ancient Egyptian Text*. London: British Museum, 1987.

Roochnik, David. *Retrieving the Ancients: An Introduction to Greek Philosophy*. London: Blackwell, 2004.

Rossi, Corinna. *Architecture and Mathematics in Ancient Egypt*. Cambridge, UK: Cambridge University Press, 2004.

Rudman, Peter S. *The Babylonian Theorem: The Mathematical Journey to Pythagoras and Euclid*. New York: Prometheus Books. 2010.

Saito, Kenneth. "Book II of Euclid's *Elements* in Light of the Theory of Conic Sections." *Historia Scientarium*. Tokyo (Japan). Vol. 28 (1985): 31–60.

———."The Diagrams of Book II and III of the Elements in Greek Manuscripts." In *Diagrams in Greek Mathematical Texts*, 39–80. Report version 2.03 (April 3, 2011). http://www.greekmath.org/diagrams/diagrams_index.html.

———. Diagrams in Greek Mathematical Texts. Report version 2.03 (April 3, 2011). www.greekmath.org/diagrams/diagrams

Schädler, Ulrich. "Der Kosmos der Artemis von Ephesos." In *Griechische Geometrie im Artemision von Ephesos* (ed. Ulrike Muss), 2001: 279–87.

Struve, W. W. *Papyrus des Staatlichen Museums der Schönen Künste in Moskau*. In *Quellen und Studien zur Geschichte der Mathematik*. Abteilung A, I (1930).

Szabó, Árpád. "ΔΕΙΚΝΥΜΙ als mathematischer Terminus für 'beweisen.'" *Maia* 10 (1958): 106–31.

———. *The Beginnings of Greek Mathematics*. Dordrecht, the Netherlands: D. Reidel Publishing Co., 1978.

Taisbak, Christian Marinus. *Euclid's Data: The Importance of Being Given*. Gylling, Denmark: Narayana Press, 2003.

Tannery, Paul. "De la solution geometrique des problemes du second degree avant Euclide." *Memoires scientifique* 1 (Paris, 1912): 254–80.

Taylor, A. E. *The Timaeus of Plato*. Oxford, UK: Clarendon Press, 1928.

Taylor, C. C. W. *From the Beginning to Plato*. London: Routledge, 1997.

Unguru, Sabetai. "On the Need to Rewrite the History of Greek Mathematics." *Archive for the History of Exact Sciences* 15 (1975): 67–114.

Van der Waerden, Bas. *Science Awakening*. Translated by Arnold Dresden. Dordrecht, the Netherland: Kluwer Academic Publishers, 1954.

———. "Eupalinos and His Tunnel." *Isis* 59, no. 1 (1968): 82–83.

———. "Defense of a 'Shocking' Point of View." *Archive for the History of Exact Sciences* 15 (1975): 199–205.

Vitruvius. *Ten Books on Architecture*. Translated by Ingrid D. Rowland, Commentary and illustrations by Thomas Noble Howe. Cambridge, UK: Cambridge University Press. 1999.

Von Fritz, Kurt. "Die ἀρχαί in der griechischen Mathematik." *Archiv für Begriffsgeschichte (ABG)* I (1955): 13–103.

———. "Gleichheit, Kongruenz und Ähnlichkeit in der antiken Mathematik bis auf Euklid." *Archiv für Begriffsgeschichte (ABG)* IV (1959): 7–81.

———. "The Discovery of Incommensurability by Hippasus of Metapontum." *Annals of Mathematics* 46 (1945): 242–64.

Waschkies, H. J. *Anfänge der Arithmetik im Alten Orient und bei den Griechen*. Amsterdam: B. R. Grüner, 1989.

Waterhouse, William C. "The Discovery of the Regular Solids." *Archive for History of Exact Sciences* 9, no. 3 (30, XII,1972): 212–21.

Werli, Fritz, ed. Die Schule des Aristoteles. Basle: Schwabe, 1944–1960; vol. VIII, *Eudemos von Rhodos*, 1955.

Wesenberg, Burkhardt. "Das Paradeigma des Eupalinos." *Deutschen Archäologischen Instituts* 122 (2007): 33–49.

WolframMathWorld: http://mathworld.wolfram.com/PlatonicSolid.html.

Wright, G. R. H. *Ancient Building Technology*. Vol. 1. Leiden: Brill, 2000.

Young, Louise B. *The Mystery of Matter*. Oxford, UK: Oxford University Press, 1966.

Zeller, Edward. *Grundriss der Geschichte der Griechischen Philosophie*, translated as *Outlines of the History of Greek Philosophy*. S. F. Alleyne (2 vols, 1866).

Zeuthen, H. *Die Lehre von den Kegelschnitten im Altertum*. Kopenhagen: Forh. Vid. Selskab,1885.

———. *Geschichte der Mathematik im Altertum und Mittelalter*. Kopenhagen: Verlag A.F. Hoest, 1896: 32–64.

Zhmud, Leonid. *Pythagoras and the Early Pythagoreans*. Oxford, UK: Oxford University Press, 2012.

———. *The Origin of the History of Science in Classical Antiquity*. Berlin: Walter de Gruyter, 2006.

———. "Pythagoras as a Mathematician," *Historia Mathematica* 16 (1989): 249–68.

Image Credits

Figure D.1: Photo by the author.

Figure I.4: Photo by the author.

Figure I.16: Photo by the author.

Figure I.20: Photo by Ulrich Schädler. "Der Kosmos der Artemis von Ephesos." In *Griechische Geometrie im Artemision von Ephesos*, ed. Ulrike Muss, p. 280, fig. 1.

Figure I.21: Drawing by Ulrich Schädler. "Der Kosmos der Artemis von Ephesos." In *Griechische Geometrie im Artemision von Ephesos*, ed. Ulrike Muss, p. 280, fig. 2.

Figure I.22: Photo by Ulrich Schädler. "Der Kosmos der Artemis von Ephesos." In *Griechische Geometrie im Artemision von Ephesos*, ed. Ulrike Muss, p. 281, fig. 3.

Figure I.23: Drawing by Ulrich Schädler. "Der Kosmos der Artemis von Ephesos." In *Griechische Geometrie im Artemision von Ephesos*, ed. Ulrike Muss, p. 281, fig. 4.

Figure I.24: Photo illustration by Hermann Kienast.

Figure 1.15: Photo by the author.

Figure 2.4: Photo by the author.

Figure 3.22: Drawing by Hermann Kienast.

Figure 3.23: Drawing by Hermann Kienast.

Figure 3.24: Photo illustration by Hermann Kienast.

Index

Aaboe, 242, 263
Abraham, xv, 11, 259, 263
Acute angle, 136, 138–139, 196
Ahmose, 97, 250
Ajades, 158
algebra, vii, x, 7–9, 45, 78, 86–87, 139, 142, 175, 242–243, 247, 249, 266–267
Allman, 29, 31, 118, 120, 123, 125, 244, 253, 263
Amasis, 13, 99, 107
Amenemmes, 250
Amenenhat, 250
anagrapheus, 247
analogia, 164, 178, 251
analogous, 37, 102, 109, 136, 168–169, 180, 202
analogy, 69, 168, 178, 261
Anangke, 155
Anaxagoras, 5–6, 41, 88
Anaximander, ix, 2, 4, 27–28, 37, 42, 88, 100–101, 103, 151, 153, 158–160, 167, 179, 182, 201, 206–207, 229–230, 251, 264–266
Anaximenes, 2–3, 201
anthyphairetic, 9
Apollodorus Logistikos (Cyzikos), 135–136
Apostol, 256, 263
Application of Areas theorems. *See also*
    Pythagoras's Other Theorem
    Euclid I.42, 232 [fig. 4.30]
    Euclid I.43, 232 [fig 4.31]
    Euclid I.44, 232 [fig. 4.32]
    Euclid I.45, 233 [fig. 4.33]
    Euclid VI.25, 233–235 [fig. 4.34, 4.35, 4.36, 4.37]
Apries, 16
architects, ix, 6, 12, 32, 34, 35, 37, 41–42, 82, 109, 148, 151, 157–158, 167, 181, 247, 252, 257, 264–265, 268–269

architecture, ix, xii, 6, 35, 37, 41–43, 104, 108, 158, 167, 202
Archytas, 136, 155, 178
areas, xi–xiv, 3, 9, 13–14, 19, 22, 42, 45–46, 51, 53, 61, 64, 66, 68–72, 74, 76–78, 80, 82, 84, 86–88, 92, 133, 140, 149–150, 154, 158, 168–169, 172–173, 176, 178, 181, 189, 191–192, 193, 195–198, 200, 202–203, 205, 209, 231–233, 235, 242, 248
Aristarchus, 266
Aristophanes, 93, 249, 263
Aristotle, xi, 1–3, 5–6, 28, 41, 116–117, 119, 121, 128, 140–141, 152–153, 155–156, 241–242, 244, 251–252, 254–256, 263, 265–266, 268
arithmetic, vii, 8, 40, 46, 70, 72–73, 92, 126, 135–136, 148, 154–156, 161, 168, 175, 178, 181, 199–200, 242, 258
Arnold, 263, 269
aroura, 243
arouras, 14, 116, 244
Artemis, xiii, 5, 32, 42, 268
Artemision, 32, 37, 42, 256, 267–268
Artmann, 242, 248, 263

Babylon, 7, 83, 92, 151
Babylonian, vii, x–xii, 7–11, 25, 92–93, 101, 103, 202, 242–243, 268
banausic, 6, 42
Barker, xv, 154, 256, 263
Barnes, 95, 250, 263
Becker, 27, 93, 144, 244, 249, 255, 263
Berggren, 9, 243, 263
Berkeley, 265, 267
bisecting, 25, 64–65, 117–118, 169, 211
Bogomolny, 248, 264

Borzacchini, 256, 264
Bridges, 114, 264
Broadie, 248, 264
Brunes, 244, 264
Bryan, 243, 264
Bryne, 261
Brynes, 264
Burkert, ix, 4, 25, 27, 42, 93–94, 103, 135, 182, 238, 241, 244, 249, 251, 255, 259, 264

Caelo, 256, 263
caliper, 182
Callimachus, 32, 94, 136, 249, 251, 264
Casselman, 260–261, 264
Categories, 152–153, 242, 263
Chefren, 98
Chersiphron, 42
Circle, vii, xiii, 10, 25–26, 31, 34, 39–40, 61, 75, 91, 93, 100–101, 108, 115, 117–119, 121–126, 129, 131, 136–137, 139–141, 153, 169, 171, 186, 188, 216–217, 219, 223–225, 230, 253–254
    every circle is built out of right triangles, 224–225 [fig. 4.14, 4.16]
Clagett, 20, 22, 244, 264
Classen, 259, 268
Cloudkookooland, 93
Commandino, 68, 215
commensurable, 142–143, 145, 147–148, 179, 255, 257–258
Componendo, 248
congruence, 54–55, 95, 108, 133, 140, 178, 249
Conica (Apollonius's), 244, 253, 265
Copernicus, 266
Cornford, 241, 259, 263–264
cosmos, ix, xi–xii, xiv, 2, 4, 6, 9, 12, 15, 29, 37, 42, 71, 77–78, 81, 88, 107, 115, 117, 126, 128, 133, 154, 158, 167–168, 178, 182, 189, 195–196, 198–200, 203–206, 209, 212, 214, 217, 231, 235–236, 241, 253
Coulton, 42, 158, 247, 256, 264
Couprie, xv, 88, 249, 264, 266
Creese, 256, 264
cubits, 21, 24, 98–99, 103–105, 148–149, 154, 157, 180–181, 243–244, 250
Cuneiform, xii, 7, 11, 83, 241

Curd, 241, 264
Cyzicus (Apollodorus), 135–136

Delphi, 259
Derrida, 95, 250, 265
Descartes, 260
diagrams. *See* geometrical diagrams
diairesis, 248
diameter, 13, 25–26, 30, 34, 37, 64–65, 91, 93, 108, 115, 117–119, 123, 125–128, 131, 140–142, 151, 158, 167, 169, 186, 190–191, 216–217, 219, 224, 230, 232, 249, 252–254
Dicks, 92, 249, 265
Didyma, ix, 94, 109, 136, 158, 251
Didymaion, 37, 256
Diodorus, 21, 136
Diogenes, 27, 31, 61, 97, 103, 117, 120, 135, 244, 251, 259, 265
dioptra, 160
Dipteros, 37, 157, 256
dittography, 223
dodecahedron, 201, 203, 206, 209, 211, 237–238, 257, 259, 261
duplicate (ratio), 69–72, 78–81, 85, 87–89, 117, 126–128, 150, 154, 173–175, 195, 217–218, 222–223, 235, 248, 257

Egypt, vii–viii, x, xii, xv, 1, 3–4, 11–16, 18, 20–22, 25–26, 28, 32, 42, 87, 91–92, 94–95, 97, 102–103, 114, 116, 118, 128, 157, 180, 202, 206, 213, 243, 263–264, 267–268
engineer, 158, 165–167, 257
engineering, iii, xii, 1, 6, 41, 43, 148, 157, 163, 165, 167, 257, 263, 267
enlargement (of Pythagorean theorem, [VI.31]), vii, x–xi, 46, 66, 70, 72–73, 79–80, 124, 138, 171, 196, 200, 204, 214–215
Ephesos, ix, xiii, 1, 5, 32, 42, 83, 158, 267–268
equations, 8, 242, 247
equinox, 25, 27, 42, 93, 103, 107, 251
Euclid, vii–viii, x–xiii, xv, 3, 7, 9, 14, 19, 26, 32, 45–46, 48, 51, 54, 56–58, 61, 63–66, 69–72, 74–76, 78, 81–83, 87, 91, 95–96, 101, 108–109, 115–121, 123, 127–128, 131–132, 138–142, 144, 151, 171, 175, 178, 180–183, 189–190, 193, 196–201, 204–206, 209–211,

214–216, 219–220, 231–232, 242, 247–249, 253–254, 259–260, 263–268
5th Postulate, 58 [fig. 1.16, 1.17]
theorems (book and proposition)
  I.4, 55 [fig. 1.12], 48, 51, 62, 108
  I.5, 119 [fig. 2.24]
  I.8, 55 [fig. 1.13]
  I.26, 55 [fig. 1.14]
  I.27, 58–59 [fig.1.18]
  I.28, 59 [fig. 1.19]
  I.29, 59 [fig. 1.20]
  I.30, 60 [fig. 1.21]
  I.31, 60 [fig. 1.22]
  I.32, 60 [fig. 1.23], 121 [fig. 2.25]
  I.33, 61 [fig. 1.24]
  I.34, 61–63 [fig. 1.25, 1.26, 1.27, 1.28, 1.29]
  I.37, 63–66 [fig. 1.30, 1.31]
  I.41, 64–65 [fig. 1.32, 1.33]
  I.42, 190 [fig. 3.57], 231 [fig. 4.30]
  I.43, 190 [fig. 3.56]
  I.44, 191 [fig. 3.59, 3.60]
  I.45, 77 [fig. 1.46], 192–193 [fig. 3.61, 3.62, 3.63]
  I.47, 47ff
  II.4, 8 [fig. 1.1]
  II.5, 183–184 [fig. 3.42, 3.43, 3.44, 3.45]
  II.11, 8 [fig. 1.1], 184–185
  II.14, 186–189 [fig. 3.48, 3.49, 3.50, 3.51, 3.52, 3.53, 3.54, 3.55, 3.56]
  III.31, 121 [fig. 2.26, 2.27], 219 [fig. 4.9]
  VI.1, 71 [fig. 1.38]
  VI.2, 74 [fig. 1.42]
  VI.3, 184
  VI.8, 75 [fig. 1.42], 216
  VI.11, 216
  VI.16, 75–76 [fig. 1.43, 1.44]
  VI.17, 76 [fig. 1.47], 220 [fig. 4.10], 217
  VI.19, 79 [fig. 1.48], 218 [fig. 4.6], 217, 218 [fig. 4.6]
  VI.25, 193–195 [fig. 3.64, 3.65, 3.66, 3.67]
  VI.20, 79 [fig. 1.49], 88 [fig. 1.55], 217 [fig. 4.5], 217
  VI.30, 184–185 [fig. 3.46, 3.47]
  VI.31, 215 [fig. 4.2, 4.3]
  X. Appendix 27, 142–147 [fig. 3.7, 3.11, 3.12, 3.14], 254–255 [Greek text] n.17
  XI.21, 204–205
  XIII.18 Remark, 206
Eudemus, 2, 5, 12–13, 25–28, 32, 42, 55–56, 92–93, 95, 103, 116, 118, 120, 124, 136, 139, 156, 197, 205, 207, 230, 251, 253–254, 257, 261, 267, 269
Eudoxus, theory of proportions, 9, 140, 178, 155, 251, 258, 265
Eupalinion (tunnel of Eupalinos)
  imagined as two right triangles, 166 [fig. 3.28], 231 [fig. 4.29]
Eupalinos, viii, xiii, 5–6, 35–41 [fig. I.24, I.25, I.26, I.27, I.28, I.29], 149, 157–167 [fig. 3.21, 3.22, 3.23, 3.24, 3.25 3.26, 3.27, 3.28], 229–231, 245–246, 256–257, 265–267, 269
Eutocius, 254
Evans, 256, 265
experiment, 91, 94, 96, 154, 158, 167, 198, 204, 236

Fabricius, 260
felting, 6, 166, 180, 252, 257, 259, 265
figurate numbers, viii, xii–xiii, 148–150, 154, 167, 203
finite, 184, 197
fire, 3, 32, 35, 117, 153, 157, 168, 196, 211
form, iv, xiv, 2, 5–6, 9, 25, 28, 37–38, 45, 71, 82, 85, 88, 92, 94, 99, 126, 128, 142, 152, 168, 170, 175, 179, 189, 191, 195, 198, 203–204, 211–212, 217, 224–225, 235, 242, 245, 247, 251, 260
Fowler, 9, 11, 242, 265
fraction, 143, 147–148, 251
Fried, xv, 199, 244, 248–249, 254–255, 257–258, 260, 265
Friedlein, 244, 247–248, 251–253, 256–257, 259–260, 268
Fritz, 6, 27, 68, 93, 96, 107, 144, 242, 244, 247, 249–251, 255, 269
Furley, 265

Geminus, 31 [fig. I.19], 124 [fig. 2.30], 214 [fig. 4.1]
geometrical diagrams
  Egyptian, 12–25
    grid with proportional rules for standing and seated figures, 17 [fig. I.7]

geometrical diagrams (*continued*)
    MMP Problem 6, 18 [fig. I.8]
    ostrakon of Senemut, 17 [fig. I.6]
    red grid drawing of Pharaoh Apries, 15 [fig. I.14]
    RMP Problem 51, 19–23 [fig. I.9, I.10, I.11, I.12]
    RMP Problem 56, 22–23 [fig. I.13]
    RMP Problem 57, 23–24 [fig. I.14]
    tomb of Menna, 16 [fig. I.5]
    tomb of Ramose, 15 [fig. I.4]
    white board with red grid lined drawing, 17 [fig. I.6]
  Greek, 32–41
    architecture, 32–41
    curve for roof yiles, archaic Artemision in Ephesus, 33–34 [fig. 1.20, 1.21, 1.22, 1.23]
  Mesopotamian, 7–12
    geometric algebra, Euclid II.4 and II.11, 8 [fig. I.1]
    Pythagorean theorem, 10–11 [fig. I.2, I.3]
  problems, 4–7
    geometry of measuring tunnel lengths, 36 [fig. I.25]
    scaled-measured diagram, 39–40 [fig. I.26, I.27, I.28]
    staking the hill, 36 [fig. I.24]
    PARADEGMA, 41 [fig. 1.29]
geometric mean
  visualization diagrams, 85 [fig. 1.52], 173 [fig. 3.34], 174 [fig. 3.35], 221 [fig. 4.11], 222, [fig. 4.12], 223 [fig. 4.13]
Gillings, 265
Giventhal, 80, 248, 265
gnomon, 28, 35, 151, 100, 102, 105–108, 113, 143, 147, 242
  carpenter's square in a diagram, 150–153 [fig. 3.18]
Goodfield, 256, 265
Gow, 29, 31, 82–83, 118, 120, 123, 244, 248, 265
Grady, 95, 250, 267
Graham, 241, 265
Graves, 266
Gregory, 253, 265
Guthrie, 201, 250, 260, 263, 265

Hahn, iii, 243, 246–248, 252, 255–257, 259, 264–266

hammers, 154
harmonic, 126, 154–155, 168, 178, 263–264, 267
Hartshorne, 266
Heath, 7, 29, 31, 46, 54, 68, 83, 91, 94, 96, 109, 112–113, 117–118, 120–121, 123–124, 137, 139, 141, 175, 178–179, 190, 193, 198–199, 201, 210–211, 215, 233, 242, 244–245, 247–261, 266
Hecateus, 136
Hegel, 95
Heisel, 32, 245, 266
Heraclitus, 182, 252, 259, 265
Heraion, 37, 158, 239, 256
Herda, 259, 266
Herodotus, 13–14, 25, 35, 157–158, 243–245, 252, 256–257, 266–267
Hippasus, viii, 136–137, 141–142, 147, 153–154, 178, 253, 269
Hippias (of Elis), 28
Hippocrates, viii, 5–6, 41, 82, 253–254, 259, 267
  squaring the Lunes, 136–141 [fig. 3.1, 3.2, 3.3, 3.4, 3.5]
Hippon, 28
Howe, 269
Hyksos, 250
hylozoistic, 133, 253
hypotenuse, viii, x–xiii, 2, 4–5, 10–11, 25, 30, 32, 46, 48, 50–51, 53–54, 57, 68–69, 73–77, 79–80, 82–87, 96, 105–106, 116–118, 120, 124, 126–129, 132–133, 135–141, 143–144, 147–148, 154, 157, 160, 167–168, 171, 173, 175–181, 188–189, 193, 196, 198, 200–201, 212, 216–217, 219–220, 222, 225–227, 230, 241, 251, 253, 255

Iambi, 94, 249, 251, 264
Iamblichus, 178, 266
icosahedron, 117, 201, 203, 206, 208–212, 257
Imhausen, 20, 244, 266
incommensurability, viii, 8, 115, 141–142, 148, 255, 258–260
inscribing a right triangle in a (semi-)circle
  Heath's proposal, 91
intuition, 15, 28, 46, 54, 57, 64–66, 69, 75–76, 79–80, 84, 87–88, 115, 120, 128, 143, 169, 178, 180, 191, 195, 200, 204–205, 209, 211, 226, 235, 247, 266
intuitive, 10

Ionia, x, 1, 4, 82, 96, 107, 117, 167, 182, 230, 243, 265
Ionians, 107, 230
irrational, 26, 68, 136, 147–148, 181, 242, 251
Isler, 243, 266
isosceles, 4, 18, 24–27, 29–31, 37–38, 40, 61, 68, 87, 91, 93, 95, 100–101, 105–106, 112, 115, 117–124, 126, 128–129, 131–133, 138, 140–141, 143, 145, 147, 164–165, 169, 171, 180–181, 196, 209, 211–216, 219, 222, 224, 230, 238, 249, 251–254
Iverson, 243, 266
Ivins, 266

Kahn, 266
Käppel, 41, 165
Kastanis, 242, 266
Katz, 247, 266
Khaefre, 97
Khufu, 97
Kienast, 37–42, 161–162, 164–165, 167, 245–246, 256–257, 265–266
Kirk, 95, 250, 259, 266–267
Klein, 267
Knorr, 259
Kocik, xv, 249
kouros, 182, 230, 239
Kranz, 265
kyma, 158

Laertius, 27, 31, 61, 97, 103, 120, 135, 244, 251, 259, 265
Laurion, 166
Lloyd, 243, 267
Lobachevskian, 249
Lockwood, xv
Lune, viii, 5, 137–140

macrocosmic, viii, 15, 71, 81, 87, 107, 153–154, 156, 158, 164, 169, 178, 184, 203, 205
Madden, 70, 248, 267
Manuel, 91, 253, 266–267
McKirahan, 95, 250, 267
mean proportion
    visualization, 85 [fig. 1.52], 127 [fig. 2.34], 129–132 [fig. 2.36 2.37, 2.38, 2.39], 221 [fig. 4.11], 223 [fig. 4.13]
Megara, 158–159, 257, 263

Meno (Plato's), 142, 254
Mesopotamia, xii, 1, 4, 7, 11–12, 41, 83, 92, 94
Metagenes, 42
metaphysics
    how I am using the term, 1–4
Meton, 93
microcosmic, 71, 81, 84, 87, 117, 153–154, 156, 158, 164, 168–169, 178, 184
Milesian, xi–xii, 2–3, 13, 37–38, 40, 42, 94–95, 101–102, 128, 133, 159–160, 164–165, 251–253, 257
monism, xii–xiii, 1–3, 153, 168, 241, 252, 265
    source vs. substance monism, 2–4
monochord, 256, 264
Morrow, 199, 244, 247–248, 251–253, 256–257, 259–261, 268
Moscow Mathematical Papyrus (MMP), 13–14, 18, 97, 124
Mueller, 9, 200, 243, 260, 267

Netz, 5–7, 32, 41–43, 82, 242, 247–249, 267
Neugebauer, 7–8, 11, 83, 92, 241–242, 249, 267
Nicomachus, 154, 178, 256, 267
Nile, 13–14, 94, 243
Niloxenus, 99, 108
NOAA, 251, 267
number, 7, 23, 51, 57, 70, 72, 83, 104–105, 115, 136, 142–156, 168, 175, 180–181, 202–203, 216, 248, 250–251, 255, 258–260
numeration, 38, 164–165

odd numbers (sandwiched between the even on a gnomon), 152 [fig. 3.19]
ostrakon, 17–18, 243

Pamphile, 31, 61, 91, 117, 120, 124, 219, 252–253
Pappus, 249, 252
PARADEGMA, 40–41, 165, 246–247
parallel, 47–50, 52–53, 57–66, 71, 75, 82–83, 93, 120, 133, 169, 178, 190–191, 194, 197, 248, 251, 257, 264
parallelogram, 47, 57, 61–65, 71, 76–77, 81–82, 151, 189–193, 194–195, 195–198, 231–235, 258–259
parallels, 47–48, 64–66, 253, 258
pebbles, 87, 144, 149–150, 255
peremus, 149
perimeter, 57, 65–66, 82, 104

perpendicular, 30–31, 48, 67–69, 73–77, 79–80, 84, 111, 123–124, 126–129, 131, 138–139, 153, 171, 186, 196, 212, 214, 216–220, 222, 226–227, 261

Pharaoh, 13, 16, 99, 107, 243, 265

philosophy, science, geometry
the missed connection, 1–4

Pheidippides, 93

Pithetaerus, 93

Plato, viii, xi, xiv, 2, 29, 105, 117, 142, 155, 195–196, 198–201, 203, 210–214, 236, 238, 241, 249, 254–256, 258–261, 263–265, 267–268

platonic solid, 261, 269

Playfair, 58, 83, 199

Polycrates, 35

polygons, 70, 79, 84, 86, 88–89, 128, 133, 198, 213, 217–218, 236–237, 260

polyhedron, 236, 260–261, 264

Polykrates, 159

porism, 67, 69, 74–75, 88, 127, 131, 205

postulate, 57–58, 82–83, 96, 133, 248, 253

Proclus, 25, 29, 55, 57–58, 61, 66, 81, 83, 92–93, 95, 105, 115–116, 118, 120–121, 124, 135, 156, 169, 178, 196–205, 209, 230, 236, 244, 247–248, 251–253, 256–257, 259–261, 268

proof, viii, x, xiii, 6, 8–9, 12–14, 26–27, 32, 46, 48, 54, 56–57, 61–62, 64–66, 68–71, 73, 75–79, 81, 83, 85, 89, 94–96, 101, 108, 115, 118–124, 132, 136, 139–147, 169–170, 175, 178–180, 182–186, 189–193, 195–196, 202–205, 207, 210, 214, 216, 222, 224, 226, 232–234, 238, 248, 250–252, 254–255, 258–259

proportions, viii, ix–xi, xiii–xiv, 8–9, 17–18, 22, 31–32, 34–35, 37, 41, 46, 69–78, 81, 83, 84–85, 86–87, 96, 99, 106–108, 113, 117, 126, 138, 141, 153–156, 158, 164, 167–168, 171, 173, 175, 178–180, 182–186, 193, 195, 197, 199–200, 203–204, 213–214, 221, 225–226, 230, 235, 243, 247–248, 254, 257–258, 260, 264

Psammetichus, 13, 243

Psamtik, 13, 243

pyramid, x, xiii, xv, 4, 6, 12, 14, 18–19, 22–24, 26–27, 32, 56, 93–94, 96–109, 113, 115–116, 123, 128–129, 132, 143, 147–149, 154, 169, 180–181, 201, 204, 206, 216, 227–228, 230, 250–251

Pythagoras, iii, v, viii–xiii, xv, 2, 4–5, 10, 25, 29, 42, 45–46, 68, 74–75, 81–84, 88, 101, 116–117, 126, 132, 135–139, 141, 143–145, 147–149, 151–157, 159–161, 163, 165–171, 173, 175, 177–183, 185–187, 189–191, 193, 195–199, 201–207, 209–213, 215, 223, 230–231, 233–234, 236, 238–239, 241–242, 249, 251, 256–257, 259–261, 264, 266, 268–269

figurate numbers, 148–153 [fig. 3.15, 3.16, 3.17]

a geometric formulation of Pythgoras's proof, 176–179 [fig. 3.37, 3.38, 3.39, 3.40]

geometry of the tunnel of eupalinos, 157–167 [fig. 3.21, 3.22, 3.23, 3.24, 3.25, 3.26, 3.27, 3.28]

Heath's algebraic formulation of Pythagoras' proof of the theorem, 175 [Fig. 3.36]

Hippasus and the proof of incommensurability, 141–148 [fig. 3.7, 3.11, 3.12, 3.14]

the immediacy of areal equivalences in isosceles triangle proof, 181 [fig. 3.41]

lengths, numbers, musical intervals, 153–156 [fig. 3.20]

mean proportional and the famous theorem, 168–182

problems connecting to him to theorem, 135–137

Pythagoras's other theorem (VI.25), 193–195 [fig. 3.64, 3.65, 3.66, 3.67]

Application of Areas, 189–195 [fig. 3.57, 3.58, 3.59, 3.60, 3.61, 3.62, 3.63, 3.64, 3.65, 3.66, 3.67, 3.68, 3.69]

Exceeding, 197 [fig. 3.68]

Falling Short, 198 [fig. 3.69]

Pythagoras and the regular solids, 198–212
cosmic figures, 198–212
folding up the regular solids around a point, 207–211 [fig. 3.71, 3.72, 3.73, 3.74, 3.75, 3.76, 3.77, and Table 3.1]

Pythagoras's regular solids and Plato's *Timaeus*, 210–212

construction of the cosmos from isosceles and scalene right triangles, 212 [fig. 3.78]

complements of parallelogram about a diameter are equal to one another, 190 [fig. 3.58]

constructing any rectilinear figure that applied to a given straight line, 192 [fig. 3.61, 3.62, 3.63]

constructing a parallelogram equal in area to a given triangle in a specified angle, as a complement of a parallelogram, on a given straight line, 191 [fig. 1.60]
  constructing a parallelogram equal to a given rectangle, 190 [fig. 3.57]
  in the bigger metaphysical picture, 195–198
Pythagorean proof that there are two right angles in every triangle, 169–170 [fig. 3.29, 3.30]
  comparison of "every triangle in (semi-)circle is right" diagram and ancient diagram of VI.31, 171 [fig. 3.31]
  similar triangles, 172 [fig. 3.32]
  square on perpendicular equals rectangle, 172–174 [fig. 3.33, 3.34, 3.35]
  theorems about odd and even numbers, 144–146 [fig. 3.9, 3.10, 3.13]
Pythagorean Proof of Mean Proportional, Euclid II, 182–189. *See also* Pythagoras: Pythagoras's other theorem (VI.25)
Pythagorean, i, iii, vii–xiv, 2, 4, 6, 8, 10–12, 14, 16, 18, 20, 22, 24, 26, 28–32, 34, 36, 38, 40, 42, 45–89, 92, 94, 96, 98, 100, 102, 104, 106, 108, 110, 112, 114, 116, 118, 120, 122, 124, 126, 128–130, 132–133, 135–136, 138, 140, 142, 144, 146, 148, 150, 152–156, 158, 160, 162, 164, 166, 168, 170–172, 174, 176, 178–180, 182–186, 188, 190, 192, 194, 196, 198–200, 202–204, 206, 208, 210, 212–216, 218, 220, 222, 224, 226, 228, 230, 232, 234, 236, 238, 241–242, 248–249, 253–256, 258–259, 264–265, 267–268
Pythagorean theorem
  Book I, Proposition 47
    Angle-Side-Angle [ASA] equality, 55 [fig. 1.14]
    comparison of squares on isosceles right triangles and scalene right triangles, 216 [fig. 4.4]
    geometrical intuitions, 54–56
    parallel postulate and parallel theorems, 58–66
      Playfair's Axiom, 58 [fig. 1.17]
      parallel lines and equal angles, 59–66
      shearing, 65 [fig. 1.34]
      Proclus: perimeter is not a criterion of area, 66
    proofs by superimposition, 56
    reflections on strategies, 48–51 [fig. 1.2–1.11]
    Side-Angle-Side [SAS] equality, 55 [fig. 1.12]
    Side-Side-Side [SSS] equality, 55 [fig. 1.13]
    theorem in Euclid, 46–47 [fig. 1.1]
  Book VI, Proposition 31
    all right triangles divide into two similar right triangles ad infinitum, 80 [fig. 1.50, 1.51]
    all triangles divide into two right triangles, 80 [fig. 150]
    arithmetical and geometrical means, 72–77
    colored diagrams of proof, 225–227 [Fig. 4.17, 4.18, 4.19, 4.20, 4.21, 4.22, 4.23]
    diagram from Medieval manuscript, 68 [fig. 1.37], 215 [fig. 4.3]
    "enlargement" figures of Pythagorean theorem, 86 [fig. 1.53]
    geometrical intuitions, 69–70
    illustration of similar triangles, 73 [fig. 1.39, 1.40, 1.41]
    ratios, proportions, and mean proportions, 70–72
    reflections on strategies, 69
    theorem in Euclid, 66–67 [fig. 1.35, 1.36], 215 [fig. 4.2]
  Metaphysics of the Pythagorean theorem (overview and summary), 81–89

quadratic, 7–8, 242
quadrature, 137–139, 242
quarry, 158–159

radius, 34, 39–40, 91, 100, 119, 123, 126, 129, 140, 171, 188, 224
Ramose, 15
ratio, vii, x–xi, xiii–xiv, 9, 22, 31–32, 46, 54, 69–72, 78–81, 83–85, 87–88, 106–108, 111, 115, 117, 126–128, 131, 142–143, 145–147, 150, 154–156, 158, 164, 167–168, 173–176, 178, 184–186, 195, 199–200, 213–214, 217–218, 220, 222–223, 225–226, 227, 230, 235, 242, 246, 247–248, 251, 257–259, 265, 267
rectilinear, xi, xiii, 12, 14, 31, 56–57, 61, 77–81, 83–84, 115–117, 133, 138–140, 168, 180, 186, 189, 192–198, 203, 211, 213, 217, 233–236, 242, 249, 25
regular solids, viii, xi–xiv, 3, 77, 81, 117, 128, 168–169, 178, 184, 198–211, 236–238, 259, 269
Reidemeister, 144, 255, 268

reincarnation, 155, 182, 238–239
resurveyed, 14–15, 24
Rhind Mathematical Papyrus (RMP), xii, 13–14, 19, 22, 31, 87, 97–99, 101, 104, 109, 116, 124, 149, 181, 202, 213, 231, 233, 250, 267–268
Rhoikos, 157, 256
Riedweg, 153, 255, 268
Riemannian, 249
right angles
　two right angles in every species of triangle, 30–31 [fig. I.17, I.18, I.19], 124 [fig. 2.30], 214 [fig. 4.1]
Robins, 16, 18–19, 243–244, 250, 268
Robson, 11
Roochnik, 95, 250, 268
Rudman, 10, 242–243, 268

Saito, 185, 247, 252, 259, 261, 268
Samian, 21, 37, 157, 166–167, 239, 256
samians, 157–158
scale, 42, 69, 73, 81, 87, 111, 143, 156, 164–165, 178, 229, 244–246
scalene, 29–31, 61, 105, 124, 131, 196, 211–214, 216, 219, 222, 224, 237–238
scaling, 69, 72, 87–88, 150, 154, 178, 248
Schädler, 32–34, 42, 245, 268
schematic (diagrams), 19–20, 68, 98, 122, 215, 250
Schofield, 95, 250, 267
Scholium, 201–202, 209, 260
semicircle, 31, 86, 91, 115, 117, 119, 120–123, 125–127, 131, 138, 140–141, 169–171, 186–187, 217, 219, 222, 224, 249, 252, 255
Senenmut (alternate spelling Senemut), 17, 243
shadows, 6, 14, 18–19, 26, 28, 97, 101, 104, 114, 148–149, 151
ship, x, xiii, 6, 26, 56, 94–96, 101, 108–113, 115, 118, 154, 169, 181, 204, 229–230, 247
similarity, x–xi, xiii–xiv, 12, 15, 24, 26, 31, 54, 56, 69, 71, 74, 77–82, 83, 96–97, 106–108, 115–117, 126–128, 131–132, 143, 151–152, 154, 156, 168–169, 171, 173, 178, 186, 205, 211, 213–220, 226–229, 230, 233–235, 244, 251, 254, 258–259
Simplicius, 5, 42
Simson, 199
sirens, 155

skiatheria, 100
solid angles
　Euclid XI.21, 236 [fig. 4.38]
solids. *See* regular solids
solstice, 25, 27, 42, 93, 103, 107, 251
sophos, 182
sphairos, 229
spindle, 155
square, 23–24, 149, 151–152
　on perpendicular equals rectangle, 172–174 [fig. 3.33, 3.34, 3.35]
　when does a square equal a rectangle, 127 [fig. 2.34], 222 [fig. 4.12]
squaring, viii, xii–xiii, 5, 9, 78, 81, 87, 137, 139, 182, 189, 217, 242
staking, 162–163, 166–167
stylobate, 144
sundial, 27, 42, 100, 102–103, 153, 251
superimposition, 55
ἐφαρμόζειν, 27, 56, 57, 93, 108, 118
surveying, xii, 13–14, 104, 148, 202, 233
surveyors, 12–16, 22, 87, 97, 99, 104, 161, 180, 192, 202
Szabó, 8–9, 27, 155, 242, 244, 247, 249, 255–256, 268

Tannery, 7–8, 242, 268
tessellation, 87, 249
tetrahedron, 117, 203, 206, 208–212, 257
Thales
　alternate technique for measurement, 110 [fig. 2.13]
　comparison diagrams of pyramid height and distance of ship, 110 [fig. 2.12]
　diagram of proportional measurement of pyramid height, when shadow does not equal height, 114 [fig. 2.20], 228 [fig. 4.25]
　ancient wheel-making, 118 [fig. 2.22]
　angles in the (semi-)circle, 118
　Euclid, Book I, 5, 119, [fig. 2.24]
　Euclid, Book I, 32, 121 [fig. 2.25]
　isosceles triangle theorem, 119–121
　photo of Giza plateau pyramid shadows, 114 [fig. 2.21]
　diagrams of theorems, 26 [fig. I.15]
　measurement of pyramid when shadow equals height, 97–107

photo of experiment measuring pyramid
    height, 23 May 2013, 99 [fig.2.4]
pyramid measurement as exercise in geometric
    similarity of triangles, 100 [fig. 2.5], 228 [fig.
    4.24]
pyramid measurement of isosceles triangle in
    square, 102 [fig. 2.6, 2.7]
RMP problem, 57, 98 [fig. 2.2 and 2.3]
testimonies of Hieronymous and Pliny, 97
dividing a Circle
    proof by mixed angles, 119 [fig. 2.23]
every triangle in the (semi-)circle is right
    ancient Manuscript diagram of Euclid VI.31,
        124 [fig. 2.29]
    comparison of isosceles and scalene right
        triangles in (semi-)circle, 122 [fig. 2.28], 224
        [fig. 4.1]
    Euclid III.31, 121 [fig. 2.26]
    Geminus report: every species of triangle has
        two right angles, 124 [fig. 2.30], 214 [fig. 4.1]
    geometric loci: all right triangle vertices form
        a circle, 125 [fig. 2.31, 2.32], 224 [fig. 4.14,
        fig. 4.16]
geometrical diagram of pyramid measurement
    with numbers, 105 [fig.2.9]
geometrical diagram of pyramid measurement in
    terms of squares, 106 [fig. 2.10]
Heath's diagram of distance measurement, 113
    [fig. 2.19]
mean proportionals
    comparison pyramid measurement diagram
        with medieval manuscript diagram of VI.31,
        128 [fig. 2.35]
    mean proportional in isosceles right triangle,
        129 [fig. 2.36]
    mean proportional in scalene right triangle,
        130 [fig. 237]
    mean proportionals and Pythagorean theorem,
        130 [fig. 238]
    mean proportions and continuous proportions
        colored diagram, 221 [fig. 4.11]
    showing when a square equals a rectangle, 127
        [fig. 2.34], 222 [fig. 4.12]
    visualizing the Mean Proportions and
        Pythagorean Theorem, 131 [fig. 2.39,
        2.40], 223 [fig. 4.13]

measurement of distance of ship at sea by similar
    triangles
    shoreline measurement, 109 [fig. 2.11], 229
        [fig. 4.26]
measurement of pyramid when shadow does not
    equal height, 107–116
medieval manuscript diagrams for III.31, 122
    [fig. 2.27]
principle of proportion in measurement, 107
Royal, Knotted Cubit Cord, 104 [fig. 2.8]
scaled-model measurement, 111 [fig. 2.15], 230
    [fig. 4.28]
solar calculation to measure pyramid height, 103
    [Table 2.1]
Thales' theorem, 111 [fig. 2.14]
tower measurement, 112 [fig. 2.16, 2.17, 2.18],
    229 [fig. 4.27]
Theodorus, 42, 157, 256
Theogonis, 153
Tile Standard, Athenian Agora, 27 [fig. I.16], 56
    [fig. 1.15]
Timaeus, viii, xi, xiv, 2, 29, 81, 105, 117, 195–196,
    198–201, 203, 211–214, 236, 238, 241,
    259–261, 263–264, 267–268
    all regular solids are reducible to right triangles,
        isosceles and scalene, 238 [fig. 4.42]
Toulmin, 256, 265
transformational equivalences, xiii, 2–3, 28–29,
    41, 84, 115, 117, 126, 128, 156, 168, 183, 189–
    190, 192, 195–197, 205, 231–233, 235, 259
trapezium, 14, 21, 62–63, 93, 124, 139, 214
triangles
    all triangles divide into two right triangles, 80
        [fig. 1.50], 84 [fig. 1.51], 218 [fig. 4.7]
    all right triangles divide into two similar right
        triangles ad infinitum, 80 [fig. 1.50], 84 [fig.
        1.51], 218 [fig. 4.7]
    every triangle in a (semi-)circle is right, 121–123
        [fig. 2.26, 2.27, 2.28], 224 [fig. 4.15]
    similar triangles, 73 [fig. 1.39], 74 [fig. 1.40, 1.41],
        219 [fig. 4.8], 172 [fig. 3.32]
triangulation, 12, 14, 202
tunnel of Eupalinos, 35–41 [fig. I.24, I.25, I.26,
    I.27, I.28, I.29], 257–267 [fig. 3.21, 3.22, 3.23,
    3.24, 3.25 3.26, 3.27, 3.28]
turtles, 214

Tuthmosis, 243

Unguru, 8–9, 242, 244, 247, 265, 269

vegetarian, 182, 238–239
vegetarianism, 238–239, 259
visualization, 10, 85, 144, 152, 182, 186, 206, 222–223, 243

Waerden, 8–9, 13, 92–94, 144, 182, 202, 238, 242, 247, 249–250, 255–256, 269

Waschkies, 144, 255, 269
water, xi, 1, 3, 117, 157–161, 166, 168, 196, 211, 238, 246, 257, 259, 263
Waterhouse, 201–203, 260, 269
Wesenberg, 246–247, 269
wheel, 37, 42, 118, 229

Zeller, 95, 250, 269
Zeuthen, 7–8, 242, 269
Zhmud, xv, 13, 135–137, 139–141, 144, 147, 153–154, 203, 241, 243, 252–257, 269

Greek Terms

ἀδύνατον, 147, 255
ἀήρ, 2, 195, 202
αἰτία, 1
αἰτίας, 241
ἀλόγων, 204, 259–260
ἀνάγκη, 195
ἀνάγκης, 155, 196
ἀναλογία, 69–70, 199
ἀνάλογον, vii, viii, 70, 72–74, 76, 78, 106, 107, 155, 168, 204 259–260
ἄπειρον, 2, 202
ἀποδεῖξαι, 25–26, 118, 120
ἁρμονίαν, 156
ἄρτιος, 144, 254, 255
ἀρχή, 1–2, 28, 117, 133, 153, 196, 203, 241
ἀσύμμετρος, 145, 254

βάθος, 195–196, 236
βοῦν, 117

γνώμων, 151, 153
γράμμα, 136
γωνία, 27, 29, 66, 115, 195, 196, 204

δείκνυμι, 26–27, 56, 268
δεικνύναι, 25–26, 93, 96, 118
δεῖξαι, 254–255
δεσμῶν, 199
διαβήτης, 93
διπλάσιον, 155, 254
διπλασίονι, 78
διπλάσιος, 255
διπλασίων, 255
δυνατὸν, 254

εἶδος, 66, 68, 195
εἰπεῖν, 25–27, 118, 120
ἔλλειψις, 197
ἔννοιαι, 96
ἐπιβάλλων, 12
ἐπίπεδον, 195–196, 236
ἐπιστῆσαι, 25–27, 118, 120–121
Εὔδημος, 26, 118, 197
εὗρεν, 12, 25, 121
εὑρημένον, 25–26, 118, 120
ἐφαρμόζειν, 27, 56, 57, 93, 108, 118

Θαλῆς, 12, 26, 93, 118

θεωρησάντων, 29
θῦσαι, 117

ἰσημερίας, 27
ἰσοσκελεῖ, 29

καθολικώτερον, 12, 25
κάλλιστος, 199
Κοιναὶ, 96
κοσμικὰ, 201, 204, 259–260
κύκλου, 117

λόγοι, 106, 108
λόγον, 196, 254–255, 260
λόγος, 70, 107–108, 111, 113, 155, 203
λόγῳ, 78

μέση ἀνάλογον, vii–viii, 70, 72–74, 76, 78, 168
μεταβαλλούσης, 28
μεταγενέστεροι, 29

ὅμοια, 68
ὁμοίας, 27, 115
ὁμοίοις, 66, 68
ὁμοίως, 66, 68
ὀρθή, 29, 46, 66, 196
ὀρθογώνιον, 46, 66, 117

παραβολὴ, 234
περιειληφέναι, 195
περισσός, περισσοί, 144, 254, 255
πολυμαθίη, 259
Πυθαγόρης, 136
πῦρ, 195
πυρὸς, 196

σκαληνῷ, 29
συμμετρία, 145

σύμμετρος, 143, 145, 254–255
συνεχής ἀναλογία, 199
σύστασις, 178, 197–198, 203–204, 206, 236, 257, 259–260
σχῆμα, 68, 94, 201, 249

τετραγώνοις, 46
τετράγωνον, 46, 68, 252, 254, 255
τρίγωνα, 29, 46, 66, 117, 195

ὕδωρ, xi, xiv, 1–3, 28, 77, 84, 107, 115, 117, 128, 180, 195, 202, 213
ὑπερβολή, 197
ὑποτείνουσα, 105, 46, 66, 147

φθείρεται, 28

χωροβάτης, 160–161

Made in the USA
Lexington, KY
23 August 2019